Lecture Notes in Computer Science 6807

Commenced Publication in 1973
Founding and Former Series Editors:
Gerhard Goos, Juris Hartmanis, and Jan van Leeuwen

W0227571

Béatrice Bouchou-Markhoff
Pascal Caron
Jean-Marc Champarnaud
Denis Maurel (Eds.)

Implementation and Application of Automata

16th International Conference, CIAA 2011
Blois, France, July 13 – 16, 2011
Proceedings

 Springer

Volume Editors

Béatrice Bouchou-Markhoff
Université François Rabelais Tours, LI
4100 Blois, France
E-mail: beatrice.bouchou@univ-tours.fr

Pascal Caron
LITIS, Université de Rouen
76801 Saint-Étienne du Rouvray Cedex, France
E-mail: pascal.caron@univ-rouen.fr

Jean-Marc Champarnaud
LITIS, Université de Rouen
76801 Saint-Étienne du Rouvray Cedex, France
E-mail: jean-marc.champarnaud@univ-rouen.fr

Denis Maurel
Université François Rabelais Tours, LI
37000 Tours, France
E-mail: denis.maurel@univ-tours.fr

ISSN 0302-9743   e-ISSN 1611-3349
ISBN 978-3-642-22255-9   e-ISBN 978-3-642-22256-6
DOI 10.1007/978-3-642-22256-6

Springer Heidelberg Dordrecht London New York

Library of Congress Control Number: 2011930727

CR Subject Classification (1998): F.2, F.1, G.2, F.3, E.1, F.4

LNCS Sublibrary: SL 1 – Theoretical Computer Science and General Issues

Typesetting: Camera-ready by author, data conversion by Scientific Publishing Services, Chennai, India

Printed on acid-free paper

Springer is part of Springer Science+Business Media (www.springer.com)

Preface

The 16th International Conference on Implementation and Application of Automata (CIAA 2011) was held at the Université François Rabelais Tours, in Blois, France, during July 13–16, 2011. It was co-located with the 9th International Workshop on Finite-State Methods and Natural Language Processing (FSMNLP 2011) that was held during July 12–15, 2011. The previous CIAA conferences were held in London, Ontario (1996 and 1997), Rouen (1998), Potsdam (1999), as WIA workshops, and then in London, Ontario (2000), Pretoria (2001), Tours (2002), Santa Barbara (2003), Kingston (2004), Nice (2005), Taipei (2006), Prague (2007), San Francisco (2008), Sydney (2009), and Winnipeg (2010).

The CIAA meeting is a cornerstone forum for researchers, application developers, and users of automata-based systems. It includes applications of automata in, for example, computer-aided verification, natural language processing, pattern matching, data storage and retrieval, document engineering and bioinformatics, as well as foundational work on automata theory. The editors would like to pay homage to Derick Wood who has been one of the founders of the CIAA conference and one of its most devoted promotors. The invited talk of Sheng Yu recalls the great contribution of Derick to theoretical computer science.

This volume of *Lecture Notes in Computer Science* contains revised versions of papers presented at CIAA 2011. The 20 full papers and 4 short papers were selected from 38 submissions. Each submitted paper was evaluated by at least three Program Committee members, with the help of external referees. We warmly thank the invited speakers, the authors of contributed papers, as well as the reviewers and the Program Committee members for their valuable work.

The authors of the papers included in these proceedings come from the following countries: Argentina, Belgium, Canada, Czech Republic, France, Germany, Israel, Italy, Republic of Korea, Poland, Russian Federation, United Arab Emirates, UK and USA.

We thank EATCS and ACL for their scientific sponsorship and Université François Rabelais Tours, CNRS, Région centre, Ville de Blois, Université de Rouen, Agglopolys Blois, Ministère de l'enseignement supérieur et de la recherche, Université Paris-Est Marne-la-Vallée, Entreprise Humanis, Université d'Orléans, and MAIF for their generous financial support.

We are indebted to Alfred Hofmann and Anna Kramer from Springer for the help in producing this volume.

May 2011   B. Bouchou-Markhoff
P. Caron
J.-M. Champarnaud
D. Maurel

Organization

Invited Speakers

Markus Holzer   Justus-Liebig-Universität Giessen, Germany
Joachim Niehren   INRIA, Lille, France
Sheng Yu   University of Western Ontario, Canada

Program Committee

Marie-Pierre Béal   Université Paris Est, France
Béatrice Bouchou-Markhoff
  (Co-chair)   Université François Rabelais Tours, France
Cezar Câmpeanu   University of Prince Edward Island, Canada
Pascal Caron   Université de Rouen, France
Jean-Marc Champarnaud
  (Co-chair)   Université de Rouen, France
Michael Domaratzki   University of Manitoba, Canada
Dora Giammarresi   University of Rome, Italy
Yo-Sub Han   Yonsei University, South Korea
Tero Harju   University of Turku, Finland
Jan Holub   Czech Technical University in Prague,
                                 Czech Republic
Markus Holzer   Justus-Liebig-Universität Giessen, Germany
Oscar Ibarra   University of California, Santa Barbara, USA
Masami Ito   Kyoto Sangyo University, Japan
Stavros Konstantinidis   University of Halifax, Canada
Martin Kutrib   University of Giessen, Germany
Andreas Maletti   University of Stuttgart, Germany
Sebastian Maneth   NICTA and University of New South Wales,
                                 Australia
Denis Maurel (Co-chair)   Université François Rabelais Tours, France
Ian McQuillan   University of Saskatchewan, Canada
Mehryar Mohri   Courant Institute of Mathematical Sciences,
                                 USA
Alexander Okhotin   University of Turku, Finland
Andrei Paun   Louisiana Tech University, USA; University of
                                 Bucharest, Romania
Giovanni Pighizzini   Università degli Studi di Milano, Italy
Bala Ravikumar   Sonoma State University, USA
Rogerio Reis   Universidade do Porto, Portugal

Kai Salomaa   Queen's University, Kingston, Canada
Colin Sterling   University of Edinburgh, UK
Bruce Watson   University of Pretoria; Stellenbosch University,
                                   South Africa
Hsu-Chun Yen   National Taiwan University, Taiwan
Sheng Yu   University of Western Ontario, Canada
Djelloul Ziadi   Université de Rouen, France

Additional Referees

Cyril Allauzen   Katja Meckel
Octavian Babus   Antoine Meyer
Thomas Braibant   Ludovic Mignot
Sabine Broda   Nelma Moreira
Yu-Fang Chen   Madhavan Mukund
Salimur Choudhury   Ernest Ngassam
Marek Chrobak   Kim Nguyen
Loek Cleophas   Xiaoxue Piao
Zhe Dang   Martin Poliak
Krystian Dudzinski   Damien Pous
Jacques Farré   Karin Quaas
Szilard Fazekas   Narad Rampersad
Dominik D. Freydenberger   Adam Roman
Yuan Gao   Johan Schalkwyk
Hermann Gruber   Richard Sproat
Lukas Holik   Tinus Strauss
Sebastian Jakobi   Avraham Trahtman
Maria Madonia   Fang Yu
Andreas Malcher

Sponsors

Table of Contents

Short Papers

Derick Wood: Always in Our Hearts

Sheng Yu

Department of Computer Science, The University of Western Ontario
London, Ontario, Canada N6A 5B7
syu@csd.uwo.ca

Professor Derick Wood passed away on October 4, 2010. He was only seventy year old. His death was a great loss to his family, friends, colleagues and students, especially to his beloved wife Mary. He left us many interesting research results, more than three hundred publications [1], including three monograph and textbooks [3,4,5], and a lot of vivid memories of an energetic, thoughtful, humorous, careful, and decisive Derick Wood.

Derick was a world-known outstanding researcher. His research was in a number of areas of theoretical computer science, including automata and formal language theory, theory of parsing, data structure and algorithms, computational geometry, and document processing.

In his early research career, he was a member of the well-known MSW (Maurer-Salomaa-Wood) club. Together they published about thirty papers on grammar forms and other topics in language theory.

He was one of the most important initiators of the Grail project, which is a computational system for automata and formal language objects. He was also one of the creators of this conference series, which was formerly called *Workshop on Implementing Automata* (WIA) and is now called *International Conference on Implementation and Application of Automata* (CIAA) since 2000.

I have known Derick for almost thirty years. I first met Derick in 1983 when I was a PhD student at the University of Waterloo. I had already heard about him through Arto Salomaa and was considering him as a potential supervisor of my PhD thesis. Although he did not become my supervisor, we were pretty close both academically and socially. He was on my thesis advisory committee and I lived in a room of his house for a number of years. Later he moved to the University of Western Ontario. We became colleagues at the Department of Computer Science there.

During the period of time when Derick was with the University of Waterloo, his main research area was in algorithm and computational geometry. He was also writing his book in the theory of computation (entitled *Theory of Computation*) at that time. Quite often he talked to me about his book and asked me to provide examples for certain problems in the book. He spent quite much time in writing the book, which was very well written and was popular as a textbook in a number of universities.

When Derick was with the University of Western Ontario, his research interests gradually shifted back to automata and formal language theory. We applied successfully for an NSERC Strategic Grant for automata implementation, especially, for the development of the Grail system. The grant was very important

B. Bouchou-Markhoff et al. (Eds.): CIAA 2011, LNCS 6807, pp. 1–2, 2011.
© Springer-Verlag Berlin Heidelberg 2011

for the development of the Grail project at that time and the Grail+ project later.

Another important collaboration between Derick and myself was the creation of the conference series of CIAA (WIA). We successfully organized the first conference in 1996 at Western [2]. Since then, the conference series has taken place in many different countries in North America, Europe, Africa, Asia, and Australia. The conference provides a forum for researchers to present their results in automata application and implementation, which is an important new direction in automata and formal language research. In recent years, many new applications of automata have appeared in natural language and speech processing, software engineering, parallel processing, etc. Automata used in those applications can be very large. Similar to the situation in the 1960's and 1970's, automata theory has been again motivated heavily by applications. The implementation of automata has become an important issue due to the large size of automata used in new applications. Derick was the initial steering committee chair of the CIAA (WIA) conference series. He showed his great foresight in automata research and studies.

After Derick moved to Hong Kong, we still visited each other at least once a year for research collaboration.

Derick was very creative in research in general. He was quick to come out new ideas. He was very conscious about the meaningfulness of his research topics. His contribution in research was significant and had a great influence.

He supervised successfully a number of PhD students who became excellent researchers. Those students included Greg Rawlins, Tony Lai, Helen Cameron, Vladimir Estivill-Castro, Xinxin Wang, and Yo-Sub Han.

Personally, he was warm hearted, straightforward, and easy-going. I consider him a great teacher and a role model in research, and a good friend in life.

Although Derick has left this world, his remarkable vision in research, his great sense of humour, and his joyful demeanor will always live in our hearts.

References

1. Jürgensen, H., Maurer, H., Salomaa, A., Yu, S. (eds.): Special issue of Journal of Universal Computer Science, for Derick Wood's 70th birthday, vol. 16(5) (2010)
2. Raymond, D.R., Yu, S., Wood, D. (eds.): WIA 1996. LNCS, vol. 1260. Springer, Heidelberg (1997)
3. Wood, D.: Grammar and L Forms: An Introduction. Springer, Heidelberg (1980)
4. Wood, D.: Theory of Computation. Wiley, Chichester (1998)
5. Wood, D.: Data Structures, Algorithms, and Performance. Addison-Wesley, Reading (1993)

Streamable Fragments of Forward XPath

Olivier Gauwin[3] and Joachim Niehren[1,2]

[1] Mostrare project, INRIA & LIFL (CNRS UMR8022)
[2] INRIA, Lille
[3] University of Mons

Abstract. We present a query answering algorithm for a fragment of Forward XPath on XML streams that we obtain by compilation to deterministic nested word automata. Our algorithm is earliest and in polynomial time. This proves the finite streamability of the fragment of Forward XPath with child steps, outermost-descendant steps, label tests, negation, and conjunction (aka filters), under the reasonable assumption that the number of conjunctions is bounded. We also prove that finite streamability fails without this assumption except if P=NP.

Keywords: tree automata, pushdown automata, query answering, XML streams, XPath, temporal logics for unranked trees.

1 Introduction

Query answering algorithms for XPath on XML streams received much interest in the database and document processing communities [2,24,3,22,5,12,29,6,21] and are currently in the focus of the W3C working groups on XSLT and XPROC [14]. A little surprisingly, the topic is far from being settled given the large remaining gap between known streamable and non-streamable fragments. The objective of this paper is to narrow this gap by providing new positive and negative results for fragments of Forward XPath. Our approach relies on the relationship between temporal logics for unranked trees [16], which abstracts from the concrete syntax of XPath, and tree automata for XML streams [1,20,18].

Streaming is particularly relevant for data collections that are too large to be stored in main memory. Instead, incremental processing is needed in order to buffer only small parts of the data collection at every time point. In the easiest case, a stream is a word over some finite alphabet and a query selects some elements of this word, for instance all a-positions with two subsequent b's. Usually, a query is considered streamable if there exists a one pass algorithm (see e.g. [26]) that computes the set of query answers with constant memory, independently of the input stream [28,27]. Note however, that streaming algorithms for element selection queries need to buffer all *alive* elements, i.e. those positions which might be selected in some continuations of the stream but not in others. In the above example, there exists at most one alive a-element at every time point, so this query can indeed be answered with bounded memory for all possible input streams.

B. Bouchou-Markhoff et al. (Eds.): CIAA 2011, LNCS 6807, pp. 3–15, 2011.
© Springer-Verlag Berlin Heidelberg 2011

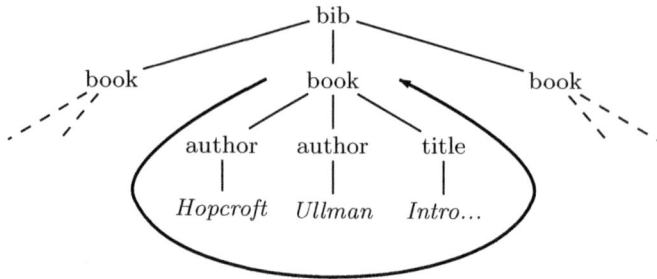

Fig. 1. Sample XML document describing a bibliography

The case of XML streams is similar except that they contain linearizations of unranked data trees and that queries select nodes in such trees. Consider for instance the XPath query */bib/book[author="Ullman"]/author* that selects all co-authors of Ullman (including himself) in all books of some bibliography (as illustrated in Fig. 1), or more precisely, all *author*-children of *book*-nodes that have at least one *author*-child with data value "Ullman". An *author*-child of a *book*-node is alive, once the corresponding opening tag was seen on the stream, and as long as the closing *book* tag was not met and no *author*-node with data "Ullman" has been read. For bibliographies, in which all books have a bounded number of authors, the maximal number of alive nodes is bounded, so that the above query can be answered with bounded memory. For unusual bibliographies, however, the number of alive candidates may grow without any bound. As a consequence, the above query is not streamable in the usual sense even though it should be intuitively.

We propose the more liberal notion of *finite streamability* for languages of node selection queries on unranked trees. Finite streamability allows the memory to grow polynomially with the number of alive candidates, the size of the query, and the depth of the tree. In order to enable negative results, we assume in addition that the computation time per step is polynomial in the above parameters, and that the memory grows at least linearly with the number of alive candidates. The latter assumptions hold for all streaming algorithms without compression tricks for representing sets of alive candidates, an assumption that is satisfied by all streaming XPath algorithms in the literature so far.

An overview on finite streamability results for XPath fragments is given in Fig. 2. Despite of the intended weakness of this notion, only few positive results exist so far. Backward XPath (Bxp) was proved finitely streamable based on transducers networks [5]. Bxp queries never have any alive candidate since node selection is always determined at opening time. The second positive result [3] applies to Fxp $(ch, o\text{-}ch_a^*, \wedge)^{thin}$, a thin fragment of positive Forward XPath on non-recursive documents, with star-restricted child steps, label-guarded (and thus outermost) descendants steps, and conjunctions (and thus filters in official XPath syntax). The only negative result so far got established for Fxp$(ch, ns^*, \wedge, \vee)$, the fragment of positive Forward XPath with child and following-sibling axes,

	bounded number of $\wedge$	unbounded number of $\wedge$
BXP	yes [5]	yes [5]
FXP $(ch,o\text{-}ch_a^*,\wedge)^{thin}$	yes [3]	yes [3]
FXP $(ch,\wedge,\neg)$	yes	no
FXP $(ch,o\text{-}ch_a^*,\wedge,\neg)$	yes	no
FXP $(ch,o\text{-}ch_a^*,ns,\wedge,\neg)$	?	no
FXP $(ch,ns^*,\wedge,\vee)$	?	no [5]
FXP $(ch,ch^*,\wedge,\neg)$	?	no

Colored results derive from the present paper. We assume here that P $\neq$ NP.

Fig. 2. Finite streamability of fragments of XPath

conjunction, and disjunction [5]. There, a counter example from online verification [19] was adapted in order to show for a family of queries in this fragment, that every streaming algorithm answering them must produce a doubly exponential number of states, and thus be of exponential size at least. This result applies even to Boolean queries (without node selection).

In this paper we study FXP(ch, $o\text{-}ch_a^*$, $\wedge$, $\neg$), the fragment of Forward XPath with child axis, outermost descendant axis, conjunction, and negation. An outermost descendant axis $o\text{-}ch_a^*$ selects all a-descendants reachable via non-a-descendants. Outermost constraints on descendant steps are a natural restriction for streaming algorithms as noticed for instance in the XSLT 2.1 definition [15]. Our first main result is a streaming algorithm for FXP(ch, $o\text{-}ch_a^*$, $\wedge$, $\neg$) that shows that this query language becomes finitely streamable if its queries are restricted to a bounded number of conjunctions. This result is relevant for the W3C pipeline language XPROC, for instance, where Forward XPath queries with at most 3 filters (and thus conjunctions) appear to be enough. Our second main result is the failure of finite streamability for FXP(ch, $\wedge$, $\neg$) except if P=NP. It shows the necessity to bound the number of conjunctions theoretically.

We obtain our streaming algorithm by compiling FXP(ch, $o\text{-}ch_a^*$, $\wedge$, $\neg$) to deterministic nested word automata (dNWAs) [1]. These are tree automata processing linearizations of unranked trees in preorder in a single pass, while mixing top-down and bottom-up determinism. For queries with a fixed number of conjunctions, our compiler is in polynomial time. Otherwise it is in exponential time, while still avoiding the usual doubly-exponential blow-up for translating XPath to deterministic automata [7]. Since the query language defined by dNWAs is finitely streamable [11], the finite streamability follows for all fragments of FXP(ch, $o\text{-}ch_a^*$, $\wedge$, $\neg$) with a bounded number of conjunctions.

Outline. Section 2 introduces FXP and Section 3 recalls dNWAs. In Section 4 we present our compiler from FXP to dNWAs. Section 5 introduces the notion of finite streamability and states our main results, positive and negative. Further related work is discussed in Section 6. The short CIAA version contains only sketches or ideas of proofs. Complete proofs are available in the long version [9].

$$[\![F_1 \wedge F_2]\!]_{t,\mu} = [\![F_1]\!]_{t,\mu} \cap [\![F_2]\!]_{t,\mu} \qquad [\![d(F)]\!]_{t,\mu} = \{\pi \mid \exists \pi' \in [\![F]\!]_{t,\mu}. \ (\pi, \pi') \in d^t\}$$
$$[\![\neg F]\!]_{t,\mu} = nod(t) - [\![F]\!]_{t,\mu} \qquad\qquad [\![a(F)]\!]_{t,\mu} = \{\pi \mid a = lab^t(\pi)\} \cap [\![F]\!]_{t,\mu}$$
$$[\![true]\!]_{t,\mu} = nod(t) \qquad\qquad\qquad [\![x]\!]_{t,\mu} = \{\mu(x)\}$$

Fig. 3. Semantics of $\mathrm{F}_{\mathrm{XP}}(ch, ch^*, \wedge, \neg)$ formulas

2 FXP

We present FXP temporal logics for unranked trees, which abstract from various aspects of the Forward XPath concrete syntax. More general temporal logics are reviewed by Libkin in [16] for instance (except for variables that we use for node selection here such as in hybrid logic).

For a finite label set Σ, we define the set of unranked trees $\mathcal{T}_\Sigma$ to be the least set such that $a(t_1, \ldots, t_k) \in \mathcal{T}_\Sigma$ if $a \in \Sigma$, $k \geq 0$ and $t_i \in \mathcal{T}_\Sigma$ for all $1 \leq i \leq k$. We write $nod(t)$ for the set of nodes of the tree t, ϵ for its root node, and $lab^t(\pi)$ for the label of node π of t. By ch^t and ch^{*t} we denote the child and descendant relations of t respectively. We will also use the outermost descendant relation $(o\text{-}ch_a^*)^t$ which navigates to all a-descendants reachable over non-a-descendants. A monadic node selection query Φ over Σ is a total function that maps trees $t \in \mathcal{T}_\Sigma$ to set of tuples of nodes $\Phi(t) \subseteq nod(t)$.

The temporal logic $\mathrm{F}_{\mathrm{XP}}(ch, o\text{-}ch_a^*, \wedge, \neg)$ is a query language for node selection in unranked trees, in which one can talk about outermost a-descendants and children while using negation and conjunction. The expressions of this logic are terms with a single fixed free variable x (for the selecting position) over the ranked signature $\Delta = \{\wedge, \neg, true, x\} \cup \mathcal{D} \cup \Sigma$ where $\mathcal{D} = \{ch\} \cup \{o\text{-}ch_a^* \mid a \in \Sigma\}$. These terms have the following form where $d \in \mathcal{D}$ and $a \in \Sigma$.

$$F ::= F_1 \wedge F_2 \mid \neg F \mid true \mid d(F) \mid a(F) \mid x$$

$\mathrm{F}_{\mathrm{XP}}(ch, o\text{-}ch_a^*, \wedge, \neg)$ corresponds to a natural class of Forward XPath expressions in the official XPath syntax modulo linear time transformations. The XPath expression $/ch^*{::}a[ch{::}b]/ch{::}*$ for instance becomes $ch^*(a(ch(x) \wedge ch(b(true))))$. Note that XPath filters are mapped to conjunctions in FXP.

Given a tree t and a variable assignment $\mu : \{x\} \rightarrow nod(t)$, we define a set valued semantics $[\![F]\!]_{t,\mu} \subseteq nod(t)$ for all formulas in Fig. 3. Path expression F defines the monadic query $[\![F]\!]$ that selects the following nodes for $t \in \mathcal{T}_\Sigma$:

$$[\![F]\!](t) = \{\mu(x) \mid \epsilon \in [\![F]\!]_{t,\mu}, \ \mu : \{x\} \rightarrow nod(t)\}$$

The size $|F|$ is the usual size of term F and its (conjunction) *width* is the number of leaves in F.

Smaller fragments of $\mathrm{F}_{\mathrm{XP}}(ch, o\text{-}ch_a^*, \wedge, \neg)$ can be obtained by removing some of the operators. For instance, we will write $\mathrm{F}_{\mathrm{XP}}(ch, \wedge, \neg)$ for the fragment using only the ch axis, conjunction and negation. The dialect of $\mathrm{F}_{\mathrm{XP}}(ch, ch^*, \wedge, \neg)$ is obtained by allowing for arbitrary descendant axis instead of only outermost a-descendants.

3   Deterministic Automata for XML Streams

We recall the notion of deterministic nested word automata (dNWAs) [1] following their presentation as streaming tree automata [8], and illustrate how to run them on XML streams. Similar kinds of tree automata were proposed for processing XML streams already in [20,18,17]. Note that these tree automata provide an explicit "visual" stack in contrast to standard tree automata.

XML streams are linearizations of unranked trees. The unranked tree $a(b, c)$ for instance becomes the XML stream `<a><b></b><c></c></a>` where `<a>` is an opening tag and `</a>` a closing tag. The events of the preorder traversal of a tree t are defined as follows (where `op` marks opening and `cl` closing events):

$$eve(t) = \{\texttt{start}\} \cup (\{\texttt{op}, \texttt{cl}\} \times nod(t))$$

Hence, $eve(a(b, c)) = \{\texttt{start}, (\texttt{op}, \epsilon), (\texttt{op}, \pi_1), (\texttt{cl}, \pi_1), (\texttt{op}, \pi_2), (\texttt{cl}, \pi_2), (\texttt{cl}, \epsilon)\}$, where π_i denotes here the ith child of the root. All events in $eve(t)$ except for `start` can be identified with a precise position in the XML stream for t. The event set is totally ordered with `start` as least element. We denote this order by $\preceq$ and for an event $\eta \neq \texttt{start}$ we write $pr_\prec(\eta)$ for the immediately preceding event wrt. $\preceq$.

Definition 1. *A dNWA is a tuple $(\Sigma, Q, \Gamma, i, F, \delta)$ where Σ is a finite alphabet, Q a finite set of states with a distinguished initial state $i \in Q$ and final states $F \subseteq Q$, Γ a finite set of stack symbols, and δ a set of rules. For each state $q_0 \in stat$ and letter $a \in \Sigma$, there is at most one rule $q_0 \xrightarrow{op\ a:\gamma} q_1$ in δ, and for each $q_0 \in Q$, $a \in \Sigma$, and $\gamma \in \Gamma$, it contains at most one rule $q_0 \xrightarrow{cl\ a:\gamma} q_1$.*

A configuration of a dNWA A on a tree t consists of an event of t, a state of Q, and a stack of elements in Γ. An opening rule $q_0 \xrightarrow{op\ a:\gamma} q_1$ can be applied to a configuration that opens some a-node in state q_0. In this case, the subsequent configuration is reached by pushing γ to the current stack, changing the state to q_1, and advancing to the next event. A closing rule $q_0 \xrightarrow{cl\ a:\gamma} q_1$ can be applied to a configuration that closes some a-node in state q_0. The symbol γ is then popped from the stack, the current state is changed to q_1, and the current event is advanced by one. It should be noticed that transitions on configurations are always deterministic.

There is exactly one initial configuration: its event is `start`, its state i, and its stack is empty. Furthermore, note that the current stack is always the sequence of symbols that were pushed to the stack by the ancestors of the current node and itself. A configuration is accepting if the current event is the closing event of the root, the current state is final, and the current stack is empty.

More formally, a run r of an dNWA A on a tree t is a pair of functions $r_e : eve(t) \to Q$ and $r_n : nod(t) \to \Gamma$, such that $r_e(\texttt{start}) = i$ and that δ contains the following rules for all $\pi \in nod(t)$ with $a = lab^t(\pi)$, $\alpha \in \{\texttt{op}, \texttt{cl}\}$ and $\eta = (\alpha, \pi)$:

$$r_e(pr_\prec(\eta)) \xrightarrow{\alpha\ a:r_n(\pi)} r_e(\eta)$$

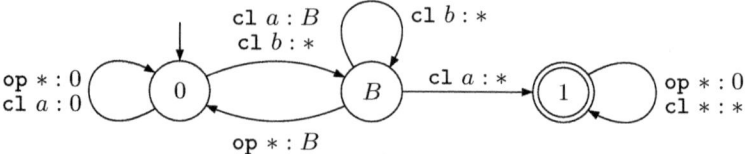

Fig. 4. A dNWA over $\Sigma = \{a, b\}$ with $Q = \{0, B, 1\}$ and $\Gamma = \{0, B\}$

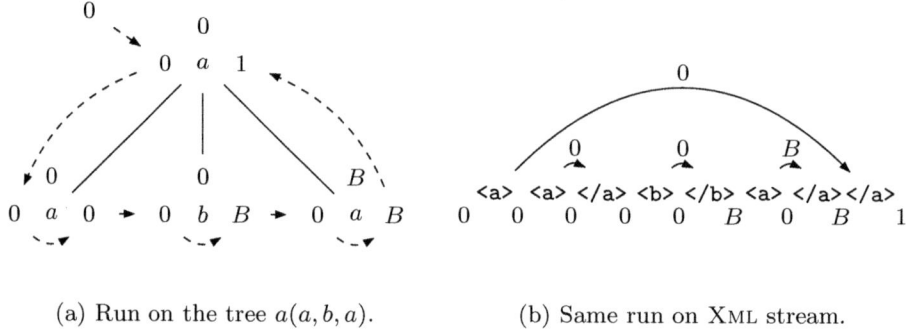

(a) Run on the tree $a(a, b, a)$.   (b) Same run on XML stream.

Fig. 5. Run of the dNWA of Fig. 4 on an input XML document

A run r is successful if $r_e((\text{cl}, \epsilon)) \in F$. The recognized language $L(A)$ is the set of trees on which A has a successful run. We call an dNWA *pseudo-complete* if there is a run on every tree $t \in \mathcal{T}_\Sigma$.

For illustration, consider the dNWA in Fig. 4, which recognizes all trees containing some a-node with some b-child. This Boolean query is $ch^*(a(ch(b(true))))$ in FXP or $[//a/b]$ in XPath syntax. We will freely use the symbol $*$ to stand either for an arbitrary letter or an arbitrary stack symbol. The idea of this automaton is to move to state B when ever closing some b-node and to propagate this state by passing B to all closing events of following-siblings (except if some of them contains some a-descendant with some b-child, so that the automaton can safely go into the successful state 1). The automaton can move to the successful state 1 when closing some a-node from state B, since state B can only be assigned to closing events of children with a previous b-sibling. The run of this dNWA on tree $a(a, b, a)$ is illustrated in Fig. 5. Stack symbols can be either annotated to nodes of trees or to edges from opening to corresponding closing events on XML streams. The horizontal propagation of B works as follows: at opening time B is pushed onto the stack and at closing time it is popped from there.

In order to compute the run of a dNWA A on an XML stream with tree t, the current configuration of A needs to be stored at each event of t. This configuration contains the state of the current event and the sequence of states annotated to the ancestors of the current node, i.e., the current stack. Note that the size of the stack is at most $depth(t)$, so that membership to $L(A)$ can be decided by a streaming algorithm with a memory of size $O(|A| + depth(t))$.

Evaluation of dNWAs encoding DTDs or other XML schemas performs streaming schema validation. A weakness of naive evaluation for testing membership $t \in L(A)$ is the laziness of A in streaming mode: it only detects a-nodes with b-children when closing the a-node, but could already do so when opening the b-child. For tree $a(a, b, a)$ for instance, the earliest event is (op, π_2) when reading the first tag <b>. The streaming algorithm from [11] improves on this situation: it decides membership $t \in L(A)$ for dNWAs A at the earliest possible event of tree t while remaining in PTIME. In order to find this earliest event, this algorithm needs to inspect the whole configuration at every event, not only the state.

Automata can also be used to define monadic queries. As before, we fix a variable x. For every tree $t \in \mathcal{T}_\Sigma$ and node $\pi \in nod(t)$, we define the canonical tree $t * \pi \in \mathcal{T}_{\Sigma \times 2^{\{x\}}}$ obtained from t by relabeling π with $(lab^t(\pi), \{x\})$ and all other nodes π' with $(lab^t(\pi'), \emptyset)$. More generally, a tree $t \in \mathcal{T}_{\Sigma \times 2^{\{x\}}}$ is canonical if exactly one of its nodes has a label in $\Sigma \times \{x\}$. A dNWA A with signature $\Sigma \times 2^{\{x\}}$ defines the query $[\![A]\!]$ on trees over Σ with $[\![A]\!](t) = \{\pi \in nod(t) \mid t * \pi \in L(A)\}$.

4   FXP to Deterministic Automata

In this section, we propose a translation of $\textsc{Fxp}(ch, o\text{-}ch_a^*, \wedge, \neg)$ to dNWAs. It runs in polynomial time if we assume a bound on the number of conjunctions. Our translation works by induction on the structure of formulas.

In order to avoid exponential blowups, our dNWAs will evaluate at most one subformula at every time point. Consider for instance the formula $ch(F')$. As all axes in F' are downwards (this would fail with the next-sibling axis), the algorithm can always know when closing a child, whether F' holds there or not. Thus, when opening the next child, the test for the previous child is finished. Therefore F' is tested for at most one child at a time. Note that an unbounded number of overlapping tests would end up in an exponential blowup. The same invariant also holds for $o\text{-}ch_a^*(F')$ formulas: no nested a-descendants need to be tested simultaneously for F'; considering outermost a-descendants is enough.

Proposition 1. *For every formula F of $\textsc{Fxp}(ch, o\text{-}ch_a^*, \wedge, \neg)$, we can build a dNWA A such that $[\![A]\!] = [\![F]\!]$ in time $O(|F|^{2 \cdot width(F)} \cdot |\Sigma|^{width(F)+1} \cdot 45^{width(F)})$.*

The automaton construction is by induction on the structure of formulas. Here we only highlight the main trick necessary that makes the construction polynomial when fixing $width(F)$. Conjunctions are mapped to automata intersection and negations to automata complementation, by swapping final states while assuming pseudo-complete dNWAs. Determinism is essential here. Note that the compilation of conjunctions might produce dNWA of size exponential in $width(F)$. The translations of label tests and variables is straightforward.

The main point, where we avoid an important blow up, appears already in the construction of the automaton for $ch(F')$ and similarly for $o\text{-}ch_a^*(F')$. The idea of the dNWA for $ch(F')$ is to run the dNWA A' testing F' on every child of

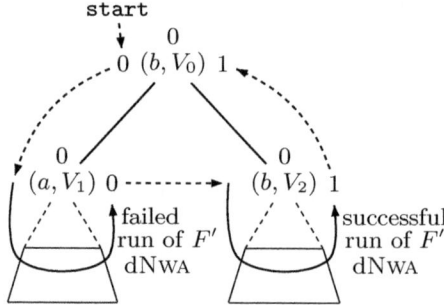

Fig. 6. Successful run of the dNWA recognizing $F = ch(F')$

the root until finding one that satisfies F'. When running on the subtree rooted by some child, the algorithm must know when the child will be closed. In order to do so, it must push a special symbol to the stack when opening the child. It could do so by pushing a tagged version of the stack symbol γ pushed by A'. However, this would double the number of node states at each ch operator (as we also have to use γ below), leading to a global size increase of 2^n for formula $ch^n(true)$. The trick here, is to push a single new symbol 0, and to recompute node state γ corresponding to the current run due to determinism: knowing the initial state of A' and the label of the child, we can infer the rule of A' applied to open this child, and thus γ.

Let $A' = (\Sigma \times 2^{\{x\}}, Q', \Gamma', i', F', \delta')$ be the automaton built for F'. Automaton $A = (\Sigma \times 2^{\{x\}}, Q, \Gamma, i, F, \delta)$ for F will produce runs of the form in Fig. 6. It has three new states $Q = Q' \uplus \{\text{start}, 0, 1\}$ and one additional stack symbol $\Gamma = \Gamma' \uplus \{0\}$.

1. State start is only used as initial state, to open the root node: $i = \{\text{start}\}$ and a rule $\text{start} \xrightarrow{\text{op } (a,V):0} 0$ is added to δ for all possible $(a, V) \in \Sigma \times 2^{\{x\}}$.
2. State 0 is used when closing a child of the root, if no matching for F' has been found so far. When a child is opened from 0, we start testing F' and assign node state 0 to this child. We have to add new rules, from rules starting from the initial state of A' (note that stack symbol γ are lost):

$$\frac{q_1 \in i' \qquad q_1 \xrightarrow{\text{op } (a,V):\gamma} q_2 \in \delta'}{0 \xrightarrow{\text{op } (a,V):\ 0} q_2 \in \delta}$$

3. State 1 is universally accepting, so we always stay there once a matching has been found: $1 \xrightarrow{\alpha\ (a,V):0} 1 \in \delta$ for all $(\alpha, a, V) \in \{\text{op}, \text{cl}\} \times \Sigma \times 2^{\{x\}}$, and $F = \{1\}$.
4. Then a test of F' is launched: the set of new rules δ subsumes δ'.
5. When closing a child of the root, we have to check whether the test of F' succeeded or not. As argued before, A pushes state 0 when oping a child, so

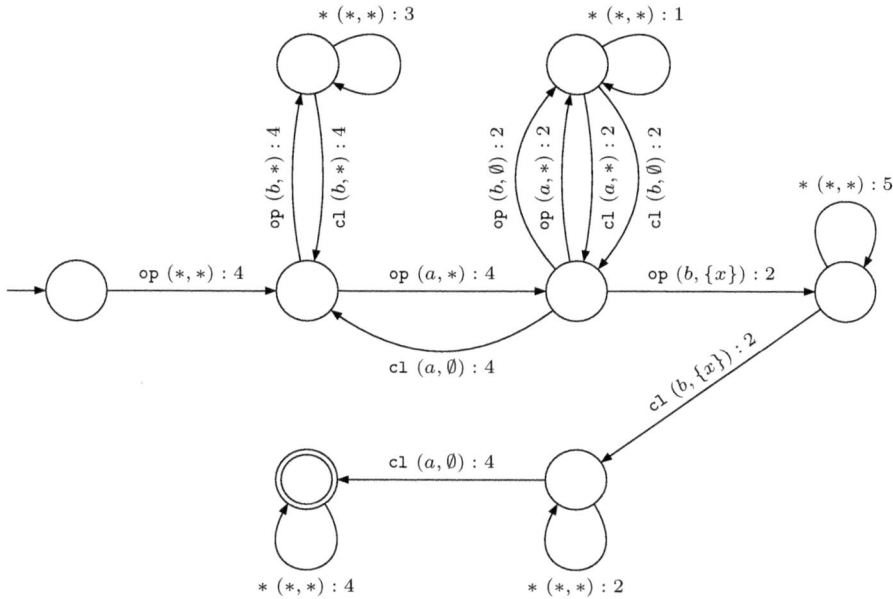

Fig. 7. dNwa constructed for $ch(a(ch(b(x))))$ with $\Sigma = \{a, b\}$

that stack symbol γ pushed by A' is lost temporarily. But A can recompute this symbol when closing the child. In case of success, A closes in state 1, otherwise in state 0.

$$\frac{q_1' \in i' \quad q_1' \xrightarrow{\text{op } (a,V):\gamma} q_2' \in \delta' \quad q_1 \xrightarrow{\text{cl } (a,V):\gamma} q_2 \in \delta' \quad q_2 \in F'}{q_1 \xrightarrow{\text{cl } (a,V):\ 0} 1 \in \delta}$$

$$\frac{q_1' \in i' \quad q_1' \xrightarrow{\text{op } (a,V):\gamma} q_2' \in \delta' \quad q_1 \xrightarrow{\text{cl } (a,V):\gamma} q_2 \in \delta' \quad q_2 \notin F'}{q_1 \xrightarrow{\text{cl } (a,V):\ 0} 0 \in \delta}$$

6. Finally, to remain pseudo-complete, we have to propagate state 0 when closing the root node: $0 \xrightarrow{\text{cl } (a,V):0} 0 \in \delta$ for all $(a, V) \in \Sigma \times 2^{\{x\}}$.

Even though the ideas of the constructions are rather simple, it should be noticed that dNwas obtained by this construction are often hard to understand. This is mainly due to the recomputation trick. See Fig. 7 for an example.

5   Streamability of Query Languages

We present the notion of finite streamability of query languages, and apply it to the query languages defined by dNwas and fragments of Forward XPath.

Definition 2. *A monadic query language for unranked trees in $\mathcal{T}_\Sigma$ is a triple $(E, [\![.]\!], |.|)$ that consists of a set E whose elements are called query definitions, a function from definitions $e \in E$ to monadic query $[\![e]\!]$, that we call the query defined by e, and a mapping of query definitions $e \in E$ to natural numbers $|e| \in \mathbb{N}$, that we call the size of e.*

How many candidates must be buffered when answering a query Φ on a tree t? Intuitively, at least all alive candidates need to be stored, where a candidate $\pi \in nod(t)$ is called *alive* at an event $\eta \in eve(t)$ if it can be selected in some continuation of the stream and rejected in other ones. The concurrency $concur_\Phi(t)$ of Φ on t is the maximal number of alive candidates at all events.

The main idea of finite streamability is to require that the number of buffered candidates must be polynomially bounded in the concurrency. In order to do so, aliveness of some candidates must be decided at some point. Doing this in PTIME in the size of query definitions imposes a serious restriction, that all finitely streamable query languages must satisfy. In order to obtain lower bounds we assume that candidate sets are always stored without compression. This property is satisfied by all streaming algorithms in the literature.

Definition 3. *We call a query language $(E, [\![.]\!], |.|)$ finitely streamable if there exists polynomials p_0, p_1, p_2 such that for all query definitions $e \in E$ one can compute in time $p_0(|e|)$ a RAM machine $\mathcal{M}_e$ computing $[\![e]\!]$, such that*

- *the space used by $\mathcal{M}_e$ per step on $t \in \mathcal{T}_\Sigma$ is at most $p_1(|e|, concur_{[\![e]\!]}(t), depth(t))$ and at least $concur_{[\![e]\!]}(t)$, and*
- *the time used by $\mathcal{M}_e$ per step on $t \in \mathcal{T}_\Sigma$ is at most $p_2(|e|, concur_{[\![e]\!]}(t), depth(t))$.*

Prior work on earliest query answering provides our first positive result on streamability for dNWAs.

Theorem 1 ([11]). *The language of monadic queries defined by dNWAs over $\Sigma \times 2^{\{x\}}$ is finitely streamable.*

Proof. For monadic queries, the streaming algorithm in [11] has the following costs per step: $O(c \cdot |A|^2)$ in time and $O(c \cdot d \cdot |A|)$ in space, where $c = concur_{[\![A]\!]}(t)$ and $d = depth(t)$. This algorithm requires the dNWA A to accept only canonical trees, which can be obtained by intersecting it with a dNWA checking canonicity (this can be done in polynomial time). A RAM machine implementing this algorithm can be built in PTIME.

We define the query language $\text{FxP}(ch, o\text{-}ch_a^*, \wedge^{(k)}, \neg)$ which expressions are formulas F of $\text{FxP}(ch, o\text{-}ch_a^*, \wedge, \neg)$ with less than k conjunctions, i.e. such that $width(F) \leq k$. For this fragment, the translation provided in Section 4 is in polynomial time, and thus avoids more general doubly exponential compilation schemas of XPath expressions into deterministic tree automata [7].

Theorem 2. *For every fixed $k \geq 0$ and alphabet Σ, $\text{FxP}(ch, o\text{-}ch_a^*, \wedge^{(k)}, \neg)$ is finitely streamable.*

Proof. Let k be fixed. For every formula F in $\text{FXP}(ch, o\text{-}ch_a^*, \wedge^{(k)}, \neg)$, $width(F) \leq k$, so, according to the translation proposed in Section 4 (Proposition 1), there exists a polynomial p such that for all formulas F of $\text{FXP}(ch, o\text{-}ch_a^*, \wedge^{(k)}, \neg)$ we can build in time $O(p(|F|))$ a dNWA A such that $[\![A]\!] = [\![F]\!]$. Hence, finite streamability of queries by dNWAs (Theorem 1) can be lifted to $\text{FXP}(ch, o\text{-}ch_a^*, \wedge^{(k)}, \neg)$.

The restriction on the width of formulas is necessary to remain in PTIME.

Theorem 3. $\text{FXP}(ch, \wedge, \neg)$ *is* not *finitely streamable, and remains non finitely streamable when restricted to non-recursive trees, unless $P = NP$.*

Here, we only give a brief sketch of the proof. We first show for all languages of descending queries that finite streamability implies that query satisfiability is in polynomial time. This can be shown by proving that aliveness of candidates must be decided for obtaining finite streamability, so that previous hardness results for earliest query answering carry over [6,11]. This works under the realistic assumption that the number of alive candidates is a space lower bound for streaming algorithms. We then show that satisfiability of $\text{FXP}(ch, \wedge, \neg)$ is NP-hard by strengthening results from [4]. Hence, without assuming P=NP or a bound on the number of conjunctions, $\text{FXP}(ch, \wedge, \neg)$ cannot be finitely streamable, nor any larger query language.

6   Related Work

Our compiler from $\text{FXP}(ch, o\text{-}ch_a^*, \wedge^{(k)}, \neg)$ must avoid the usual doubly exponential blow-up when translating XPath expressions into deterministic tree automata [7]. One exponential goes away by bounding the number of conjunctions and all kinds of overlapping tests, for instance when adding ns or ns^* steps. The other exponential is circumvented by the restriction to outermost descendants steps since these can be checked deterministically.

   As proved in the current paper, finite streamability of $\text{FXP}(ch, \wedge, \neg)$ continues to fail even if restricted to non-recursive documents. This shows that the memory consumption of the two algorithms of [2] and [12] cannot be polynomial in the number of alive candidates, in contrast to what is stated there[1] except if P=NP. We also note that streaming algorithms for Forward XPath in [22] and [23,24] do not claim finite streamability. The complexity results stated there count the maximal number of candidates stored simultaneously by their algorithms, rather than the maximal number of alive candidates with respect to the query.

   Space lower bounds for multi-pass streaming algorithms were shown in [13]. Previous space lower bounds for one-pass streaming algorithms for XPath were obtained by communication complexity arguments without any assumptions on compression tricks. Therefore, they remained limited to very specific fragments. In [2], wildcard-free queries in $\text{FXP}(ch, ch^*, \wedge, \neg)$ are considered under the assumption of an infinite signature. It is shown that the maximal number of closed

---

[1] Authors of [2] and [12] have been notified. The journal version of [2] will take this remark into account.

simultaneously alive answer candidates is a lower bound for "mostly all" non-recursive trees in the sense of instance complexity. In [25], it is shown that for some queries in $\mathrm{FXP}(ch, ch^*, \wedge)$ with independent ch predicates, the lower bound becomes $n \cdot c$ where n is the length of the selecting branch of the XPath expression, and c is maximal number of concurrently alive candidates. This shows that even compression tricks do not help for these query languages.

In [10] it was shown that it is decidable in polynomial time for queries defined by deterministic nested word automata, whether the maximal number of concurrently alive candidates is bounded. This result can be lifted to $\mathrm{FXP}(ch, o\text{-}ch_a^*, \wedge^{(k)}, \neg)$ by using our P-time compiler to dNwAs.

7 Conclusion

We have shown that $\mathrm{FXP}(ch, o\text{-}ch_a^*, \wedge, \neg)$ becomes finite streamability when fixing the number of conjunctions. Without such a bound, even $\mathrm{FXP}(ch, \wedge, \neg)$ is not finitely streamable. Our results reveal some errors in previous work. This illustrates that they are nontrivial even though proofs are straightforward (once the translation is set up properly). It should also be noticed that our algorithm can be extended to support schemas (defined by DTDs or dNwAs) as well as for queries selecting tuples of nodes instead of nodes.

In QuiXProc (see www.quixproc.com), a transfer project of INRIA and IN-NOVIMAX, we are currently working on highly efficient streaming algorithms for $\mathrm{FXP}(ch, o\text{-}ch_a^*, \wedge, \neg)$ based on similar dNwA constructions, which enable early node selection (not necessarily always earliest). First tests with our implementation, whose source code is freely available at fxp.lille.inria.fr, confirm this expectation. We are working on improving the integration of these algorithms into XPROC to industrial quality. We are thus confident to prove the practical relevance of the methods presented here in the near future.

References

1. Alur, R., Madhusudan, P.: Adding nesting structure to words. Journal of the ACM 56(3), 1–43 (2009)
2. Bar-Yossef, Z., Fontoura, M., Josifovski, V.: Buffering in query evaluation over XML streams. In: ACM PODS, pp. 216–227 (2005)
3. Bar-Yossef, Z., Fontoura, M., Josifovski, V.: On the memory requirements of XPath evaluation over XML streams. J. Comp. Syst. Sci. 73(3), 391–441 (2007)
4. Benedikt, M., Fan, W., Geerts, F.: XPath satisfiability in the presence of DTDs. Journal of the ACM 55(2), 1–79 (2008)
5. Benedikt, M., Jeffrey, A.: Efficient and expressive tree filters. In: FST-TCS, pp. 461–472 (2007)
6. Benedikt, M., Jeffrey, A., Ley-Wild, R.: Stream Firewalling of XML Constraints. In: ACM SIGMOD, pp. 487–498 (2008)
7. Francis, N., David, C., Libkin, L.: A direct translation from XPath to nondeterministic automata. In: 5th Alberto Mendelzon International Workshop on Foundations of Data Management (2011)

8. Gauwin, O.: Streaming Tree Automata and XPath. PhD thesis, Université Lille 1 (2009)
9. Gauwin, O., Niehren, J.: Streamable fragments of Forward XPath. Long version (2011), http://hal.inria.fr/inria-00442250/en
10. Gauwin, O., Niehren, J., Tison, S.: Bounded delay and concurrency for earliest query answering. In: Dediu, A.H., Ionescu, A.M., Martín-Vide, C. (eds.) LATA 2009. LNCS, vol. 5457, pp. 350–361. Springer, Heidelberg (2009)
11. Gauwin, O., Niehren, J., Tison, S.: Earliest query answering for deterministic nested word automata. In: Kutyłowski, M., Charatonik, W., Gebala, M. (eds.) FCT 2009. LNCS, vol. 5699, pp. 121–132. Springer, Heidelberg (2009)
12. Gou, G., Chirkova, R.: Efficient algorithms for evaluating XPath over streams. In: ACM SIGMOD, pp. 269–280 (2007)
13. Grohe, M., Koch, C., Schweikardt, N.: Tight lower bounds for query processing on streaming and external memory data. In: Caires, L., Italiano, G.F., Monteiro, L., Palamidessi, C., Yung, M. (eds.) ICALP 2005. LNCS, vol. 3580, pp. 1076–1088. Springer, Heidelberg (2005)
14. Kay, M.: Saxon diaries (2009), http://saxonica.blogharbor.com/blog
15. Kay, M.: XSLT 2.1 – W3C working draft (May 2010)
16. Libkin, L.: Logics over unranked trees: an overview. Logical Methods in Computer Science 3(2), 1–31 (2006)
17. Kumar, V., Madhusudan, P., Viswanathan, M.: Visibly pushdown automata for streaming XML. In: WWW, pp. 1053–1062 (2007)
18. Koch, C., Scherzinger, S., Schweikardt, N., Stegmaier, B.: Schema-based Scheduling of Event Processors and Buffer Minimization for Queries on Structured Data Streams. In: 30th VLDB, pp. 228–239. Morgan Kaufmann, San Francisco (2004)
19. Kupferman, O., Vardi, M.Y.: Model checking of safety properties. Form. Meth. in Syst. Design 19(3), 291–314 (2001)
20. Neumann, A., Seidl, H.: Locating matches of tree patterns in forests. In: Arvind, V., Sarukkai, S. (eds.) FST TCS 1998. LNCS, vol. 1530, pp. 134–146. Springer, Heidelberg (1998)
21. Nizar, A., Kumar, S.: Efficient Evaluation of Forward XPath Axes over XML Streams. In: COMAD, pp. 222–233 (2008)
22. Olteanu, D.: SPEX: Streamed and progressive evaluation of XPath. IEEE Trans. on Know. Data Eng. 19(7), 934–949 (2007)
23. Ramanan, P.: Evaluating an XPath Query on a Streaming XML Document. In: COMAD, pp. 41–52 (2005)
24. Ramanan, P.: Worst-case optimal algorithm for XPath evaluation over XML streams. J. of Comp. Syst. Sci. 75, 465–485 (2009)
25. Ramanan, P.: Memory lower bounds for XPath evaluation over XML streams. J. of Comp. Syst. Sci (2010) (in Press, Corrected Proof)
26. Schweikardt, N.: Machine models and lower bounds for query processing. In: ACM PODS, pp. 41–52 (2007)
27. Chomicki, J.: Consistent query answering: Five easy pieces. In: Schwentick, T., Suciu, D. (eds.) ICDT 2007. LNCS, vol. 4353, pp. 1–17. Springer, Heidelberg (2006)
28. Segoufin, L., Vianu, V.: Validating streaming XML documents. In: ACM PODS, pp. 53–64 (2002)
29. Wu, X., Theodoratos, D.: Evaluating Partial Tree-Pattern Queries on XML Streams. In: CIKM, pp. 1409–1410 (2008)

Gaining Power by Input Operations: Finite Automata and Beyond

Markus Holzer and Martin Kutrib

Institut für Informatik, Universität Giessen,
Arndtstr. 2, 35392 Giessen, Germany
{holzer,kutrib}@informatik.uni-giessen.de

Abstract. We summarize results on extended finite automata, which are basically finite state machines with the additional ability to manipulate the still unread part of the input. Well-known manipulation functions are reversal, left-revolving, right-revolving, and circular interchanging, or even biologically motivated functions as hairpin inversion. We mainly focus on the computational power of these machines and on the closure properties by standard formal language operations of the induced language families. Moreover, we also discuss several generalizations of this concept, the natural generalization to hybrid extended finite automata, which allows several input manipulation functions, and in particular, extended pushdown automata, which lead to an alternative characterization of Khabbaz hierarchy of languages. We do not prove these results but we merely draw attention to the big picture, some of the main ideas involved, and open problems for further research.

1   Introduction

Finite automata have intensively been studied and moreover, have been extended in several different ways. Typical extensions in the view of [18] are, for example, pushdown tapes [10], stack tapes [19], or Turing tapes. The investigations in [18] led to a rich theory of abstract families of automata, which is the equivalent to the theory of abstract families of languages (AFL) (see, for example, [26]). On the other hand, recently in several papers, see [4,5,6,7,8,9], finite state machines have been extended in quite a different manner. The models considered there, called extended finite automata, are (nondeterministic) finite state machines which are enriched with the ability to apply a string operation on the part of the input that has not been consumed yet. Extended finite automata are inspired by the model of flip pushdown automata [27] which can flip the contents of their pushdown stores in certain configurations. The authors in [21] showed that $k + 1$ pushdown-flips are better than k, and established an interrelation between the pushdown-flips and reversal operations on the unprocessed input of a flip pushdown automaton. This link between storage operations on the one hand and operations on the unread part of the input on the other hand brought up the idea of extended finite automata as the simplest model for investigating typical input operations for themselves. Obviously, with no further limitation on the input operations one can define devices that provide computational

B. Bouchou-Markhoff et al. (Eds.): CIAA 2011, LNCS 6807, pp. 16–29, 2011.

power beyond that of Turing machines, for example, by defining an operation to be an oracle for the halting problem. So, there is an interest in natural operations that are somehow feasible. Examples of feasible input operations are input-reversal, left-revolving, right-revolving, and circular-interchanging. In this paper we briefly survey some recent results on both extended finite automata and extended pushdown automata with some of the aforementioned input operations. We summarize what is known on the computational power of these devices with respect to each other, and to classical formal language families such as variants of context-free and context-sensitive languages. Moreover also closure properties of some language families induced by extended automata are discussed. The relation between different input operations in the context of hybrid automata, which are extended automata that are allowed to use more than one particular input operation during the computation, are considered. Finally, the focus is set to extended pushdown automata, which for the input operation on reversal nicely fit to previous work done one iterated pushdown automata [28]. In turn, these iterated pushdown machines are related to controlled linear context-free languages [24], which lead to the so called geometric hierarchy of languages [23,24], that has its name from the geometric series involved in the pumping lemmas for these language families. Our tour on the subject obviously lacks completeness due to the fact that this research field is still in its infantile state—we come across of some of the open problems. More research on extended automata hopefully closes some of these gaps, and furthermore identifies other language operations that are of interest not only for theory, but also for applications. We hope that this survey stimulates further investigations on extended automata and variants thereof.

2   Extended Finite Automata

In connection with formal languages, strings are called *words*. Let Σ^* denote the set of all words over a finite alphabet Σ. The *empty word* is denoted by λ, and we set $\Sigma^+ = \Sigma^* \setminus \{\lambda\}$. For the *length of a word* w we write $|w|$; in particular, the length of the empty word is zero, that is, $|\lambda| = 0$. A *formal language* L is a subset of Σ^*. We use $\subseteq$ for *inclusions* and $\subset$ for *strict inclusions*. The uniform definition of an extended finite automaton reads as follows:

A *(nondeterministic) extended finite automaton* is a 6-tuple $(Q, \Sigma, \delta, \Delta, q_0, F)$ where Q is a finite set of states, Σ is the input alphabet, δ and Δ are mappings from $Q \times (\Sigma \cup \{\lambda\})$ to 2^Q, where δ is called the transition function, and Δ is called the input operation function, $q_0 \in Q$ is the initial state, and $F \subseteq Q$ is the set of accepting states. Furthermore, A is said to be λ-*free*, if both δ and Δ are mappings from $Q \times \Sigma$ to 2^Q.

The different operations on the input are formally distinguished by different interpretations of the mapping Δ. To this end, we consider *configurations* of extended finite automata $A = (Q, \Sigma, \delta, \Delta, q_0, F)$ to be tuples (q, w), where $q \in Q$ is the current state, and $w \in \Sigma^*$ is the yet unread part of the input. The transition of a configuration into a successor configuration can be induced by either δ or Δ:

1. Let a be in $\Sigma \cup \{\lambda\}$ and w in Σ^*. If p is in $\delta(q,a)$, then $(q, aw) \vdash_A (p, w)$. Those transitions are referred to as *ordinary transitions*.
2. An input operation is performed by applying the mapping Δ. Concise formal definitions of the different possible interpretations of the mapping Δ are given as follows (cf. Figure 1). The precise interpretation depends on the type of the automaton in question. For $a \in \Sigma \cup \{\lambda\}$, $b, c \in \Sigma$, $w \in \Sigma^*$, and p in $\Delta(q, a)$,

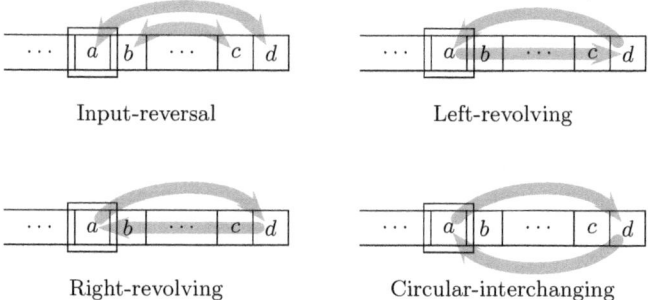

Input-reversal   Left-revolving

Right-revolving   Circular-interchanging

Fig. 1. Input operations: input-reversal, left-revolving, right-revolving, and circular-interchanging

(a) an *input-reversal transition* is defined by $(q, aw) \vdash_A (p, w^R a)$, if $a \in \Sigma$, and $(q, bw) \vdash_A (p, w^R b)$ and $(q, \lambda) \vdash_A (p, \lambda)$, if $a = \lambda$,
(b) a *left-revolving transition* is defined by $(q, a) \vdash_A (p, a)$ and $(q, awb) \vdash_A (p, baw)$,
(c) a *right-revolving transition* is defined by $(q, aw) \vdash_A (p, wa)$, if $a \in \Sigma$, and $(q, bw) \vdash_A (p, wb)$ and $(q, \lambda) \vdash_A (p, \lambda)$, if $a = \lambda$, and
(d) a *circular-interchanging transition* is defined by $(q, a) \vdash_A (p, a)$ and $(q, awb) \vdash_A (p, bwa)$, if $a \in \Sigma$, and $(q, cwb) \vdash_A (p, bwc)$ and $(q, \lambda) \vdash_A (p, \lambda)$, if $a = \lambda$.

Further interpretations of the Δ transition are possible and some will be briefly discussed in Sections 3 and 5.

The corresponding transitions are referred to as *input transitions*. Note that the formal definitions involve both λ-transitions (that is "blind" input operations) and those depending on the input symbol which is currently read. Especially, for any operation, if $p \in \Delta(q, \lambda)$, then $(q, \lambda) \vdash_A (p, \lambda)$. For any extended finite automaton, whenever there is a choice between an ordinary or an input transition, the automaton nondeterministically chooses the next move. A *deterministic* extended finite automaton is an extended finite automaton for which there is at most one choice of action for any possible configuration. This includes a unique interpretation of the mapping Δ. As usual, the reflexive transitive closure of $\vdash_A$ is denoted by $\vdash_A^*$. The subscript A will be dropped from $\vdash_A$ and $\vdash_A^*$ whenever the meaning remains clear. We define the language *accepted* by an extended finite automaton A to be

$$L(A) = \{ w \in \Sigma^* \mid (q_0, w) \vdash_A^* (q, \lambda) \text{ with } q \in F \}$$

A nondeterministic extended finite automaton whose input operation function is interpreted as input-reversal, left-revolving, right-revolving, or circular-interchanging is called an *input-reversal* (ir-NFA), *left-revolving* (lr-NFA), *right-revolving* (rr-NFA), or *circular-interchanging finite automaton* (ci-NFA). The corresponding deterministic types of automata are abbreviated ir-DFA, lr-DFA, rr-DFA, and ci-DFA, respectively. We denote the *family of languages accepted* by devices of type X by $\mathscr{L}(X)$.

Let us recall an example that can be found in [5], which shows that already input-reversal deterministic finite automata can accept context-free languages.

Example 1. The context-free language $\{ wcw^R \mid w \in \{a, b\}^* \}$ is accepted by the ir-DFA $A = (\{q_0, q_a, q_b, q_a', q_b', q_f\}, \{a, b, c\}, \delta, \Delta, q_0, \{q_f\})$, where

1. $\delta(q_0, a) = \{q_a\}$ 4. $\delta(q_b', b) = \{q_0\}$ 7. $\delta(q_0, c) = \{q_f\}$
2. $\delta(q_0, b) = \{q_b\}$ 5. $\Delta(q_a, \lambda) = \{q_a'\}$
3. $\delta(q_a', a) = \{q_0\}$ 6. $\Delta(q_b, \lambda) = \{q_b'\}$

From state q_0 automaton A tries to read matching symbol pairs one symbol from each end of the input. The transitions 1 and 2 allow A to store the currently read input letter in the finite control in order to search for a corresponding mate letter, which must be at the end of the input. Then with transitions 5 through 8 the symbol at the end of the input is brought to the left, and with transitions 3 and 4 it is verified. Then the search process is repeated. Finally, with transition 9 the sole symbol c is read while A changes to the accepting state. It is straightforward to modify the construction such that the nondeterministic context-free language $\{ ww^R \mid w \in \{a, b\}^* \}$ is accepted some ir-DFA.   □

3   Computational Capacity

The definition of deterministic extended finite automata allows λ-transitions of δ as well as of Δ. They have been included for the sake of compatibility and convenience, since often constructive proofs are much more readable if λ-transitions are used. In [5,6] it was shown that as in the case of ordinary finite automata λ-moves do not increase the computational power of extended finite automata, regardless whether the underlying device is deterministic or nondeterministic.

Theorem 2. *For a nondeterministic extended finite automaton A of any type, one can construct a λ-free extended finite automaton B of the same type, such that $L(A) = L(B)$. If A is deterministic, so is B. The statements remain true if a bounded number of input transitions is allowed.*

It is worth mentioning that if the number of non-ordinary moves is bounded to a constant, then extended finite automata of any type can only accept regular languages. Moreover, in [5] it was even shown that providing finite automata with an unbounded number of circular-interchanging operations does not increase the power beyond regular languages.

Theorem 3. $\mathscr{L}(\textit{ci-DFA}) = \mathscr{L}(\textit{ci-NFA}) = REG.$

What concerns the computational power of the remaining extended finite automata, if the number of non-ordinary moves is not restricted? Let's start with input-reversal automata, since the nondeterministic automata give rise to an alternative characterization of a well known language family. In [6] it was shown that nondeterministic ir-NFAs characterize the family of linear context-free languages (cf. Figure 2).

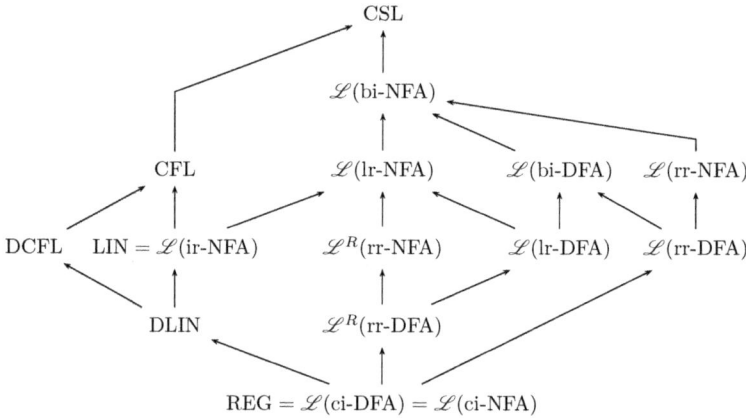

Fig. 2. Inclusion structure. All inclusions depicted are strict, and families that are not linked by a path are pairwise incomparable. CSL, CFL, LIN, DCFL, DLIN, and REG denote the families of context-sensitive, context-free, linear context-free, deterministic context-free, deterministic linear context-free, and regular languages. If X is some family of revolving automata, then $\mathscr{L}^R(X) = \{\, L^R \mid L \in \mathscr{L}(X)\,\}$.

Theorem 4. $\mathscr{L}(\textit{ir-NFA}) = LIN.$

For the language family $\mathscr{L}(\text{ir-DFA})$ it turns out that it properly includes the regular languages and is incomparable to DLIN [4]. Together with the results from [6,7] one obtains that $\mathscr{L}(\text{ir-DFA})$ is properly included in the family of linear context-free languages, that is, $\mathscr{L}(\text{ir-DFA}) \subset \mathscr{L}(\text{ir-NFA})$. In fact, for all types of extended finite automata considered the question whether deterministic machines are as powerful as nondeterministic ones is answered in similar veins [6,7].

Theorem 5. *Let* $x \in \{ir, lr, rr\}$. *Then* $\mathscr{L}(x\text{-}DFA) \subset \mathscr{L}(x\text{-}NFA)$.

Before we survey the relation between different modes in more detail, we introduce a slight generalization of extended finite automata, namely *bi-revolving* finite automata, where both left- and right-revolving steps can be performed— this is the first step towards a hybrid automata model, whose computational power will be discussed at the end of this section. The corresponding nondeterministic and deterministic families of bi-revolving automata are abbreviated by bi-NFA and bi-DFA. It will turn out that this hybrid model is somehow the most general one that one can get for the input operations in question. Then in [4,5,9] the following (strict) chains of inclusion relations were shown.

Theorem 6. *The following statements remain valid also for deterministic extended finite automata:*

1. $\mathscr{L}(\text{bi-DFA}) \subset \mathscr{L}(\text{bi-NFA}) \subset CSL$.
2. $REG = \mathscr{L}(\text{ci-NFA}) \subset \mathscr{L}(\text{rr-NFA}) \subset \mathscr{L}(\text{bi-NFA})$
3. $\mathscr{L}(\text{ir-NFA}) \subset \mathscr{L}(\text{lr-NFA}) \subset \mathscr{L}(\text{bi-NFA})$.

Among the depicted inclusions in Figure 2 also new language families of the form $\mathscr{L}^R(X) = \{ L^R \mid L \in \mathscr{L}(X) \}$ were introduced and investigated. The remaining inclusions can be found in [5]. Here it is interesting to note hat there is a certain asymmetry in the statements when reversal of language families are involved. Observe, that in [5] it was shown that the language families induced by both deterministic and nondeterministic bi-revolving automata are closed under reversal and hence $\mathscr{L}(\text{bi-DFA}) = \mathscr{L}^R(\text{bi-DFA})$ and $\mathscr{L}(\text{bi-NFA}) = \mathscr{L}^R(\text{bi-NFA})$.

Theorem 7. $REG \subset \mathscr{L}^R(\text{rr-DFA}) \subset \mathscr{L}^R(\text{rr-NFA}) \subset \mathscr{L}(\text{lr-NFA})$.

Most of these inclusions are shown by straightforward simulations, while the strictness mostly relies on involved pumping arguments. Comparing to pumping arguments on ordinary finite automata, in the case of extended finite automata one has to face the problem that although the automaton has a one-way input head, an input symbol can be read several times due to the application of a input operations. Nevertheless, by appropriately designing the witness languages one can overcome this subtle problem in most cases. We give a small example, which we literally take from [5]:

Theorem 8. *The linear context-free language $L = \{ a^n b^n \mid n \geq 0 \}$ cannot be accepted by any rr-NFA.*

Proof. Assume that L is accepted by a λ-free nondeterministic right-revolving finite automaton $A = (Q, \{a, b\}, \delta, \Delta, q_0, F)$.

The proof is done in two steps: Let $n = |Q|$. First we show that the number of ordinary steps reading a sequence of a's between two consecutive revolving moves is bounded by n. This is obvious, because otherwise one state is repeated at least once due to the pigeon hole principle. Thus, cutting this loop leads to a valid computation. Therefore, whenever the original word is accepted, also the new word induced by the cut loop is also accepted. Since after the cutting the number of a's is not equal to the number of b's on the input the automaton accepts a word not of the appropriate form. Therefore, in the forthcoming we may assume that the automaton A fulfills the above mentioned property.

Second, consider an accepting computation of the right-revolving automaton A on input $w = a^{n(n+1)+1}b^{n(n+1)+1}$. Because of the above mentioned fact, there are at least $n+1$ positions where a right-revolving move is started by reading a letter a. Because of the pigeon hole principle we find a state, say p, which appears at least twice. Thus, starting the computation in state q_0 with input w, the first appearance of state p is reached by i ordinary moves and j right-revolving moves (inter-winded), with $0 \le j < n+1$. Hence we have

$$(q_0, w) = (q_0, a^{n(n+1)+1}b^{n(n+1)+1}) \vdash_A^* (p, a^{n(n+1)+1-i-j}b^{n(n+1)+1}a^j).$$

Then from the latter configuration state p is reached a second time by k ordinary moves and ℓ right-revolving moves (inter-winded) with $1 \le \ell \le (n+1) - j$. Therefore we find

$$(p, a^{n(n+1)+1-i-j}b^{n(n+1)+1}a^j) \vdash_A^* (p, a^{n(n+1)+1-i-j-k-\ell}b^{n(n+1)+1}a^j a^\ell).$$

Since we are considering an accepting configuration, there is a state $q_f \in F$ such that

$$(p, a^{n(n+1)+1-i-j-k-\ell}b^{n(n+1)+1}a^{j+\ell}) \vdash_A^* (q_f, \lambda).$$

Observe, that $j + \ell \le (n+1)$ and $i + j + k + \ell \le n(n+1)$. Now we can fool the automaton A by constructing an accepting computation for the word

$$w' = a^{n(n+1)+1-k-\ell}b^{n(n+1)+1}a^\ell$$

by cutting out the above considered loop in the computation. For this word we have the accepting computation

$$(q_0, w') = (q_0, a^{n(n+1)+1-k-\ell}b^{n(n+1)+1}a^\ell) \vdash_A^*$$
$$(p, a^{n(n+1)+1-i-j-k-\ell}b^{n(n+1)+1}a^\ell a^j) =$$
$$(p, a^{n(n+1)+1-i-j-k-\ell}b^{n(n+1)+1}a^{j+\ell}) \vdash_A^* (q_f, \lambda)$$

of A. Since the constructed word w' is not a member of L we obtain a contradiction.   □

In order to complete the picture drawn in Figure 2 we list some of the known incomparability results. Observe, that the previous theorem already shows that there is a language accepted by an lr-NFA but not by any rr-NFA. On the other hand, in [5] it was shown that, for example, the language

$$L = \{\, a^{2n}bv \mid n \ge 0, v \in \{a,b\}^*, \text{ and } n + |v|_a = 1 + |v|_b \,\}$$

is accepted by some rr-DFA but cannot be accepted by any lr-NFA. Thus, the family $\mathscr{L}(\text{lr-NFA})$ is incomparable with $\mathscr{L}(\text{rr-NFA})$. For the language families under consideration the following more general result holds:

Theorem 9. *Let X and Y be any type of extended finite automata. Whenever the two language families $\mathscr{L}(X)$ ($\mathscr{L}^R(X)$, if $X \in \{rr\text{-}NFA, rr\text{-}DFA\}$) and $\mathscr{L}(Y)$ are not linked by a (directed) path in Figure 2, then these languages families are incomparable.*

The comparisons of the language families in question with some well-known families of the Chomsky-hierarchy read as follows [5]:

Theorem 10. *The language families $\mathscr{L}(lr\text{-}NFA)$, $\mathscr{L}(rr\text{-}NFA)$, $\mathscr{L}^R(rr\text{-}NFA)$, and $\mathscr{L}(bi\text{-}NFA)$, are incomparable with DLIN, LIN, DCFL, and CFL. A similar statement is valid for the corresponding language families induced by deterministic extended finite automata.*

Besides inclusions and incomparability results also closure and non-closure properties under standard language operations were discussed for some families of extended finite automata. It turns out that, for instance, right-revolving deterministic finite automata form a non-reversal and non-intersection closed anti-AFL, what is surprising for a language family defined by a deterministic automaton model. Although anti-AFLs are sometimes referred to an "unfortunate family of languages" there is linguistical evidence that such language families might be of crucial importance, since in [11] it was shown that the family of natural languages is an anti-AFL. Hence the question for uncommon automata models such as, for example, extended finite automata with revolving, that induce anti-AFLs seem to be worth to consider. Particularly, deterministic language families were considered [5]. Here we list some closure properties of these language families in Table 1.

Table 1. Closure properties of families of deterministic revolving automata languages; entry + means the the language family is closed under the operation under consideration, − means that it is *not* closed, and ? means that the answer is not known

$\mathscr{L}(\cdot)$	Operation								
	$\cup$	$\cap$	$\sim$	$\cap_{reg}$	R	$\cdot$	$*$	h^{-1}	h_λ
lr-DFA	−	−	−	?	−	−	−	−	−
rr-DFA	−	−	−	−	−	−	−	−	−
bi-DFA	−	−	−	?	+	−	−	−	−

Nevertheless, one can also find some partial results on nondeterministic extended finite automata with revolving operations [5,7,9]. Unfortunately, here the picture is far from being complete.

In order to gain a better understanding of the computational power of extended finite automata with different input operations, also a generalization to so-called *hybrid* extended automata was considered [9]. We have already seen an example of a hybrid machine, namely bi-revolving automata. More generally, instead of having a *single* operation the automaton may choose from a finite set $\mathcal{O}$ of operations during the computation. In [9] the set $\mathcal{O} = \{\text{ir}, \text{lr}, \text{rr}, \text{ci}, \text{cs}\}$ is considered, where the *circular-shift transition* (cs) is a generalization of left- and right-revolving defined for $a \in \Sigma \cup \{\lambda\}$ and $w \in \Sigma^*$ and $p \in \Delta(q, a)$ by $(q, a) \vdash_A (p, a)$ and $(q, aw) \vdash_A (p, vau)$, for all u and v with $w = uv$. For the dependencies of modes the following picture emerges. Note that a similar statement as Theorem 2 also applies for hybrid finite automata [9].

Theorem 11. *Let $X \subseteq \{\text{ir}, \text{lr}, \text{rr}, \text{ci}, \text{cs}\}$. Then $\mathscr{L}(X\text{-}NFA) \subset CSL$.*

Moreover, it turns out that hybrid extended finite automata without the circular-interchanging operation, but with at least *two* operations from the set $\{\text{ir}, \text{lr}, \text{rr}\}$ characterize the family of languages accepted by bi-revolving finite automata [7], which we came across already above.

Theorem 12. *For any $X \subseteq \{\text{ir}, \text{lr}, \text{rr}, \text{cs}\}$ with $|X \cap \{\text{ir}, \text{lr}, \text{rr}\}| > 1$, we have*

$$\mathscr{L}(X\text{-}NFA) = \mathscr{L}(bi\text{-}NFA).$$

Observe, that a bi-revolving nondeterministic finite automaton is an $\{\text{lr}, \text{rr}\}$-NFA in the terminology of hybrid extended finite automaton. Thus, this family of languages is the most general family that can be obtained by operations from the set $\mathcal{O} \setminus \{\text{ci}\}$. What concerns the computational power of hybrid automata with the circular-interchanging operation? Although circular-interchanging finite automata characterize the family of regular languages, combining circular-interchanging with some other operation may increase the computational power of the underlying device. The following results were obtained in [9]:

Theorem 13.   *1. $\mathscr{L}(ir\text{-}NFA) = \mathscr{L}(\{ir, ci\}\text{-}NFA)$.*
2. $\mathscr{L}(rr\text{-}NFA) \subset \mathscr{L}(\{rr, ci\}\text{-}NFA)$.
3. $\mathscr{L}(cs\text{-}NFA) \subset \mathscr{L}(\{cs, ci\}\text{-}NFA)$.

The question on the exact power of hybrid finite automata which are allowed to perform, among others, the circular-interchanging operation, in particular in combination with the left-revolving operations was left open in [9]. As a side result in that paper the location of the family $\mathscr{L}(cs\text{-}NFA)$ within the hierarchy of hybrid language families was obtained.

Theorem 14. *For $x \in \{\text{lr}, \text{rr}\}$, we have $REG \subset \mathscr{L}(cs\text{-}NFA) \subset \mathscr{L}(x\text{-}NFA) \subset \mathscr{L}(\{x, cs\}\text{-}NFA)$.*

Not much is known either on the computational power of hybrid finite automata. Only a few incomparability results show up in [9] with some classes from the Chomsky hierarchy. Moreover, a wide open field of further research is that of deterministic variants of hybrid automata. It seems that here the picture of the language families induced maybe much more diverse, since some of the techniques used in [9] seem to be not applicable for deterministic devices. Nevertheless, an exact picture is still missing. Furthermore, research on extended and hybrid automata hopefully identifies other languages operations that are of interest not only for theory, but also for applications.

4 Extended Pushdown Automata

Next we consider input operations for pushdown automata and some computational models beyond. The theory of extended pushdown automata is much less developed than that for extended finite automata. In fact, only the input-reversal operation was investigated for pushdown automata and generalizations [6]. Extended pushdown automata are analogously defined as ordinary pushdown automata with the additional ability to perform certain input operations on the unread part of the input. Such a definition generalizes to iterated pushdown automata [28] that come into play later. These definitions are in perfect correspondence to the definition of extended finite automata. We give a small example literally taken from [6]:

Example 15. Let $A = (\{q_0, q_1\}, \{a, b\}, \{X, Y, Z_0\}, \delta, \Delta, q_0, Z_0, \emptyset)$ be an input-reversal pushdown automaton, where

1. $\delta(q_0, a, Z_0) = \{(q_0, Z_0 X)\}$
2. $\delta(q_0, b, Z_0) = \{(q_0, Z_0 Y)\}$
3. $\delta(q_0, a, X) = \{(q_0, XX)\}$
4. $\delta(q_0, b, X) = \{(q_0, XY)\}$
5. $\delta(q_0, a, Y) = \{(q_0, YX)\}$
6. $\delta(q_0, b, Y) = \{(q_0, YY)\}$
7. $\delta(q_1, a, X) = \{(q_1, \lambda)\}$
8. $\delta(q_1, b, Y) = \{(q_1, \lambda)\}$
9. $\delta(q_1, \lambda, Z_0) = \{(q_1, \lambda)\}$

and $\Delta(q_0) = \{q_1\}$ that accepts by empty pushdown the non-context-free language $L = \{ ww \mid w \in \{a, b\}^* \}$. This is seen as follows.

The transitions (1) through (6) allow A to store the input on the pushdown. If A decides that the middle of the input string has been reached, then the input reversal operation specified by $\Delta(q_0) = \{q_1\}$ is selected and A goes to the state q_1 and tries to match the remaining input symbols with reversed input. This is done with the transitions (7) and (8). If successful, A will empty its pushdown with transition (9) and therefore accept the input string (by empty pushdown) if and only if the guess of A was correct and the input is of the form ww.

First observe, that by a simple adaption of the proof for ordinary pushdown automata (see, for example, [22]) one can show that for nondeterministic input-reversal pushdown automata acceptance by empty pushdown is equally powerful as acceptance by final state (even with exactly the same number of input reversals). Moreover, the presented example nicely contrasts the results on extended finite automata with a constant number of revolving (or input-reversal) steps, since input-reversal pushdown automata with a *single* input reversal already accept non-context-free languages. In fact, it was shown in [6] that input-reversals on pushdown automata induce a strict hierarchy of language families.

Theorem 16. *Let k be a natural number. Then there is a language L, which is accepted by a $(k+1)$-input-reversal (deterministic) pushdown automaton, but cannot be accepted by any k-input-reversal pushdown automaton.*

The proof of the hierarchy follows closely the proof of a hierarchy on the number of pushdown flips for flip-pushdown automata, which is based on the so called "flip-pushdown input-reversal" theorem [21]. There it is shown that pushdown flips can be undone by considering a more complicated language, where certain suffixes of the original language are reversed. Hence trading pushdown flips by input reversals. A corresponding statement to the "flip-pushdown input-reversal" theorem is also valid for input-reversal pushdown automata and can be found in [6]. The statement reads as follows:

Lemma 17. *Let k be a natural number. Language L is accepted by an input-reversal pushdown automaton $A = (Q, \Sigma, \Gamma, \delta, \Delta, q_0, Z_0, \emptyset)$ with $k+1$ reversals if and only if the language*

$$L_R = \{\, w\$v^R \mid (q_0, w, Z_0) \vdash^*_{A_1} (q_1, \lambda, Z_0\gamma) \text{ with no reversals, } q_2 \in \Delta(q_1),$$
$$\text{and } (q_2, v^R, Z_0\gamma) \vdash^*_{A_1} (q_3, \lambda, \lambda) \text{ with } k \text{ reversals} \,\}$$

is accepted by an input-reversal pushdown automaton B with k reversals, where $\$$ is a new symbol not contained in Σ.

Finally the proof of the hierarchy stated in Theorem 16 runs along the following lines: Assume to the contrary, that a specific language L_{k+1} is accepted by some input-reversal pushdown automaton A with exactly k input reversals. Then applying Lemma 17 exactly k times, results in a context-free language L. Now the idea is to pump an appropriate word from the context-free language and to undo the input reversals, in order to obtain a word that must be in L_{k+1}. If the pumping is done such that no input reversal boundaries which are marked by appropriate symbols in the word are pumped, then the input reversals can be undone. By applying a generalization of Ogden's lemma, which is due to Bader and Moura [3] and incooperates excluded positions, one can succeed with this task. For the exact definition of the languages L_k and more details on the proof we refer to [6].

When we turn to an unbounded number of input-reversal operations on pushdown automata a nice and unexpected link to Khabbaz geometric hierarchy of languages [24], which has its name from the geometric series involved in the pumping lemmas for these language families, shows up. The levels of the geometric hierarchy of languages are characterized by, for example, controlled linear context-free grammars [23], context-free based finite-reversal checking-stack automata [20], or alternatively by iterated one-turn pushdown automata where the innermost pushdown is unrestricted [28]. Intuitively, for some family of languages $\mathscr{L}$, an $\mathscr{L}$-controlled linear context-free grammar consists of a linear context-free grammar G and a control language L in $\mathscr{L}$, where the terminals of L are interpreted as labels of rules of G. Then the language generated by G under L-control is the set of all terminal words that can be generated by a derivation such that the labels of the sequence of rules applied form a word in L. The control of linear context-free grammars can be iterated by starting with $\mathscr{L}$ and by taking the result of the kth step as family of control languages for the $(k + 1)$st step. For $k \geq 1$, let $\mathrm{CTRL}_k(\mathscr{L})$ refer to the kth level of this hierarchy and define $\mathrm{CTRL}_0(\mathscr{L}) = \mathscr{L}$. In this way, we obtain two hierarchies, namely $\mathrm{CTRL}_k(\mathrm{LIN})$

and $CTRL_k(CFL)$. Observe, that $CTRL_1(CFL)$ and $CTRL_1(LIN)$ are equal to the families of linear context-free and restricted indexed languages [17]. Originally, indexed languages were introduced in [1,2] as a generalization of context-free grammars. Then these base levels of this hierarchy of controlled language families are characterized by input-reversal (one-turn) pushdown automata [6].

Theorem 18. *The language L is accepted by some input-reversal (one-turn, respectively) pushdown automaton if and only if L is a linear context-free (restricted, respectively) indexed language.*

This correspondence extends further, on the basis of the previously mentioned iterated pushdown automata. Note that a language L belongs to $CTRL_k(LIN)$ ($CTRL_k(CFL)$, respectively), for $k \geq 1$, if and only if L is accepted by a k-iterated one-turn pushdown automaton (where the innermost pushdown is unrestricted, respectively) [28]. By the relation of input-reversal automata and controlled linear context-free languages we can show that a $(k+1)$-iterated one-turn pushdown automaton (where the innermost pushdown is unrestricted, respectively) can be simulated by an input-reversal k-iterated one-turn pushdown automaton (where the innermost pushdown is unrestricted, respectively) and *vice versa*, thus trading one-turn pushdown iteration by input-reversal [6].

Theorem 19. *Let k be some natural number. Then language L is accepted by some input-reversal k-iterated one-turn pushdown automaton if and only if L is accepted by some $(k+1)$-iterated one-turn pushdown automaton. The statement remains true in case the first storage is an unrestricted pushdown.*

Not much further is known for input-reversal pushdown automata, except for some computational complexity considerations on the fixed membership performed in [6]. The following completeness results were obtained, which nicely fit to known results for the complexity of fixed membership for linear context-free and context-free languages.

Theorem 20. *The following problems are complete with respect to deterministic logspace many-one reductions: Let $k \geq 1$ be some natural number. (1) The fixed membership problem for k-input-reversal one-turn pushdown languages, where the first storage is an unrestricted pushdown, is $\mathsf{LOG}(CFL)$-complete and (2) the fixed membership problem for k-input-reversal one-turn pushdown automata languages is NL-complete.*

For extended finite automata also some partial complexity results are known. In particular, the fixed membership for X-NFA, for $X \subseteq \{ir, lr, rr, ci, cs\}$ is strictly contained in $CSL = \mathsf{NSPACE}(n)$ [25] and belongs to the complexity class NP. A closer look on Figure 1 reveals that for instance the fixed membership problem for ir-NFA is complete for NL. The complexity of other types of extended finite automata is untouched yet. Research into the direction of other automata and formal language relevant problems such as, for example, general membership non-emptiness, infiniteness, universality, etc., still lack investigations. Also the whole field of extended pushdown automata seems to be worth to be considered further. The study of input-reversal pushdown automata and their iterated versions is just the first step into this direction.

5   Conclusions

We have surveyed some recent results on extended finite automata and extended pushdown automata and variants thereof. Here the extension of the underlying machine is the ability to perform operations on the unread part of the input. Mostly we have focused on the operations of reversal, left-revolving, right-revolving, and circular-interchanging. The results presented are far from being complete, but a large interesting picture of the power and limitations of these new devices has already emerged yet. Finally, it is worth mentioning that extended automata (in general) are a host of natural problems due to the freeness of the interpretation of the input transition function Δ. For instance, recently in [8] so called hairpin finite automata were investigated as a simple model for the biological process to manipulate molecules. Formally the *hairpin inversion* (hairpin loop with pointers) which reverses a substring between a pointer a is defined for $w \in \Sigma^+$ as follows

$$hi(w) = \{\, xay^R az \mid w = xayaz, \text{ for } x, y, z \in \Sigma^* \text{ and } a \in \Sigma \,\}.$$

Besides hairpin inversion, also other simple operations such as *ld* (loop with directed repeat of pointers) which deletes a substring between two occurrences of a pointer, and *dlad* (double loop with alternating direct repeat of pointers) which swaps two substrings marked by pointer-pairs, were already investigated from a purely language theoretical perspective (see, for example, [12,13,14,15,16]). Nevertheless, in combination with finite automata in the framework of extended machines, these natural operations are still untouched from a theoretical as well as a practical point of view.

References

1. Aho, A.V.: Indexed grammars—An extension of context-free grammars. J. ACM 15, 647–671 (1968)
2. Aho, A.V.: Nested stack automata. J. ACM 16, 383–406 (1969)
3. Bader, C., Moura, A.: A generalization of Ogden's lemma. J. ACM 29, 404–407 (1982)
4. Bensch, S., Bordihn, H., Holzer, M., Kutrib, M.: Deterministic input-reversal and input-revolving finite automata. In: Martín-Vide, C., Otto, F., Fernau, H. (eds.) LATA 2008. LNCS, vol. 5196, pp. 113–124. Springer, Heidelberg (2008)
5. Bensch, S., Bordihn, H., Holzer, M., Kutrib, M.: On input-revolving deterministic and nondeterministic finite automata. Inform. Comput. 207, 1140–1155 (2009)
6. Bordihn, H., Holzer, M., Kutrib, M.: Input reversals and iterated pushdown automata: A new characterization of khabbaz geometric hierarchy of languages. In: Calude, C.S., Calude, E., Dinneen, M.J. (eds.) DLT 2004. LNCS, vol. 3340, pp. 102–113. Springer, Heidelberg (2004)

7. Bordihn, H., Holzer, M., Kutrib, M.: Revolving-input finite automata. In: De Felice, C., Restivo, A. (eds.) DLT 2005. LNCS, vol. 3572, pp. 168–179. Springer, Heidelberg (2005)
8. Bordihn, H., Holzer, M., Kutrib, M.: Hairpin finite automata. In: Harju, T., Karhumäki, J., Lepistö, A. (eds.) DLT 2007. LNCS, vol. 4588, pp. 108–119. Springer, Heidelberg (2007)
9. Bordihn, H., Holzer, M., Kutrib, M.: Hybrid extended finite automata. Internat. J. Found. Comput. Sci. 18, 745–760 (2007)
10. Chomsky, N.: Formal Properties of Grammars. In: Handbook of Mathematic Psychology, vol. 2, pp. 323–418. Wiley & Sons, Chichester (1962)
11. Culy, C.: Formal properties of natural language and linguistic theories. Linguistics and Philosophy 19, 599–617 (1996)
12. Daley, M., Ibarra, O., Kari, L.: Closure properties and decision questions of some language classes under ciliate bio-operations. Theoret. Comput. Sci. 306, 19–38 (2003)
13. Daley, M., Kari, L., McQuillan, I.: Families of languages defined by ciliate bio-operations. Theoret. Comput. Sci. 320, 51–69 (2004)
14. Dassow, J., Holzer, M.: Language families defined by a ciliate bio-operation: Hierarchies and decision problems. Internat. J. Found. Comput. Sci. 16, 645–662 (2005)
15. Dassow, J., Mitrana, V., Salomaa, A.: Operations and language generating devices suggested by genome evlution. Theoret. Comput. Sci. 270, 701–738 (2002)
16. Dassow, J., Păun, G.: Remarks on operations suggested by mutations in genomes. Fund. Inform. 36, 183–200 (1998)
17. Duske, J., Parchmann, R.: Linear indexed languages. Theoret. Comput. Sci. 32, 47–60 (1984)
18. Ginsburg, S.: Algebraic and Automata-Theoretic Properties of Formal Languages. North-Holland, Amsterdam (1975)
19. Ginsburg, S., Greibach, S.A., Harrison, M.A.: One-way stack automata. J. ACM 14, 389–418 (1967)
20. Greibach, S.A.: One way finite visit automata. Theoret. Comput. Sci. 6, 175–221 (1978)
21. Holzer, M., Kutrib, M.: Flip-pushdown automata: $k + 1$ pushdown reversals are better than k. In: Baeten, J.C.M., Lenstra, J.K., Parrow, J., Woeginger, G.J. (eds.) ICALP 2003. LNCS, vol. 2719, pp. 490–501. Springer, Heidelberg (2003)
22. Hopcroft, J.E., Ullman, J.D.: Introduction to Automata Theory, Languages and Computation. Addison-Wesley, Reading (1979)
23. Khabbaz, N.A.: Control sets and linear grammars. Inform. Control 25, 206–221 (1974)
24. Khabbaz, N.A.: A geometric hierarchy of languages. J. Comput. System Sci. 8, 142–157 (1974)
25. Kuroda, S.Y.: Classes of languages and linear bounded automata. Inform. Control 7, 207–223 (1964)
26. Salomaa, A.: Formal Languages. Academic Press, New York (1973)
27. Sarkar, P.: Pushdown automaton with the ability to flip its stack. Report TR01-081,ECCC (2001)
28. Vogler, H.: Iterated linear control and iterated one-turn pushdowns. Math. Systems Theory 19, 117–133 (1986)

Weak Inclusion for XML Types⋆

Joshua Amavi, Jacques Chabin, Mirian Halfeld Ferrari, and Pierre Réty

LIFO - Université d'Orléans, B.P. 6759, 45067 Orléans cedex 2, France
{joshua.amavi,jacques.chabin,mirian,pierre.rety}@univ-orleans.fr

Abstract. Considering that the *unranked* tree languages $L(G)$ and $L(G')$ are those defined by given *non-recursive XML types* G and G', this paper proposes a *simple and intuitive* method to verify whether $L(G)$ is "approximatively" included in $L(G')$. Our approximative criterion consists in weakening the father-children relationships. Experimental results are discussed, showing the efficiency of our method in many situations.

1 Introduction

Today, XML is the *lingua franca* for data exchange on the web. To allow interoperability among systems, one usually needs to obtain partial information from another system file. In the context of tree-modeled data, this operation corresponds to the retrieval of sub-trees according to some given application requests. This retrieval may be approximative, trying to find the XML document that best fit some given constraints. The situation is more complex when the problem consists in comparing (or retrieving) XML types (or schemas) defining approximate sub-trees of the trees generated by a given XML type.

Example 1. Suppose an application where we want to replace an XML type G by a new type G' (*eg.*, a web service composition where a service replaces another, each of them being associated to its own XML message type). We want to analyse whether the XML messages supported by G' contains (in an approximate way) those supported by G. XML types are regular tree grammars where we just consider the structural part of the XML documents, disregarding data attached to leaves. Thus, to define leaves we consider rules of the form $A \rightarrow a[\epsilon]$.

Now let us suppose that both of our grammars contain the following rules: F → firstName[ϵ], L → lastName[ϵ] , T → title[ϵ], Y → year[ϵ] and C → conference[ϵ]. However, G defines a publication by using the following rule PUB → publication[$(F.L)^+.T.Y.C$]; while in G' the definition is done by the set of rules: PUB → publication[$A^*.P$]; A → authors[$F.L$] and P → paper[$T.Y.C$]. We want to know whether messages valid with respect to G can be accepted (in an approximate way) by G'. Notice that G accepts trees such as t in Figure 1 that are not valid with respect to schema G' but that represent the same kind of information G' deals with. Indeed, in G', the same information would be organised as the tree t' in Figure 1. □

⋆ Partially supported by: Codex ANR-08-DEFIS-04.

B. Bouchou-Markhoff et al. (Eds.): CIAA 2011, LNCS 6807, pp. 30–41, 2011.

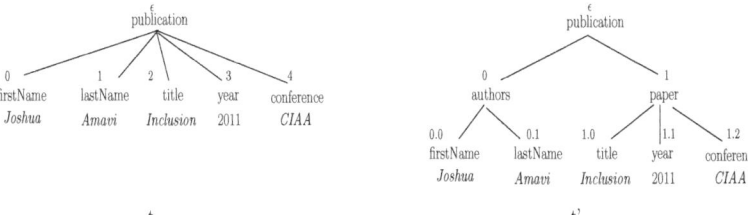

Fig. 1. Examples of trees t and t' valid with respect to G and G', respectively

The approximative criterion for comparing trees that is commonly used consists in weakening the father-children relationships (*i.e.*, they are implicitly reflected in the data tree as only ancestor-descendant). In this paper, we consider this criterion in the context of tree languages. We denote this relation *weak inclusion* to avoid confusion with the *inclusion* of languages (*i.e.*, the inclusion of a set of trees in another one).

Given two types G and G', we call $L(G)$ and $L(G')$ the set of XML documents valid with respect to G and G', respectively. Our paper proposes a method for deciding whether $L(G)$ is weakly included in $L(G')$, in order to know if the substitution of G by G' can be envisaged. The unranked-tree language $L(G)$ is weakly included in $L(G')$ if for each tree $t \in L(G)$ there is a tree $t' \in L(G')$ such that t is weakly included in t'. Intuitively, t is weakly included in t' (denoted $t \lhd t'$) if we can obtain t by removing nodes from t' (a removed node is replaced by its children, if any). For instance, in Figure 1, t can be obtained by the removal of the nodes *authors* and *paper* from t'.

To decide whether $L(G)$ is weakly included in $L(G')$, we consider the set of trees $WI(L(G')) = \{t \mid \exists t' \in L(G'), t \lhd t'\}$. Note that $L(G)$ is weakly included in $L(G')$ iff $L(G) \subseteq WI(L(G'))$.

Assuming that $L(G')$ is bounded in depth (which holds for most XML types), we propose a direct and simple approach that deals with unranked trees, using hedge grammars. The intuition of our method is to change types by allowing the deletion of XML tree levels. Roughly speaking, according to this new type, a given node in an XML tree can have as children those imposed by the original XML type or any of its descendants. With this simple idea we can compute a grammar capable of generating all the weakly included trees of a original non-recursive type G'. We prove that our algorithm is correct and complete.

Example 2. Let us consider G' from Example 1. We start from this tree grammar and use our algorithm to obtain a tree grammar which generates the language containing all the trees weakly-included in $L(G')$. The obtained grammar is:

PUB $\rightarrow$ publication$[(A \mid ((F|\epsilon).(L|\epsilon)))^*.(P|((T|\epsilon).(Y|\epsilon).(C|\epsilon)))]$

A $\rightarrow$ authors$[(F|\epsilon).(L|\epsilon)]$ P $\rightarrow$ paper$[(T|\epsilon).(Y|\epsilon).(C|\epsilon)]$

F $\rightarrow$ firstName$[\epsilon]$ L $\rightarrow$ lastName$[\epsilon]$

T $\rightarrow$ title$[\epsilon]$ Y $\rightarrow$ year$[\epsilon]$

C $\rightarrow$ conference$[\epsilon]$.

Given this new grammar G'' we can verify that $L(G)$ is included in $L(G'')$.   □

However, if $L(G')$ is not bounded in depth, computing $WI(L(G'))$ may be difficult as illustrated by the following example.

Example 3. Let G_1' be a grammar containing the rule $A \rightarrow a[B.(A|\epsilon).C]$ where non-terminals B and C generate leaves b and c respectively. In this simple case, it is easy to imagine an extension of our basic algorithm for computing $WI(G_1')$. This new grammar replaces the first rule by $A \rightarrow a[B^*.(A|\epsilon).C^*]$. However, one can take G_2' with a more complex rule such as $A \rightarrow a[B.(A \mid A.A \mid \epsilon).C]$. The solution here should be given by replacing this rule by $A \rightarrow a[(A|B|C)^*.(A|\epsilon).(A|B|C)^*]$. Notice, for instance, that in $WI(L(G_2'))$ we can have trees where nodes a, b or c appear on the left of a node labelled a while according to G_2' this was not possible. We can remark that the method needed to obtain $WI(G_2')$ is more sophisticated than the one used for $WI(G_1')$. The situation becomes worse if we suppose G_3' similar to G_2' except for the rule concerning B, which is now $B \rightarrow b[B|\epsilon]$. In this case, we should guarantee that in $WI(G_3')$ nodes labelled b will have at most one child. Thus, in $WI(G_3')$, the rule $B \rightarrow b[B|\epsilon]$ stays unchanged. This represents another special case to be treated. □

It seems difficult to define a general and simple algorithm for treating all the recursive cases. To obtain simple methods we believe that different classes of recursivity should be considered. A generic approach may need sophisticated tools.

In this paper, given *non-recursive regular tree grammars*[1] G and G', to check if $L(G)$ is weakly included in $L(G')$, we proceed according to the following steps:

1. Starting from G', we compute a grammar $WI(G')$ that generates $WI(L(G'))$.
2. Then we check whether $L(G) \subseteq WI(L(G'))$, i.e. the inclusion of regular tree languages. The runtime of this step is exponential in the worst case [18]. However, if G' satisfies some deterministic-like restrictions, we show that so does $WI(G')$ and thus the runtime of this step becomes polynomial [15,6].

Paper organisation: Section 2 gives some theoretical background. Section 3 presents how to compute $WI(G)$ for a given non-recursive grammar G, while Section 4 analyses some experimental results of our method. Section 5 considers the special case of deterministic DTDs. Due to the lack of space, missing proofs are given in [1].

Related work: Several works deal with the (weak) tree inclusion problem in the context of ordered trees: different improvements (e.g. [2,7,17]) have been presented to the initial proposal in [13]. Our proposal differs from these approaches because it considers the weak inclusion with respect to *tree languages* (and not with respect to trees only). Given a pattern query, to select the answers, [11] proposes a polynomial algorithm which verifies whether a sub-tree

---

[1] Notice that although Example 2 deals with local tree grammars (DTDs), our algorithm can be applied to any non-recursive regular tree grammar.

belongs to the language defined by the pattern and by: (i) weakening the father-children relationship and (ii) disregarding the ordering of children. Contrary to us, they do not compare XML types, and, thus, are not concerned by horizontal constraints in general. Testing precise inclusion of XML types is considered in [6,8,9,15]. In [15], the authors study the complexity of the inclusion, identifying tractable cases. In [6] we find a new polynomial algorithm for checking whether $L(A) \subseteq L(D)$, where A is an automaton for unranked trees and D is a *deterministic* DTD.

2 Preliminaries

An XML document is an unranked tree, defined in the usual way as a mapping t from a set of positions $Pos(t)$ to an alphabet Σ. Thus for $v \in Pos(t)$, $t(v)$ is the label of t at the position v, and $t|_v$ denotes the sub-tree of t at position v. Positions are sequences of integers in $\mathbb{N}^*$ and the set $Pos(t)$ satisfies: $j \geq 0, u.j \in Pos(t), 0 \leq i \leq j \Rightarrow u.i \in Pos(t)$. As usual, ϵ denotes the empty sequence of integers, i.e. the root position. In the following definition, let t, t' be unranked trees. The char "." denotes the concatenation of sequences of integers. Figure 1 illustrates trees with positions and labels: we have, for instance, $t(1) = lastName$ and $t'(1) = paper$. The sub-tree $t'|_0$ is the one whose root is *authors*.

Definition 1. Relationships on a tree: Let $p, q \in Pos(t)$. Position p is an *ancestor* of q (denoted $p < q$) if there is a non-empty sequence of integers r such that $q = p.r$. Position p is *to the left* of q (denoted $p \prec q$) if there are sequences of integers u, v, w, and $i, j \in \mathbb{N}$ such that $p = u.i.v$, $q = u.j.w$, and $i < j$. □

Definition 2. Resulting tree after node deletion: For a tree t' and a non-empty position q of t', let us note $Rem_q(t') = t$ the tree obtained after the removal of the node at position q in t' (a removed node is replaced by its children, if any). We have:

1. $t(\epsilon) = t'(\epsilon)$,
2. $\forall p \in Pos(t')$ such that $p < q$: $t(p) = t'(p)$,
3. $\forall p \in Pos(t')$ such that $p \prec q$: $t|_p = t'|_p$,
4. Let $q.0, q.1..., q.n \in Pos(t')$ be the positions of the children of position q, if q has no child, let $n = -1$. Now suppose $q = s.k$ where $s \in \mathbb{N}^*$ and $k \in \mathbb{N}$. We have:
   - $t|_{s.(k+n+i)} = t'|_{s.(k+i)}$ for all i such that $i > 0$ and $s.(k + i) \in Pos(t')$ (the siblings located to the right of q shift),
   - $t|_{s.(k+i)} = t'|_{s.k.i}$ for all i such that $0 \leq i \leq n$ (the children go up). □

Definition 3. Weak inclusion for unranked trees: The tree t is *weakly included in* t' (denoted $t \lhd t'$) if there exists a series of positions $q_1 \ldots q_n$ such that $t = Rem_{q_n}(\cdots Rem_{q_1}(t'))$. □

Example 4. In Figure 1, we have tree $t \lhd t'$. Notice that for each node of t, there is a node in t' with the same label, and this mapping preserves vertical order and left-right order. However a tree t_1 such as $publication(lastName, firstName)$ is not weakly included in t' since the left-right order is not preserved. □

Definition 4. Regular Tree Grammar: A *regular tree grammar* (RTG) (also called hedge grammar) is a 4-tuple $G = (NT, T, S, P)$, where: NT is a finite set of *non-terminal symbols*; T is a finite set of *terminal symbols*; S is a set of *start symbols*, where $S \subseteq NT$ and P is a finite set of *production rules* of the form $X \rightarrow a\,[R]$, where $X \in NT$, $a \in T$, and R is a regular expression over NT. We recall that the set of regular expressions over $NT = \{A_1, \ldots, A_n\}$ is inductively defined by: $R ::= \epsilon \mid A_i \mid R|R \mid R.R \mid R^+ \mid R^* \mid R^? \mid (R)$. □

Definition 5. Derivation: For an RTG $G = (NT, T, S, P)$, we say that a tree t built on $NT \cup T$ derives (in one step) into t' iff (i) there exists a position p of t such that $t|_p = A \in NT$ and a production rule $A \rightarrow a\,[R]$ in P, and (ii) $t' = t[p \leftarrow a(w)]$ where $w \in L(R)$ ($L(R)$ is the set of words of non-terminals generated by R). We write $t \rightarrow_{[p, A \rightarrow a\,[R]]} t'$. A derivation (in several steps) is a (possibly empty) sequence of one-step derivations. We write $t \rightarrow_G^* t'$. Let $Tree_T$ be the set of all trees that contain only terminal symbols. The language $L(G)$ generated by G is defined by: $L(G) = \{t \in Tree_T \mid \exists A \in S, A \rightarrow_G^* t\}$. □

Remark 1. As usual, in this paper, *we only consider regular tree grammars such that*: (A) every non-terminal generates at least one tree containing only terminal symbols and (B) distinct production rules have distinct left-hand-sides (*i.e.*, tree grammars in the normal form [14]). □

Remark 2. Given an RTG $G = (NT, T, S, P)$, for each $A \in NT$, there exists in P a unique rule of the form $A \rightarrow a[E]$, i.e. whose left-hand-side is A. □

Example 5. Grammar $G_0 = (NT, T, S, P_0)$, where $NT = \{X, A, B\}$, $T = \{f, a, c\}$, $S = \{X\}$, and $P_0 = \{X \rightarrow f\,[A.B], A \rightarrow a[\epsilon], B \rightarrow a[\epsilon], A \rightarrow c[\epsilon]\}$ does not respect the conditions stated in this paper since it is not in the normal form. The conversion of G_0 into normal form gives the set $P_1 = \{X \rightarrow f\,[(A|C).B], A \rightarrow a[\epsilon], B \rightarrow a[\epsilon], C \rightarrow c[\epsilon]\}$.

Among regular tree grammars we are particularly interested in local tree grammars which have the same expressive power as DTDs[2]. We recall their definition from [16]:

Definition 6. Local Tree Grammar: A *local tree grammar* (LTG) is a regular tree grammar that does not have competing non-terminals. Two non-terminals A and B (of the same grammar G) are said to be *competing with each other* if $A \neq B$ and G contains production rules of the form $A \rightarrow a[E]$ and $B \rightarrow a[E']$ (i.e. A and B generate the same terminal symbol). A *local tree language* (LTL) is a language that can be generated by at least one LTG. □

To finish this section we recall some definitions and results concerning the regular expressions that will be important for us in Section 5.

Firstly we recall that, as W3C standard, only 1-unambiguous regular expressions are allowed in DTDs. A regular expression is 1-unambiguous if every symbol

---

[2] Note that converting an LTG into normal form produces an LTG as well.

in any input string can be uniquely matched to one occurrence of the symbol in the regular expression, without looking ahead in the string. As an example, consider the regular expression $E = (A|B)^*.A.A^*$. and the word $w = BAA$ in $L(E)$. The word w can be parsed in two different ways: (i) the first and the second A in w match the first and the second A in E, respectively; (ii) the first and the second A in w match the second and the third A in E, respectively. The regular expression E is therefore *not* 1-unambiguous. We refer to [4] for a formal definition of this concept. It is also known that a regular expression E is 1-unambiguous if and only if its corresponding Glushkov automaton is deterministic [4,5,19].

Definition 7. Monadic and strict regular expression: A regular expression E is *monadic* if each non-terminal of E occurs only once in E. It is *strict* if it does not contain operators + (positive closure) nor ? (optional). A grammar is monadic (resp. strict) if all its regular expressions are monadic (resp. strict). □

The following lemma is an immediate consequence of the previous notions.

Lemma 1. *A monadic regular expression is 1-unambiguous. Consequently, a strict and monadic LTG is deterministic[3].*   □

It may happen that algorithm for testing tree language inclusion (second step of our proposal) are built by considering strict regular expressions only. In this case, recall that it is always possible to make a regular expression strict, by replacing each $E^?$ by $E|\epsilon$ and each E^+ by $E.E^*$. Unfortunately, removing operator + does not preserve monadicity. However if $\epsilon \in L(E)$ then $L(E^+) = L(E^*)$ and in this case we can just replace each + by $*$, which preserves monadicity.

3   Weak Inclusion for Regular Tree Grammars

Given a non-recursive regular tree grammar G, in this section we present how to generate a grammar G_1 such that $L(G_1) = WI(L(G))$. To do that, we introduce some definitions and results.

Definition 8. Relation $\leadsto_G$ over non-terminals: Let $G = (NT, T, S, P)$ be an RTG and A, B be non-terminals. We write $A \leadsto_G B$ if there exists a rule $A \rightarrow a[E]$ in G s.t. $B \in NT(E)$ (where $NT(E)$ denotes the set of non-terminals occurring in E). We say that $A_0, \ldots, A_n$ ($A_i \in NT$) is a chain for $\leadsto_G$ if $A_0 \leadsto_G \cdots \leadsto_G A_n$. The relation $\leadsto_G$ is *noetherian* if $\leadsto_G$ does *not* have an infinite chain $A_0 \leadsto_G \cdots \leadsto_G A_n \leadsto_G \cdots$. Grammar G is *recursive* if there exists a non-terminal A s.t. $A \leadsto_G^+ A$ (where $\leadsto_G^+$ is the transitive closure of $\leadsto_G$).   □

Lemma 2. *If G is non-recursive then $\leadsto_G$ is noetherian.*   □

To compute $WI(G)$, the idea is: for each non-terminal A that generates terminal a, either we generate a, or a is not generated and we generate its children instead. First, we extend $\leadsto_G$ to regular expressions. Moreover, to each non-terminal A, we associate a new non-terminal denoted $A^\sharp$ (called *marked non-terminal*).

---

[3] An LTG or DTD is deterministic if all its regular expressions are 1-unambiguous [4].

Definition 9. Relation $\leadsto_G$ over regular expressions: Let G be a grammar and E be a regular expression appearing in one of its production rules. Suppose that A is a non-terminal appearing at some position in E and that there is a rule $A \rightarrow a[E'']$ in G. Let E' be the regular expression defined by $E' = E[A \leftarrow A^{\sharp}|E'']$ (i.e. this occurrence of A is replaced by $A^{\sharp}|E'']$). Then we say that $E \leadsto_G E'$. $\square$

Lemma 3. *If G is non-recursive then $\leadsto_G$ (over reg. exp.) is noetherian.* $\square$

Definition 10. Substitutions in the context of $\leadsto_G$: Let G be a grammar. We define a substitution σ over non-terminals as follows. Due to the assumptions, for each non-terminal A there exists in G a unique rule whose left-hand-side is A, say $A \rightarrow a[E]$. Then $\sigma(A) = A^{\sharp}|E$. We extend σ to regular expressions: if E contains at least one non-marked non-terminal, $\sigma(E)$ is the regular expression obtained by replacing each non-marked non-terminal A in E by $\sigma(A)$. Otherwise $\sigma(E)$ is not defined. Note that $E \leadsto_G^+ \sigma(E)$ (where $\leadsto_G^+$ is the transitive closure of $\leadsto_G$). $\square$

Example 6. In grammar G' of Example 1, let us consider the rule $PUB \rightarrow publication[A^*.P]$. Let $E = A^*.P$ be its regular expression. Then, according to Definition 10, we have $\sigma(E) = (A^{\sharp} | (F.L))^*.(P^{\sharp} | T.Y.C)$. $\square$

In the following definition we present an algorithm to produce grammar $WI(G)$ for a given grammar G. By σ^n we denote n successive applications of σ, i.e. $\sigma^n = \sigma \circ \cdots \circ \sigma$ (n times).

Definition 11. Algorithm for computing $WI(G)$: Let G be a non-recursive grammar. As $\leadsto_G$ and $\leadsto_G^+$ are noetherian, for any regular expression E, there exists $n \in \mathbb{N}$ s.t. $\sigma^n(E)$ is defined and $\sigma^{n+1}(E)$ is not, which means that $\sigma^n(E)$ contains only marked non-terminals. We define $E{\uparrow} = \sigma^n(E)$. The grammar $G{\uparrow}$ is the one obtained from G by replacing each regular expression E in G by $E{\uparrow}$. $\square$

Example 2 shows the resulting grammar after applying Definition 11. Notice that the marks inserted by our algorithm are just to follow substitutions already done. The resulting grammar is one where every non terminal is marked, *i.e.*, all substitutions have been applied. We can then rewrite the grammar as usual, disregarding the marks used during the algorithm processing. This is why, when talking about $WI(G)$ we do not consider the marks anymore.

Theorem 1. *Given a non-recursive grammar G, we have $L(G{\uparrow}) = WI(L(G))$ (with common roots).* $\square$

4   Experimental Results

Given a grammar G', the computation of $WI(G')$ (Definition 11) considers each non-terminal of each production rule. Our implementation avoids repeating computation (which may lead to an exponential blow-up in the worst case) by computing each $A{\uparrow}$ only once. Thus, supposing that G' has n non-terminals (and thus n production rules), the computation of $WI(G')$ can be seen as the traversal of a graph having n nodes and $n \times l$ edges (where l is the max. length of reg. exp.). Notice that $n \times l$ equals the *number of non-terminal occurrences*, denoted by $|G'|$, the size of G'. Thus, the complexity of our algorithm is $O(n + |G'|)$.

Our prototype is implemented in Java and our experiments are done on an Intel Dual Core T2390 with 1.86GHz and 2GB of memory. The first phase of our tests concerns the generation of $WI(G')$. Results shown in Figure 2 correspond to 400 synthetic DTDs whose size ranges from 50 to 10000 non-terminal (NT) occurrences. These experiments concern DTDs with simple regular expressions composed by the concatenation of $A_1 \ldots A_n$; where we vary the number n of non-terminals, allowing as maximal value $n = 9$. Notice that our algorithm does not exceed $100ms$ for DTDs having less than 10000 NT-occurrences. We have also considered 10 real DTDs having about 50 NT-occurrences. The execution time was approximately $10ms$.

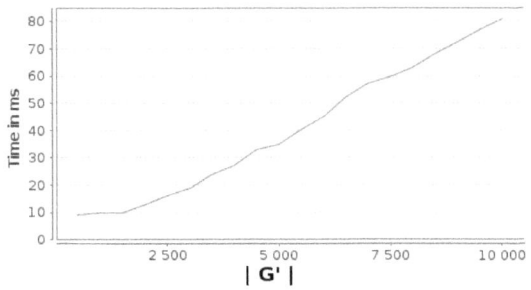

Fig. 2. Runtime for computing $WI(G')$ for grammar G'

We have run a hundred complete tests and Table 1 shows the results for 21 of them. Here we have considered more complex DTDs with $\star$, $+$, $?$, $|$ and imbrications. In this case, most regular expressions are of the form $E = E_1.E_2.E_3$ where each E_i is a disjunction involving one or more Kleene or positive closure. The DTDs are deterministic or non-deterministic. When a DTD is non-deterministic, some E_i of E are of the form $(A_j.A_{j+1})|(A_j.A_{j+2})$ or $(A_j|(A_{j+3}|A_{j+4}))^+.(A_{j+2}|(A_{j+3}|A_{j+4}))^\star$. Results on lines 1 to 9 concern synthetic non-deterministic DTDs, while those on lines 10 to 18 correspond to synthetic deterministic DTDs. On lines 19 to 21 we deal with deterministic real DTDs.

The second phase of our tests analyses the performance of the other steps of our method. Given a grammar G, to decide whether $L(G) \subseteq L(WI(G'))$, we have implemented the algorithm presented in [3]. Although the complexity of this method is exponential, the authors show that it allows very important performance improvement. Table 1 summarizes our results. Notice that, as the algorithm in [3] is proposed for ranked trees, to apply this method, we convert $WI(G')$ and G into binary grammars $bin(WI(G'))$ and $bin(G)$, respectively.

This conversion gives us grammars having more rules than their unranked counterpart. Given a grammar G, the production rules of $bin(G)$ are generated by considering each regular expression of each rule in G. The number of rules also depends on the format of the regular expressions (*eg.*, the presence of the Kleene closure). For $WI(G')$ this augmentation can be very important since in this grammar regular expressions are more complex than those in G'.

Table 1. Runtime in seconds for Phase1 (computing $WI(G')$) and Phase2 (converting unranked grammars $WI(G')$ and G to their binary counterpart and testing if $L(bin(G)) \subseteq L(bin(WI(G')))$). Result is the boolean value for the inclusion test.

	Unranked grammars					Ranked grammars		Runtime		Result
	$|G|$	$|G'|$	$\|WI(G')\|$	#Rules G	#Rules G'	#Rules bin(G)	#Rules bin(WI(G'))	Phase1 (s)	Phase2 (s)	T/F
1	32	52	123	25	40	113	5622	0	73	T
2	37	68	167	29	50	82	6420	0	139	T
3	42	98	233	33	77	93	19107	0	350	F
4	98	68	167	77	50	314	6420	0	354	F
5	86	98	233	65	77	249	19107	0	918	F
6	19	98	233	14	77	72	19017	0	14	F
7	42	86	222	33	65	93	22762	0	1455	T
8	52	98	233	43	77	168	19107	0	1890	T
9	68	86	222	50	65	200	22762	0	1729	F
10	10	62	125	9	53	30	5728	0	2	T
11	33	62	125	28	53	96	5728	0	61	T
12	42	78	183	34	62	174	7483	0	278	F
13	62	96	249	53	78	166	21808	0	522	F
14	47	96	249	40	78	210	21808	0	90	F
15	42	96	249	34	78	174	21808	0	110	F
16	20	90	224	18	74	22	11299	0	8	F
17	27	96	249	24	78	148	21808	0	18	F
18	48	96	249	40	78	167	21808	0	3217	T
19	31	31	86	25	25	35	3625	0	114	T
20	32	32	68	14	14	190	2254	0	36	T
21	32	31	86	14	25	190	3625	0	1	F

As expected, the first phase is much more faster than the second. In order to have tractable tests in Phase 2, we have chosen small examples having thus insignificant (0s) time for Phase 1 (see also Figure 2). In general, the execution time of Phase 2 is higher when the inclusion is true. However, when languages are very similar, Phase 2 can take a lot of time even for non-included languages (as in line 5, 9). On the contrary, for very different languages the inclusion test is very fast (as in lines 6, 16, 17 and 21). It is interesting to consider the case on line 18 which takes about 2-times longer than for any other examples. Notice that we have DTD with more than 90 non-terminal occurrences, and a positive result for the inclusion test. Indeed, DTD G corresponds to a subset of the rules of DTD G'. To achieve some improvement on Phase 2, we may envisage to apply techniques presented in [15] to find regular expressions for which inclusion verification is tractable or to restrict ourselves to the use of deterministic DTDs which allow us to use a polynomial time algorithm for testing language inclusion. The latter option (that we intend to implement) is discussed in the following section.

5   The Special Case of Deterministic DTDs

We finally discuss a restricted situation where the weak inclusion between XML types can be computed in polynomial time. We first define $Succ(A)$ as the set of non-terminals obtained from A by applying rules of the grammar G (including A itself). Then we consider LTGs respecting some constraints.

Definition 12. Set of successive non terminals: Let $G = (NT, T, S, P)$ be an LTG and $\leadsto_G$ the relation introduced in Definition 8. For any $A \in NT$ we define $Succ(A) = \{B \in NT \mid A \leadsto_G^* B\}$ where $\leadsto_G^*$ is the reflexive-transitive closure of $\leadsto_G$. □

Theorem 2. *Let $G = (NT, T, S, P)$ be a non-recursive monadic LTG such that*

$$\forall C \to c[E] \in P, \ \forall A, B \in NT(E), \ (A \neq B \implies Succ(A) \cap Succ(B) = \emptyset)$$

Then $G\uparrow$ is a monadic LTG. □

The following example illustrates the need of the condition imposed on non-terminals by Theorem 2. It also introduces the idea that by renaming common terminals and non-terminals one can adapt a given grammar to the condition imposed by Theorem 2.

Example 7. Consider a non-recursive monadic LTG G having the following rules:
$$R \to root[PROF^*.STUD^*] \quad PROF \to professor[F.L] \quad STUD \to stud[F.L]$$
$$F \to firstName[\epsilon] \quad\quad\quad L \to lastName[\epsilon]$$
and not respecting the condition in Theorem 2. The resulting $G\uparrow$ computed by our algorithm (Definition 11) has a production rule $R \to root[E]$ where $E = (PROF \mid ((F|\epsilon).(L|\epsilon)))^*.(STUD \mid ((F|\epsilon).(L|\epsilon)))^*$. Clearly the regular expression E is not 1-unambiguous and thus the LTG $G\uparrow$ is not deterministic □

Now we consider how to compute the weak inclusion of the language generated by a grammar G into the language generated by a grammar G', when G' is a non-recursive monadic (and maybe non-strict) LTG that respects the condition of Theorem 2. Indeed, to decide whether $L(G)$ is weakly included in $L(G')$, we compute $G'\uparrow$, which is also a monadic LTG (Theorem 2). Clearly, $G'\uparrow$ may be non-strict. However, it is interesting to remark that the construction of $G'\uparrow$ (Definition 11) gives us a grammar where each non terminal of a regular expression in G' can be replaced by ϵ. Indeed, let $E = A_1 \circ A_2 \circ \cdots \circ A_n$ be a part of a regular expression, composed of non-terminals A_i (where $\circ$ is any allowed operator). Each step of our algorithm consists in changing $E = A_1 \circ A_2 \circ \cdots \circ A_n$ into a new regular expression $E' = (A_1 \mid E_1) \circ (A_2 \mid E_2) \circ \cdots \circ (A_n \mid E_n)$ where each E_i is a regular expression in G' (see Definition 11). Then E' is modified by replacing each non terminal B_{i_j} in each expression E_i by $B_{i_j}|E_{i_j}$ and so on, until reaching some $E_{i_{j\ldots k}} = \epsilon$. It follows that all resulting regular expression have the form $E'' = A_1 \mid (B_{1_1} \mid (\cdots \mid \epsilon)) \circ \cdots \circ A_n \mid (B_{n_1} \mid (\cdots \mid \epsilon))$. In other words, $\epsilon \in L(E'')$. As explained at the end of Section 2, for a given regular expression E, when $\epsilon \in L(E)$ we have that $L(E^+) = L(E^*)$ and thus we can replace each $+$ by $*$. Based on all these points one can easily see that the obtained $G'\!\uparrow$ can be transformed into a strict grammar G'_1 by transforming operator ? and by replacing $+$ by $*$. As the LTG G'_1 is strict and monadic, it is also deterministic. Now, to decide whether the language $L(G)$ is *weakly included* into the language $L(G')$, we just need to check whether $L(G) \subseteq L(G'_1)$. Since $L(G'_1)$ is generated by a deterministic LTG, which is equivalent to a deterministic DTD, this can be done in polynomial time by using the method presented in [6].

6   Conclusion

The main contribution of this paper is a simple algorithm for computing the weak inclusion between two non-recursive XML types. It extends the weak inclusion notion, normally used for trees, to tree languages. Our approach is composed of two steps: the generation of $WI(G')$, which is linear; and precise language inclusion testing, exponential for non-recursive tree grammars (but polynomial for deterministic DTDs). Our tests show a good performance for practical cases. Weak inclusion is important for comparing types by relaxing father-children relationship and can be useful in applications such as the substitution of a web service in a composition.

To process recursive tree grammars, we envisage two directions: by defining restricted classes of recursive grammars, and trying to keep simple the generation of $WI(G')$; or by translating unranked trees into binary trees and using a complex machinery. Another idea could consist in translating the initial regular tree grammars G and G' into context-free word grammars $word(G)$ and $word(G')$ that generate the corresponding XML texts. We refer to [12,10] as examples of the translation of a DTD or a tree automaton to a context-free word grammar. By using similar techniques it is possible to compute $WI(word(G'))$. Unfortunately, checking that $L(word(G)) \subseteq L(WI(word(G')))$ (phase 2) is undecidable since it amounts to check inclusion between context-free languages.

References

1. Amavi, J., Chabin, J., Halfeld Ferrari, M., Réty, P.: Weak Inclusion for XML Types (full version). Tech. Rep. RR-2011-07, LIFO, Université d'Orléans (2011), http://www.univ-orleans.fr/lifo/prodsci/rapports/RR/RR2011/RR-2011-07.pdf
2. Bille, P., Li Gørtz, I.: The tree inclusion problem: In optimal space and faster. In: Caires, L., Italiano, G.F., Monteiro, L., Palamidessi, C., Yung, M. (eds.) ICALP 2005. LNCS, vol. 3580, pp. 66–77. Springer, Heidelberg (2005)
3. Bouajjani, A., Habermehl, P., Holík, L., Touili, T., Vojnar, T.: Antichain-based universality and inclusion testing over nondeterministic finite tree automata. In: Ibarra, O.H., Ravikumar, B. (eds.) CIAA 2008. LNCS, vol. 5148, pp. 57–67. Springer, Heidelberg (2008)

4. Brüggeman-Klein, A., Wood, D.: One-unambiguous regular languages. Information and Computation 142(2), 182–206 (1998)
5. Caron, P., Ziadi, D.: Characterization of Glushkov automata. Theor. Comput. Sci (TCS) 233(1-2), 75–90 (2000)
6. Champavère, J., Gilleron, R., Lemay, A., Niehren, J.: Efficient Inclusion Checking for Deterministic Tree Automata and DTDs. In: Martín-Vide, C., Otto, F., Fernau, H. (eds.) LATA 2008. LNCS, vol. 5196, pp. 184–195. Springer, Heidelberg (2008)
7. Chen, Y., Shi, Y., Chen, Y.: Tree inclusion algorithm, signatures and evaluation of path-oriented queries. In: Symposium on Applied Computing, pp. 1020–1025 (2006)
8. Colazzo, D., Ghelli, G., Pardini, L., Sartiani, C.: Linear Inclusion for XML Regular Expression Types. In: Proceedings of the 18th ACM Conference on Information and Knowledge Management, CIKM, pp. 137–146. ACM Digital Library (2009)
9. Colazzo, D., Ghelli, G., Sartiani, C.: Efficient Asymmetric Inclusion between Regular Expression Types. In: Proceeding of International Conference of Database Theory, ICDT, pp. 174–182. ACM Digital Library (2009)
10. Fujiyoshi, A.: Combination of context-free grammars and tree automata for unranked and ranked trees. In: Ibarra, O.H., Ravikumar, B. (eds.) CIAA 2008. LNCS, vol. 5148, pp. 283–285. Springer, Heidelberg (2008)
11. Götz, M., Koch, C., Martens, W.: Efficient algorithms for descendant-only tree pattern queries. Inf. Syst. 34(7), 602–623 (2009)
12. Hopcroft, J.E., Motwani, R., Ullman, J.D.: Introduction to Automata Theory Languages and Computation, 2nd edn. Addison-Wesley Publishing Company, Reading (2001)
13. Kilpeläinen, P., Mannila, H.: Ordered and unordered tree inclusion. SIAM J. Comput. 24(2), 340–356 (1995)
14. Mani, M., Lee, D.: XML to Relational Conversion Using Theory of Regular Tree Grammars. In: Bressan, S., Chaudhri, A.B., Li Lee, M., Yu, J.X., Lacroix, Z. (eds.) CAiSE 2002 and VLDB 2002. LNCS, vol. 2590, pp. 81–103. Springer, Heidelberg (2003)
15. Martens, W., Neven, F., Schwentick, T.: Complexity of decision problems for simple regular expressions. In: Int. Symp. Mathematical Foundations of Computer Science, MFCS, pp. 889–900 (2004)
16. Murata, M., Lee, D., Mani, M., Kawaguchi, K.: Taxonomy of XML schema languages using formal language theory. ACM Trans. Inter. Tech. 5(4), 660–704 (2005)
17. Richter, T.: A new algorithm for the ordered tree inclusion problem. In: Hein, J., Apostolico, A. (eds.) CPM 1997. LNCS, vol. 1264, pp. 150–166. Springer, Heidelberg (1997)
18. Seidl, H.: Deciding equivalence of finite tree automata. SIAM J. Comput. 19, 424–437 (1990)
19. Ziadi, D., Ponty, J.L., Champarnaud, J.: Passage d'une expression rationnelle un automate fini non-deterministe. Bull. Belg. Math. Soc. 4, 177–203 (1997)

Categorial Grammars with Iterated Types form a Strict Hierarchy of k-Valued Languages

Denis Béchet[1], Alexandre Dikovsky[1], and Annie Foret[2]

[1] LINA UMR CNRS 6241, Université de Nantes, France
Denis.Bechet@univ-nantes.fr,
Alexandre.Dikovsky@univ-nantes.fr
[2] IRISA, Université de Rennes 1, France
Annie.Foret@irisa.fr

Abstract. The notion of k-valued categorial grammars where a word is associated to at most k types is often used in the field of lexicalized grammars as a fruitful constraint for obtaining several properties like the existence of learning algorithms. This principle is relevant only when the classes of k-valued grammars correspond to a real hierarchy of languages. Such a property had been shown earlier for classical categorial grammars.

This paper establishes the relevance of this notion when categorial grammars are enriched with iterated types.

1 Introduction

The field of natural language processing includes lexicalized grammars such as classical categorial grammars [1], the different variants of Lambek calculus [11], lexicalized tree adjoining grammars [8], etc. In these lexicalized formalisms, a k-valued grammar associates at most k categories to each word of the lexicon. For a particular model of lexicalized grammars and their corresponding languages, this definition forms a (strict) hierarchy of classes of grammars when k increases. To this hierarchy of grammars, it corresponds a growing list of classes of languages that does not necessarily form a strict hierarchy.

In fact, in the field of lexicalized grammars, the concept of k-valued grammars is often used to define sub-classes of grammars and languages that satisfy some property when the whole class does not satisfy it. In particular, this notion is important for a lot of learnability results in Gold's model [7].

In the paper, we prove that the extension of classical categorial grammars with iterated types *AB form strict hierarchies of classes of languages. Since Categorial Dependency Grammars [6,3,5] use a very similar mechanism, the result is also extended to these classes of grammars. The results give a direct justification of the notion of k-valued grammars for such systems.

The paper is organized as follows. Section 2 gives some background knowledge on categorial grammars and on iterated types. Section 3 focuses on parsing or deduction structures (the two notions are closely related for type-logical or categorial grammars). Section 4 presents the proof that the class of k-valued categorial grammars with iteration form a strict hierarchy. Section 5 considers some variants. Section 6 concludes.

B. Bouchou-Markhoff et al. (Eds.): CIAA 2011, LNCS 6807, pp. 42–52, 2011.
© Springer-Verlag Berlin Heidelberg 2011

2   Background

2.1   Categorial Grammars

In categorial grammars, when a word w_1 has a type of the form $B \setminus A$, this means that w_1 can be concatenated on its left with a word of type B, so as to produce a group of words of type A. Similarly a type of the form A / B, expresses a possible concatenation on the right with a word of type B. This concatenation principle extends to groups of words. See Example 1.

Definition 1 (Types). *The types Tp, or formulas, are generated from a set of primitive types Pr, or atomic formulas, by two binary connectives[1] " / " (over) and "\ " (under):* $Tp ::= Pr \mid Tp \setminus Tp \mid Tp / Tp$

Definition 2 (Rigid and k-valued categorial grammars). *A categorial grammar is a structure $G = (\Sigma, \lambda, S)$ where:*
- *Σ is a finite alphabet (the words in the sentences);*
- *$\lambda : \Sigma \mapsto \mathcal{P}^f(Tp)$ is a function (called a lexicon) that maps a finite set of types to each element of Σ (the possible categories of each word);*
- *$S \in Pr$ is the main type associated to correct sentences.*

If $X \in \lambda(a)$, we say that G associates X to a and we write $G : a \mapsto X$.

A k-valued categorial grammar is a categorial grammar where, for every word $a \in \Sigma$, $\lambda(a)$ has at most k elements. A rigid categorial grammar is a 1-valued categorial grammar.

Definition 3 (Language). *Given a type calculus, based on a derivation relation $\vdash$ on Types, a sentence $v_1...v_n$ belongs to the language of G, written $L(G)$, provided its words v_i can be assigned types X_i whose sequence $X_1...X_n$ derives S according to $\vdash$.*

2.2   *AB Calculus

Categorial grammars usually express optional and repeatable arguments by a recursive mechanism. Here, we present a different approach that uses an extension of atomic formulas. With *AB Calculus, an atomic formula can be either a primitive type $x \in Pr$ or the iteration of a primitive type written $x^*, x \in Pr$. This extension lets naturally express optional repeatable dependencies. The calculus is very similar except that an iterated primitive type can be used zero, one or several times.

Categorial Dependency Grammars [6,3,5] use a very similar mechanism. However, in this case, types are of order one (flat)[2] , but a complex system of polarities produces non projective dependencies. Thus, CDG is not a conservative extension of *AB Calculus and the reverse does not hold either.

The iterated types originate from one of the basic principles of dependency syntax, which concerns optional repeatable dependencies (cf. [12]): all modifiers

---

[1] No product connective is used in the paper.
[2] The order o is null on primitive types s.t. $o(X/Y) = o(Y \setminus X) = max(o(X), 1 + o(Y))$.

of a noun share the noun as their governor and, similarly, all circonstants of
a verb share the verb as their governor. At the same time, the iterated dependencies are a challenge for grammatical inference [2]. For example, as in
Example 3, a repeatable circumstancial dependency A may be determined by
the type $[N \backslash S / A^*]$ attached to an intransitive verb, instead of several types :
$[N \backslash S]$, $[N \backslash S / A]$, $[N \backslash S / A / A]$ ...

Definition 4 (Types). *The types Tp, or formulas, are generated from a set of*
primitive types Pr, or iteration of primitive types $Pr^* = \{x^*, x \in Pr\}$ by two
binary connectives "/ " (over) and "\ " (under):
$$Tp ::= Pr \mid Pr^* \mid Tp \backslash Tp \mid Tp / Tp$$

The elimination rules are as follows :

$$
\begin{array}{ll}
X / Y, Y \vdash X & (\mathbf{L^r}) \\
X / y^*, y \vdash X / y^* & (\mathbf{L^{r*}}) \\
X / y^* \vdash X & (\mathbf{\Omega^r})
\end{array}
\qquad
\begin{array}{ll}
Y, Y \backslash X \vdash X & (\mathbf{L^l}) \\
y, y^* \backslash X \vdash y^* \backslash X & (\mathbf{L^{l*}}) \\
y^* \backslash X \vdash X & (\mathbf{\Omega^l})
\end{array}
$$

Remark. The AB Calculus (without iteration) derivation relation is defined by
the two rules L^r and L^l. AB grammars are equivalent to Context-free grammars.
In more details, to each ϵ-free Context-Free Grammar G in *Greibach Normal*
Form, we can associate $cg_{AB}(G)$, whose alphabet consists in the terminals of G,
whose primitive types are the non terminals of G, with the following lexicon :
    $a \mapsto ((\dots (X/X_n)/X_{n-1} \dots)/X_1)$ for each rule $X \to aX_1 \dots X_{n-1}X_n$ in G ;
G and $cg_{AB}(G)$ have the same language ($cg_{AB}(G)$ is of order 1). For the converse direction, to each AB grammar G, we associate $cf(G)$ with the same
language, having the alphabet of G as terminals, the set $Tp(G)$ of *subformu-*
las of types of G as non-terminals, with rules $\{B \to A\ A \backslash B \mid A \backslash B \in Tp(G)\}$
$\cup \{B \to B/A\ A \mid B/A \in Tp(G)\} \cup \{A \to c \mid c \mapsto A \in G\}$. These equivalences are
said *weak*, because they concern string languages, not structures.

Definition 5 (Head and arguments). *Any type X can be written in the fol-*
lowing form: $((p|A_1)|\dots|A_n)$ where $A|B$ stands for A/B or $B \backslash A$ and p has no
binary operator ; p is the head *of X, each subtype $((p|A_1)|\dots A_k)$ is a* head subtype
of X, n is the arity *of X, and each A_i is said an* argument subtype *of X.*

Example 1. Let $\lambda(John) = \lambda(Mary) = N$, $\lambda(loves) = [N \backslash S / N]$: *John loves*
Mary belongs to the language (for AB or *AB). See also Example 3 for iteration.

CDG. The *AB calculus on flat types (order 1) is the basis of Categorial Dependency grammars (CDG) used for natural language. In fact CDG involve more
complex types, we only give their supplementary rule D^l that handles distant
dependencies (rule D^r is similar on the right): $\mathbf{D^l}$. $\alpha^{P_1(\swarrow C)P(\nwarrow C)P_2} \vdash \alpha^{P_1 P P_2}$,
if the potential $(\swarrow C)P(\nwarrow C)$ satisfies the following pairing rule **FA** (*first avail-*
able): **FA :** P has no occurrences of $\swarrow C, \nwarrow C$ (see ref [5] for full details).
    The relation of CDG to automata is explained below.

2.3   Abstract Automata Equivalent to CDG

There is a class of simple abstract automata equivalent to CDG [10] (in Russian). Intuitively, these are automata with one stack and several completely independent counters. In fact, each polarized valency of a CDG corresponds to one independent counter.

Definition 6. *A* `real-time pushdown independent counters automaton` *$(RtPiCA^{(k)}, k \geq 0)$ is a system $A = (W, \Gamma, Q, q_0, k, I)$, where: W is the set of input symbols (words), Γ is the set of stack symbols containing a special symbol $\perp \in \Gamma$ (bottom), Q is a set of states, $q_0 \in Q$ is the start state, $k \geq 0$, and I is a set of instructions of the form*

$$i = (aqz \rightarrow q'\alpha v)$$

in which: $a \in W$, $q, q' \in Q$, $z \in \Gamma$, $\alpha \in \Gamma^*$ and v is an integer vector of length k (empty if $k = 0$), i.e. $v \in \mathbb{Z}^k$ if $k > 0$. k is the number of counters.

Computations of $RtPiCA^{(k)}$ are defined in terms of the following transition system over configurations. A `configuration` is a tuple (q, w, γ, V), where $w \in W^*$ (non read part of input string), $q \in Q$ (current state), $\gamma \in \Gamma^*$ (stack contents) and $V \in \mathbb{Z}^k$ (current counters' values).

A computation step is the following transition relation:

$$< q, s, \gamma, V > \vdash_A^i < q', s', \gamma', V' >,$$

where: 1) $s = as'$;
2) $\gamma = z\gamma''$, $\gamma' = \alpha\gamma''$ γ, γ' have non-negative components ;
3) $V' = V + v$ for the instruction $i = (aqz \rightarrow q'\alpha v) \in I$.
$\vdash_A^*$ is the reflexive-transitive closure of $\vdash_A^i$.

A string $s \in W^*$ is recognized by the automaton A if $< q_0, s, \perp, (0, \ldots, 0) > \vdash_A^* < q, \varepsilon, \varepsilon, (0, \ldots, 0) >$ for some q. $L(A)$ (the language recognized by A) is the set of all strings recognized by A.

Example 2. The language $L = \{ w_1^n w_2^n w_3^n \mid n = 0, 1, \ldots \}$ is recognized by the automaton $A = (W, \Gamma, Q, q_0, k, I)$ in which: $W = \{ w_1, w_2, w_3 \}$, $Q = \{ q_0, q_1, q_2 \}$, $\Gamma = \{ z_0, w_1, w_2, w_3 \}$, $k = 1$ and the set of instructions I is as follows:

$$
\begin{array}{ll}
w_1\ q_0\ \perp \ \rightarrow\ q_0\ w_1\perp\ 1 & w_1\ q_0\ w_1\ \rightarrow\ q_0\ w_1 w_1\ 1 \\
w_2\ q_0\ w_1\ \rightarrow\ q_1\ \varepsilon\ 0 & w_2\ q_1\ w_1\ \rightarrow\ q_1\ \varepsilon\ 0 \\
w_3\ q_1\ \perp \ \rightarrow\ q_2\ \perp\ -1 & w_3\ q_2\ \perp \ \rightarrow\ q_2\ \perp\ -1 \\
w_3\ q_2\ \perp \ \rightarrow\ q_2\ \varepsilon\ -1 &
\end{array}
$$

The equivalence of $RtPiCA^{(k)}$ and CDG is proved in [10].

Theorem 1. *A language L is recognized by a $RtPiCA^{(k)}$ A for some k if and only if it is generated by a CDG.*

3   Deduction Structures

In this section we focus on structures for the calculus $*AB$ (and CDG) ; in fact, these rules are extensions of the cancellation rules of classical categorial grammars that lead to the generalization of FA-structures used here.

3.1 *FA* Structures over a Set $\mathcal{E}$

We give a general definition of *FA* structures over a set $\mathcal{E}$, whereas in practice $\mathcal{E}$ is either an alphabet Σ or a set of types such as Tp.

Definition 7 (*FA* structures). *Let $\mathcal{E}$ be a set, a *FA* structure over $\mathcal{E}$ is a binary tree where each leaf is labelled by an element of $\mathcal{E}$ and each internal node is labelled by $\mathbf{L^r}$ (forward application) or $\mathbf{L^l}$ (backward application):*

$$\mathcal{F}\!A_{\mathcal{E}} ::= \mathcal{E} \mid \mathbf{L^r}(\mathcal{F}\!A_{\mathcal{E}}, \mathcal{F}\!A_{\mathcal{E}}) \mid \mathbf{L^l}(\mathcal{F}\!A_{\mathcal{E}}, \mathcal{F}\!A_{\mathcal{E}})$$

3.2 Functor-Argument Structures with Iterated Subtypes

The **functor-argument structure** and **labelled functor-argument structure** associated to a (dependency) structure proof in $*AB$ (or in CDG), are obtained as below.

Definition 8. *Let ρ be a structure proof, ending in a type t. The **labelled functor-argument structure** associated to ρ, denoted $lfa_{iter}(\rho)$, is defined by induction on the length of the proof ρ considering the last rule in ρ:*

- if ρ has no rule, then it is reduced to a type t assigned to a word w, let then $lfa_{iter}(\rho) = w$;
- if the last rule is $\mathbf{L^l}\ c^{P_1}\ [c \setminus \beta]^{P_2} \vdash [\beta]^{P_1 P_2}$, by induction let ρ_1 be a structure proof for c^{P_1} and $T_1 = lfa_{iter}(\rho_1)$; and let ρ_2 be a structure proof for $[c \setminus \beta]^{P_2}$ and $T_2 = lfa_{iter}(\rho_2)$: then $lfa_{iter}(\rho)$ is the tree with root labelled by $\mathbf{L^l}_{[c]}$ and subtrees T_1, T_2;
- if the last rule is $\Omega^l_* [c^* \setminus \beta]^{P_2} \vdash [\beta]^{P_2}$, by induction let ρ_2 be a structure proof for $[c^* \setminus \beta]^{P_2}$ and $T_2 = lfa_{iter}(\rho_2)$: then $lfa_{iter}(\rho)$ is T_2;
- if the last rule is $\mathbf{L^{l*}}\ c^{P_1}\ [c^* \setminus \beta]^{P_2} \vdash [c^* \setminus \beta]^{P_1 P_2}$, by induction let ρ_1 be a structure proof for c^{P_1} and $T_1 = lfa_{iter}(\rho_1)$ and let ρ_2 be a structure proof for $[c^* \setminus \beta]^{P_2}$ and $T_2 = lfa_{iter}(\rho_2)$: $lfa_{iter}(\rho)$ is the tree with root labelled by $\mathbf{L^l}_{[c]}$ and subtrees T_1, T_2;
- we define similarly the function lfa_{iter} when the last rule is on the right, using $/$ and $\mathbf{L^r}$ instead of $\setminus$ and $\mathbf{L^l}$;
- (in the CDG case) if the last rule is $\mathbf{D^l}$, then $lfa_{iter}(\rho)$ is taken as the image of the proof above.

The functor-argument structure $fa_{iter}(\rho)$ is obtained from $lfa_{iter}(\rho)$ (the labelled one) by erasing the labels $[c]$.

Example 3. Let $\lambda(John) = N$, $\lambda(ran) = [N \setminus S / A^*]$, $\lambda(yesterday) = \lambda(fast) = A$, then $s'_3 = \mathbf{L^l}_{[N]}(John, \mathbf{L^r}_{[A]}(\mathbf{L^r}_{[A]}(ran, fast), yesterday)$ (labelled structure) and $s_3 = \mathbf{L^l}(John, \mathbf{L^r}(\mathbf{L^r}(ran, fast), yesterday)$ are associated to ρ_1 below :

ρ_1 :

(dependency structure)

4   A Strict Hierarchy

For each $k \in \mathbb{N}$, we are interested in classes of the form $\mathcal{C}^k_{<constraint>}$ of languages corresponding to k-valued grammars with some $< constraint >$. This section proves for some $< constraint >$ that such families forms a strict hierarchy (if the lexicon has at least 2 elements):

For instance, a first very easy result when we consider the $*$AB calculus (denoted by $*$ as class constraint) is given by the fact that $\mathcal{C}^0_* \subsetneq \mathcal{C}^1_*$ because $\mathcal{C}^0_* = \emptyset$ and $\mathcal{C}^1_*$ contains the (finite) language $\{a\} = \mathcal{L}_*(G)$ for the rigid grammar $G : a \mapsto S$.

Note that the class of languages corresponding to rigid AB-grammars is a proper subset of the languages of rigid $*$AB-grammars: consider $L = \{a^+\}$ generated by $G = \{a \mapsto S \,/\, S^*\}$, which cannot be generated by a rigid AB-grammar.

4.1   Overview

We first sum up some previous work for classical categorial grammars (AB) and non-associative Lambek grammars (NL).

AB. A similar problem was solved by Kanazawa in [9] for the classes of k-valued classical categorial grammars. The proof scheme was as follows:
- Languages: for $k > 0$, $L_{AB,k} =_{def} \{a^i b a^i b a^i \mid 1 \le i \le 2k\}$
- Grammars:[3] for $k > 0$,

$$G_k = \begin{cases} a \mapsto x, \\ \quad (\cdots(S\underbrace{/x)\cdots/x)}_{i}/y)\underbrace{/x)\cdots/x)}_{i}/y)\underbrace{/x)\cdots/x)}_{i-1} \quad (1 \le i \le k) \\ b \mapsto y, \\ \quad (x\backslash(\cdots\backslash(x\backslash}_{i}(\cdots(S\underbrace{/x)\cdots/x)}_{i}/y)\underbrace{/x)\cdots/x)}_{i}\cdots)(k+1\le i\le 2k) \end{cases}$$

- The language (for AB) of G_k is $L_{AB,k}$.
- Property: for $k > 0$, $L_{AB,k}$ is a $(k+1)$-valued language but is not a k-valued language for classical categorial grammars.

NL. For Lambek non-associative calculus the proof scheme [4] is based on the previous one (for AB), but using grammars beyond order 1, $2k+1$ words and generalized AB-deductions. The proof scheme is as follows:
- Languages: for $k > 0$, $L_{NL,k} =_{def} \{abb\} \cup \{a^i b a^i b a^i \mid 1 \le i \le 2k\}$
- Grammars: $k+1$-valued grammar $G'_k = \sigma(G_k)$ where G_k is as above, with substitution $\sigma = x := (S \,/\, y) \,/\, y$.
- The language (for NL) of G'_k is $L_{NL,k}$.
- Property: for $k > 0$, $L_{NL,k}$ is a $(k+1)$-valued language but is not a k-valued language for NL.

Towards Iteration. We can easily show that the languages of grammars G_k is the same when we consider the $*$AB calculus instead of the AB rules (because G_k has not iteration). The same remark holds for grammar G'_k.

---

[3] In fact, the second type of a can be abbreviated as $S \,/\, x^i y x^i y^{i-1}$ and the second type of b can be abbreviated as $x^i \backslash (S \,/\, x^i y x^i)$.

This shows that the languages $L_{AB,k}$ are also $(k+1)$-valued languages for the $*$AB calculus. It is thus natural to ask whether they are k-valued for the $*$AB calculus as well. This is the purpose of next section.

Remark. One key point in the adaptation is that, when the language is finite ($L_{AB,k}$ is finite), an iterated argument subtype cannot be used in a proof tree for application of L^{l*} or L^{r*}.

4.2   Order 1 and Iteration

For each $k \in \mathbb{N}$, we can consider the class $\mathcal{C}^k_{*,flat}$ of languages corresponding to k-valued $*$-AB grammars with types of order at most 1. This section proves that this family forms a strict hierarchy (if the lexicon has at least 2 elements):

Theorem 2. $\forall k \in \mathbb{N}\ \ \mathcal{C}^k_{*,flat} \subsetneq \mathcal{C}^{k+1}_{*,flat}$

Before the details of proof, we introduce some definitions and remarks.

In this section, we consider the *binary deduction trees* obtained by omitting the Ω unary steps and where each node is decorated with the type that is obtained by application of the elimination rule on the immediate subtrees. These trees also correspond to the functor-argument structures previously described.

Definition 9. *We say that B is an *-context of A, when we can write:*
   $B = (G^*_{i,p'_i} \backslash ... G^*_{i,1} \backslash A / D^*_{i,1} ... / D^*_{i,p_i})$ *where the sequences of iterated types (on the left, or on the right of A) are possibly empty.*

When B is an *-context of A: if $\Delta, A, \Gamma \vdash X$ then $\Delta, B, \Gamma \vdash X$ as well (using Ω^r* and Ω^l*).

Rule patterns. We observe that each type occurring in a binary deduction tree obtained by omitting the Ω unary steps is a head subtype of some type associated to a leaf. The patterns are as follows :
(on the right - similarly on the left -)

$$\frac{(G^*_{i,p'_i} \backslash ... G^*_{i,1} \backslash A_i / C_i / D^*_{i,1} ... / D^*_{i,p_i})\ \ C'_i}{A_i}\ L^r (\textit{several } \Omega^r \textit{ and } \Omega^l)$$

$$\frac{(G^*_{i,p'_i} \backslash ... G^*_{i,1} \backslash A_i / C^*_i / D^*_{i,1} ... / D^*_{i,p_i})\ \ C'_i}{A_i}\ L^{r*} (\textit{several } \Omega^r \textit{ and } \Omega^l)$$

$$\frac{(G^*_{i,p'_i} \backslash ... G^*_{i,1} \backslash A_i / C^*_i / D^*_{i,1} ... / D^*_{i,p_i})\ \ C'_i}{A_i / C^*_i}\ L^{r*} (\textit{several } \Omega^r \textit{ and } \Omega^l)$$

where C'_i is a *-context of C_i.

Steps of proof

1. Obviously, we have $\forall k \in \mathbb{N}\ \ \mathcal{C}^k_* \subseteq \mathcal{C}^{k+1}_*$
2. For $k > 0$, we consider $L_{*,k} =_{def} \{a^i b a^i b a^i \mid 1 \le i \le 2k\}$
3. We see that $L_{*,k}$ is a $(k+1)$-valued language : because G_k is $(k+1)$-valued, without $*$ in its types, its language is as in the AB case, which is $\{a^i b a^i b a^i \mid 1 \le i \le 2k\}$ as shown in [9].

4. We prove that $L_{*,k}$ is not a k-valued language for $*$AB languages.

   Proof : suppose G is a k-valued grammar with $*$AB language $L_{*,k}$

   (a) For each element of $L_{*,k}$, there exists a *binary deduction tree* : T_i for $a^i b a^i b a^i$ $(1 \leq i \leq 2k)$

   (b) For $0 < i \leq 2k$ let A_i denote the root type of **the smallest subtree** in T_i **whose yield includes both** b. This gives two subtrees with one b with yields $a^{i_0} b a^{i_1}$ and $a^{i_2} b a^{i_3}$ $(i_1 + i_2 = i)$. Then, we consider the antecedents of A_i in T_i : C'_i and B_i such that : $B_i = (G^*_{i,p'_i} \backslash ... G^*_{i,1} \backslash A_i / C^\delta_i / D^*_{i,1} ... / D^*_{i,p_i})$ (or $B_i = (D^*_{i,p_i} \backslash ... \backslash D^*_{i,1} \backslash C^\delta_i \backslash A_i) / G^*_{i,1} ... / G^*_{i,p'_i})$) where δ is either $*$ or empty, and such that C'_i is a $*$context of C_i.

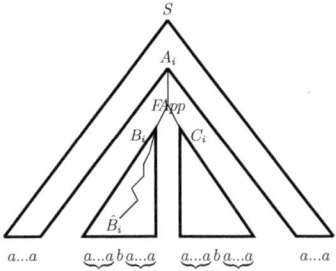

       In fact, δ cannot denote $*$, otherwise, we would get deductions involving iterations of C_i (replacing one C_i) for words with more than two b. Each B_i is thus an $*$-context of A_i / C_i or of $C_i \backslash A_i$.

       We define $\widehat{B_i}$ as the type in G "providing" B_i (following functors) in T_i.

       We define $\widehat{C'_i}$ as the type in G "providing" C'_i (following functors) in T_i.

   (c) We remark that $\forall i : B_i \neq A_i$ *and* $C'_i \neq A_i$.

   Otherwise, if $B_i = A_i$ by replacing the subtree ending in B_i (or C'_i if $C'_i = A_i$) by the subtree ending in A_i, we would get a derivation of a word with three b instead of two.

   (d) More generally : $\forall i, j : A_j$ cannot have B_i or C_i as head subtype.

   Otherwise, a subtree ending in B_i (or a $*$context of C_i) would contain the subtree ending with A_j that has two b.

   (e) We prove that: $\forall i \neq j : B_i \neq B_j$

   Let $y^i_{ce}(X_i)$ denote the center part of the yield with root X_i in T_i. (this is i_1 for the left subtree with yield $a^{i_0} b a^{i_1}$ and i_2 for the right subtree with yield $a^{i_2} b a^{i_3}$), we have $\forall i : y^i_{ce}(B_i) + y^i_{ce}(C'_i) = i$.

   - Suppose (from the contrary) (i) $B_i = B_j$, for some $i \neq j$;

   Since $i \neq j$, either $y^i_{ce}(B_i) \neq y^j_{ce}(B_j)$ or $y^i_{ce}(C'_i) \neq y^j_{ce}(C'_j)$.

   - - Suppose first (ii) $y^i_{ce}(B_i) \neq y^j_{ce}(B_j)$; from (ii) replacing in T_j, $(j \neq 0)$, B_j by B_i is a derivation of a word $w = ...ba^{j'}ba^j$ or $w = a^j ba^{j'} b...$, where $j' = y^i_{ce}(B_i) + y^j_{ce}(C'_j)$ this word w is not in $L_{*,k}$ since $j' = y^i_{ce}(B_i) + y^j_{ce}(C'_j) \neq y^j_{ce}(B_j) + y^j_{ce}(C'_j) = j$; this contradicts the assumption that G has $L_{*,k}$ as language (for $*AB$).

   - - Suppose instead (ii)' $y^i_{ce}(C'_i) \neq y^j_{ce}(C'_j)$;

   - - - if (iii) $C_i = C_j$: replacing in T_j, C'_j by C'_i yields a similar word w not in $L_{*,k}$ with $j' = y^j_{ce}(B_j) + y^i_{ce}(C'_i)$ occurrences of a between the b and $j' \neq j$, (ii)' also leads to a contradiction.

   - - - otherwise (iii) $C_i = D_{i,k}$ for some $D^*_{i,k}$ of B_i

   $B_i = (G^*_{i,p'_i} \backslash ... G^*_{i,1} \backslash A_i / C_i / D^*_{i,1} ... / D^*_{i,p_i})$ (in the right case) ;

   however in such a case, we could replace C'_i by a succession of C'_i, using

the iteration rule, producing a word with more than two b.
Therefore (i) is not possible : this means that all B_i are distinct.

(f) We prove that: $\forall i,j : \widehat{B_i} \neq \widehat{B_j}$.

We write $X|Y$ as an abbreviation for $X \,/\, Y$ or for $Y \,\backslash\, X$ (functor first).
- Suppose $\widehat{B_i} = \widehat{B_j}$. One (say B_i) is a head subtype of the other (B_j),
that is in the form: $\qquad B_j = ...(B_i|D'_1...)|D'_n$
with $B_j = (G^*_{j,p'_j} \backslash ...G^*_{j,1} \backslash (A_j\,/\,C_j)\,/\,D^*_{j,1}.../\,D^*_{j,p_j})$ (in the right case) ;

- - if B_i is a strict[4] head subtype of $A_j\,/\,C_j$, we then get A_j in a subtree
ending in B_i , which is impossible since the yield would then have three
b instead of two.

- - otherwise, B_i is a *context[5] of $A_j\,/\,C_j$ (in the right case), which entails
that $C_i = C_j$; then, replacing B_j by B_i in T_j or C'_i by C'_j in T_i gives
deduction trees: which leads to a contradiction using a reasoning similar
to that of $B_i \neq B_j$.[6]

(g) As a consequence, we get a contradiction as follows.

Let $f(i)$ denote the index s.t. $\widehat{C'_i} = \widehat{B_{f(i)}}$. By definition C_i is a head
subtype of $\widehat{C'_i}$ and $B_{f(i)}$ is a head subtype of $\widehat{B_{f(i)}}$, that is the same
type. Therefore, one of C_i and $B_{f(i)}$ is a head subtype of the other ;
because C_i is primitive and $B_{f(i)}$ is not, C_i is a head subtype of $B_{f(i)}$.
This entails that C_i is a head subtype of $A_{f(i)}$ as well, which is impossible
as shown previously.

5. Thus $\forall k > 0$ $\ C^k_{*,flat} \neq C^{k+1}_{*,flat}$ (we have also seen in the introduction to the
section that the property is also true for $k = 0$).

4.3   Order >1 and Iteration

The previous reasoning can be adapted to the *AB calculus where types are not
necessarily flat (order >1), using the same deduction rules and structures.

Theorem 3. $\qquad \forall k \in \mathbb{N}\ \ C^k_* \subsetneq C^{k+1}_*$

Sketch of proof. To this end, we use in this section the languages $L_{NL,k} =$
$\{abb\} \cup \{a^iba^iba^i \mid 1 \le i \le 2k\}$ and consider $2k+1$ proof trees instead of $2k$ in
the previous section.

- Languages: for $k > 0$, $L_{NL,k} =_{def} \{abb\} \cup \{a^iba^iba^i \mid 1 \le i \le 2k\}$
- Grammars: $k+1$-valued grammar $G'_k = \sigma(G_k)$ where G_k is as above, with
substitution $\sigma = x := (S\,/\,y)\,/\,y$. and we can show $\mathcal{L}_*(\sigma(G_k)) = L_{NL,k}$.
- Property: for $k > 0$, $L_{NL,k}$ is a $(k+1)$-valued language (using G'_k) but is not
a k-valued language (see details below) for the *AB calculus.

Details of proof. To prove that $L_{NL,k}$ is not a k valued language, we proceed as
in the previous section: we suppose the existence of a k-valued gammar G', with
language $L_{NL,k}$ and we consider a deduction tree T_i for $a^iba^iba^i$ ($1 \le i \le 2k$)

---

[4] (Not equal to).
[5] Possibly equal to.
[6] $B_i = (G^*_{j,p'_{j,q'}} \backslash ...G^*_{j,1} \backslash A_j\,/\,C_j\,/\,D^*_{j,1}.../\,D^*_{j,q})$ for some $q' \le p'_j$ and $q \le p_j$.

and $\mathcal{T}_0$ for abb. For $0 \leq i \leq 2k$, we define A_i as the root type of the smallest subtree in $\mathcal{T}_i$ with a yield including both b.

- We prove that: $\forall i \neq j : B_i \neq B_j$ (similarly to the previous subsection)
- $\forall i \neq j : \widehat{B_i} \neq \widehat{B_j}$ (details are similar to the previous subsection)
- As a consequence, we need $2k + 1$ distinct $\widehat{B_i}$.
- Contradiction: $2k + 1$ distinct $\widehat{B_i}$ are needed with a k-valued grammar with a useful lexicon of 2 words (a and b).

The advantage of this construction is to handle directly $2k + 1$ types ($2k$ in the previous one). However, a main difference is the presence of types of order 2 in the grammar.

5 Conclusion

∗AB. The paper studies variants of grammatical systems with iterated types: involving flat type (order 1) or not. We have proved that the classes of k-valued categorial grammars form a strict hierarchy of classes of languages. Thus, the notion of k-valued grammars is relevant for both systems: each $k \in \mathbb{N}$ defines a particular class of languages. The proof relies on generalized AB deductions and their corresponding functor-argument structures that enables us to define languages of structured sentences as for classical categorial grammars.

CDG. In fact, our strict hierarchy theorem also extends to categorial dependency grammars (CDG) with empty potentials, due to the following argument. A CDG-grammar G with empty potentials, has the same language, when considered as CDG-grammar or as ∗AB grammar (of order 1). Therefore the hierarchy for CDG with empty potentials cannot collapse.

Future work could concern other extensions of type logicial grammars, such as the extension of pregroups with iterated types.

References

1. Bar-Hillel, Y.: A quasi arithmetical notation for syntactic description. Language 29, 47–58 (1953)
2. Béchet, D., Dikovsky, A., Foret, A.: Two models of learning iterated dependencies. In: Proc. of the 15th Conference on Formal Grammar (FG 2010), Copenhagen, Denmark. LNCS (2010) (to appear), http://www.angl.hu-berlin.de/FG10/fg10_list_of_papers
3. Béchet, D., Dikovsky, A., Foret, A., Moreau, E.: On learning discontinuous dependencies from positive data. In: Proc. of the 9th Intern. Conf. Formal Grammar (FG 2004), Nancy, France, pp. 1–16 (2004)
4. Béchet, D., Foret, A.: k-valued non-associative lambek grammars (Without product) form a strict hierarchy of languages. In: Blache, P., Stabler, E.P., Busquets, J.V., Moot, R. (eds.) LACL 2005. LNCS (LNAI), vol. 3492, pp. 1–17. Springer, Heidelberg (2005)

5. Dekhtyar, M., Dikovsky, A.: Generalized categorial dependency grammars. In: Avron, A., Dershowitz, N., Rabinovich, A. (eds.) Pillars of Computer Science. LNCS, vol. 4800, pp. 230–255. Springer, Heidelberg (2008)
6. Dikovsky, A.: Dependencies as categories. In: Recent Advances in Dependency Grammars (COLING 2004) Workshop, pp. 90–97 (2004)
7. Gold, E.: Language identification in the limit. Information and control 10, 447–474 (1967)
8. Joshi, A.K., Shabes, Y.: Tree-adjoining grammars and lexicalized grammars. In: Tree Automata and LGS, Elsevier Science, Amsterdam (1992)
9. Kanazawa, M.: Learnable Classes of Categorial Grammars. In: Studies in Logic, Language and Information, Center for the Study of Language and Information (CSLI) and The European association for Logic, Language and Information (FOLLI), Stanford, California (1998)
10. Karlov, B.: On properties of generalized categorial dependency grammars. In: Proc. of the 12th National Conference on Artificial Intelligence, Russia, vol. 1, pp. 283–290 (2010)
11. Lambek, J.: The mathematics of sentence structure. American mathematical monthly 65 (1958)
12. Mel'čuk, I.: Dependency Syntax. SUNY Press, Albany (1988)

Bouma2 – A High-Performance Input-Aware Multiple String-Match Algorithm

Erez Buchnik

European Research Centre of Huawei Technologies Co. Ltd.,
Riesstr.25, C-3.0G, 80992 Munich, Germany
Erez.Buchnik@Huawei.com

Abstract. We present Bouma2, a new algorithm for exact multiple string-match. It is highly parallelizable, has small footprint, and can be tuned using statistics of the input stream. It uses a special hashing technique to map the pattern-set to 2-symbol sequences, allowing the match procedure to be considerably optimized. This algorithm employs a fast-path/slow-path principle at match-time, which facilitates pipelining in H/W. We also produce experimental comparative results.

Keywords: aho corasick, pattern match, hash functions, motif finding, clique partitioning, integer linear programming, fast path, slow path, dpi.

1 Introduction

We present Bouma2,[1] a new exact multiple string-match algorithm. The main idea behind this algorithm is that an optimized match procedure can compensate for multiple passes over the input stream - especially when match attempts are infrequent. The heart of the match procedure is a simple pass over the input in 2-symbol strides, in search of "hints" of matches, termed *'Motifs'* (named after *Sequence Motifs* in Computational Biology [6]). For each motif match, the input string is examined around the match location to corroborate the match. This can be compared to collision-resolving of a hash value. Bouma2 features the ability to assist the mapping of patterns to motifs at compile-time by using statistics, in order to eliminate motif false-positives, control memory footprint etc.

We show benchmark results demonstrating the superiority of Bouma2 over a basic Aho-Corasick implementation in S/W, yet we believe that the true benefit of Bouma2 may be in H/W form; inherent statelessness of the match process, small footprint, a fast-path/slow-path approach and cache-sensitivity make it a promising candidate for many optimizations that are impossible in S/W.

This paper is organized as follows: Section 2 surveys related work; Section 3 defines basic concepts and notations; Section 4 presents the compilation process as an ILP[16] problem and as a clique-partitioning problem; Section 5 describes the match structures, the match-time algorithms, and the resolving process; Section 6 provides experimental results and Section 7 describes future work.

---

[1] Bouma Shape - the outline, or contour, of a written word. Boumas were used in Cognitive Psychology for some Word-Recognition models (see [22]).

B. Bouchou-Markhoff et al. (Eds.): CIAA 2011, LNCS 6807, pp. 53–64, 2011.

2   Related Work

Bouma2 is partly inspired by motif-finding [6] as well as pairwise [7] and multiple [8] sequence alignment, adapting similar concepts to general pattern-match; our graph-based representation (Section 4.2) is a simplified variant of the one in [9]. Many automaton-based multiple pattern-match algorithms (usually based on [1], [2] or [3]) exist; most of them suffer from the state-explosion problem [4]. Hash-based pattern match is described in [5]; the main difference is that our hashed value is already part of the match. 2-symbol strides (or more) were used in [10] for multi-pattern match, but the resolving is done with Rabin-Karp [5] or similar. Using different data-structures for matching the same language has been suggested mainly for reducing memory footprint ([11], [12]), and not directly for performance. Using statistics is common for motif-finding [13], but rare for other pattern-match applications. Finally, we have no record of any other pattern-match algorithm that is stateless (like our fast-path algorithm, Section 5.2), nor of a structure similar to the *Mangled Trie* (Section 5.4).

3   Basic Definitions

An alphabet Σ is a nonempty set of symbols. A word over Σ is a finite sequence of symbols of Σ. The empty word is denoted by ϵ and the length of a word w is denoted by $|w|$. Σ^* is the set of words over Σ. A language L is a subset of Σ^*. Σ^2 is the set of all 2-symbol words. The total length of all words in a language ($\sum_{w \in L} |w|$) is denoted by $sz(L)$. *The Multiple Pattern-Match Problem* is defined as follows: given a language $L \subseteq \Sigma^*$ and a long word $W_I \in \Sigma^*$, find all occurrences of all words in L that are substrings of W_I (also refer to [14]).

3.1   Traces and Motifs

Definition 1. *Any set $T_L \subseteq \Sigma^2$ that satisfies $t \in T_L \Leftrightarrow \exists w \in L : w = w_p t w_s$ is named the Trace-Set of L. Any $t \in T_L$ is named a Trace.*

Definition 2. *The Trace-Occurrence Function $occ : L \times \Sigma^2 \times \mathbb{N} \to \{0,1\}$:*

$$occ(w, t, l) = 1 \Leftrightarrow w = w_p t w_s \wedge |w_p| = l . \tag{1}$$

Definition 3. *The functions $assoc_0, assoc_1 : L \times \Sigma^2 \to \{0,1\}$ are respectively named the Even and Odd Trace-Association Functions, and are defined as:*

$$assoc_0(w, t) = 1 \Leftrightarrow \sum_{l=0}^{\lfloor |w|/2 \rfloor} occ(w, t, 2l) > 0$$

$$assoc_1(w, t) = 1 \Leftrightarrow \sum_{l=0}^{\lfloor (|w|-1)/2 \rfloor} occ(w, t, 2l + 1) > 0 . \tag{2}$$

Definition 4. *A Motif-Set is any $M_L \subseteq T_L$ that satisfies for every $w \in L$:*

$$\sum_{t \in M_L} assoc_0(w,t) \geq 1 \ \wedge \ \sum_{t \in M_L} assoc_1(w,t) \geq 1 \ . \tag{3}$$

A Motif $\mu \in M_L$ is every trace that belongs to a motif-set.

Theorem 1. *For every L satisfying $\forall w \in L : |w| > 2$ there exists a motif-set.*

Proof. By example: consider the trace-set M_L, in which the condition in Definition 4 is inherently satisfied:

$$M_L := \bigcup_{w \in L} \{t : occ(w,t,0) = 1\} \ \cup \ \bigcup_{w \in L} \{t : occ(w,t,1) = 1\} \ . \tag{4}$$

$\square$

3.2   Resolve-Sets

Definition 5. *A Motif-Set Hash Function is any function H_{M_L} such that:*

$$H_{M_L} \ : \ L \times \{0,1\} \to M_L \ , \ \ H_{M_L} \ is \ surjective$$
$$H_{M_L}(w,i) = \mu \ \Rightarrow \ assoc_i(w,\mu) = 1 \ . \tag{5}$$

Definition 6. *For a given $H_{M_L}, \mu \in M_L$, a Motif's Resolve-Set $R_\mu \subseteq L$, is:*

$$R_\mu := \{w : H_{M_L}(w,0) = \mu \ \vee \ H_{M_L}(w,1) = \mu\} \ . \tag{6}$$

4   The Bouma2 Compilation Process

Bouma2 is a hash between the words in a language and a motif-set. The choice of motif-set and the words-to-motifs coupling is part of the compilation process. At match-time, motif occurrences in the input trigger a resolving process for words with common motifs, followed by a word-specific comparison at a fixed offset. Words map to 2 motifs, one for an even offset and one for an odd offset, so the motif search can proceed in 2-symbol strides. The selection of a suitable motif-set and efficiency of the match-time procedure are thus crucial for performance.

4.1   Cost Functions

Cost-functions for motifs are used when selecting the final motif-set. Different cost-functions serve different purposes, like improving performance, reducing memory size, or speeding up compile-time.

Definition 7. *We define a Motif Cost-Function $c : \Sigma^2 \to \mathbb{R}$, and a Maximizing Motif-Set $M_{L|c}$ solving the following Integer Linear Programming[16] problem:*

$$\text{Maximize} \ \sum_{t \in T_L} c(t)x_t \ : \ \ x_t \in \{0,1\} \ \forall t \in T_L$$
$$\text{s.t.} \ \forall w \in L : \sum_{t \in T_L} x_t \cdot assoc_0(w,t) \geq 1 \ \wedge \ \sum_{t \in T_L} x_t \cdot assoc_1(w,t) \geq 1 \ . \tag{7}$$

Example 1 (Minimizing Motif False-Positives). Cost functions allow the use of statistics gathered on the input string and language. For example, the conditional probability $P(w|t)$, i.e. the probability of the word w appearing in the input string, given that the trace t was observed, can be used as a weight:

$$c(t) = \sum_{w \in L} (assoc_0(w, t) \vee assoc_1(w, t)) \cdot P(w|t) \,. \tag{8}$$

Example 2 (Memory Cost Function). Maximizing the number of words per motif ensures smaller overall memory requirements:

$$c(t) = \sum_{w \in L} (assoc_0(w, t) + assoc_1(w, t)) \,. \tag{9}$$

Example 3 (Resolve Dimension Cost Function). Controlling the resolve dimension (see Section 5.4) can help maintain deterministic matching performance:

$$c(t) = D(\{w \in L : \; assoc_0(w, t) = 1 \; \vee \; assoc_1(w, t) = 1\}) \,. \tag{10}$$

Example 4 (Motif-Sets). Consider the following 7-word language of 8-bit characters, and its corresponding trace-set:

$$L = \{boat, book, bore, oral, cooks, core, coredump\}$$
$$T_L = \{bo, oa, at, oo, ok, or, re, ra, al, co, ks, ed, du, um, mp\} \,. \tag{11}$$

Maximizing the number of words per motif yielded $M_{L|c1}$, and applying weights based on conditional occurrence probabilities for minimizing motif false-positives yielded $M_{L|c2}$. Table 1 shows resolve-sets for the two solutions.

$$M_{L|c1} = \{bo, oa, oo, or, co, ra\}$$
$$M_{L|c2} = \{bo, oa, oo, or, co, ra, ok, al, ks, du, um\} \,. \tag{12}$$

4.2   The Bouma2 Compilation Graph

An alternative to the ILP formulation in Definition 7 is to treat the compiler as a Weighted Clique Partitioning problem [9] (see example in Figure 1):

Definition 8. *We define the Bouma2 Graph, $G_L = (V, E)$, which satisfies:*

$$V := \bigcup_{w \in L} \{v_{0|w}, v_{1|w}\}$$
$$E := \{(v_{i|w}, v_{j|w'}) : v_{i|w}, v_{j|w'} \in V \; \wedge \; i, j \in \{0, 1\} \; \wedge \; w \neq w' \; \wedge$$
$$\exists t \in T_L : assoc_i(w, t) = 1 \; \wedge \; assoc_j(w', t) = 1 \} \cup$$
$$\{(v_{0|w}, v_{1|w}) : v_{0|w}, v_{1|w} \in V \; \wedge$$
$$\exists t \in T_L : assoc_0(w, t) = 1 \; \wedge \; assoc_1(w, t) = 1 \} \,. \tag{13}$$

Theorem 2. *Every trace in L can be mapped to a maximal clique in G_L.*

Table 1. Two options of motif-sets for the same language

Min. Motif False-Positives:	Max. Words per Motif:
$R_{bo} = \{\textbf{\underline{bo}}at, \textbf{\underline{bo}}re\}$	$R_{bo} = \{\textbf{\underline{bo}}at, \textbf{\underline{bo}}re, \textbf{\underline{bo}}ok\}$
$R_{oa} = \{b\textbf{\underline{oa}}t\}$	$R_{oa} = \{b\textbf{\underline{oa}}t\}$
$R_{oo} = \{b\textbf{\underline{oo}}k\}$	$R_{oo} = \{b\textbf{\underline{oo}}k, c\textbf{\underline{oo}}ks\}$
$R_{or} = \{b\textbf{\underline{or}}e, c\textbf{\underline{or}}e\}$	$R_{or} = \{b\textbf{\underline{or}}e, c\textbf{\underline{or}}e, \textbf{\underline{or}}al, c\textbf{\underline{or}}edump\}$
$R_{co} = \{\textbf{\underline{co}}re\}$	$R_{co} = \{\textbf{\underline{co}}re, \textbf{\underline{co}}oks, \textbf{\underline{co}}redump\}$
$R_{ra} = \{o\textbf{\underline{ra}}l\}$	$R_{ra} = \{o\textbf{\underline{ra}}l\}$
$R_{ok} = \{bo\textbf{\underline{ok}}, co\textbf{\underline{ok}}s\}$	–
$R_{al} = \{or\textbf{\underline{al}}\}$	–
$R_{ks} = \{coo\textbf{\underline{ks}}\}$	–
$R_{du} = \{core\textbf{\underline{du}}mp\}$	–
$R_{um} = \{cored\textbf{\underline{um}}p\}$	–

Proof. For every trace $t \in T_L$, we identify the set $C_t \subseteq E, C_t := \{(v_{i|w}, v_{j|w'}) : assoc_i(w,t) = 1 \wedge assoc_j(w',t) = 1\}$. V_{C_t} represents[2] words having t in common (either at an even or an odd offset), and no other vertex represents such a word, so V_{C_t} is necessarily a maximal clique. □

Theorem 3. *Any clique partition of a Bouma2 graph represents a motif set.*

Proof. Given a clique partition, we denote an arbitrary clique in it by $V_{C'_t} \subseteq V_{C_t}$, where $V_{C_t} \subseteq V$ is a maximal clique representing t according to Theorem 2. We can thus specify a trace-set M_L such that for every clique $V_{C'_t}$ in the partition, $t \in M_L$. By definition, all the vertices have to be covered by the clique partition, and each vertex belongs to a single clique. Consider the 2 vertices $v_{0|w}, v_{1|w}$ corresponding to the word w. Let $v_{0|w}$ belong to $V_{C'_\mu}$, and let $v_{1|w}$ belong to $V_{C'_{\mu*}}$ in said partition. By Definition 8, $assoc_0(w, \mu) = 1$ and $assoc_1(w, \mu*) = 1$. Thus, for M_L the condition in Definition 4 is necessarily satisfied. □

Algorithm 1 BOUMA2-GRAPH-MAX-CLIQUES($L \in \Sigma^*$)

```
 1: procedure BOUMA2-GRAPH-MAX-CLIQUES(L)
 2:     for all t ∈ Σ² do
 3:         CLIQUE-REF(t)={}
 4:     end for
 5:     TRACE-SET={}
 6:     for all w ∈ L do
 7:         for all i ∈ {x :  0 ≤ x < |w|} do
 8:             TRACE-SET = TRACE-SET ∪ {t : occ(w,t,i) = 1}
 9:             CLIQUE-REF(t) = CLIQUE-REF(t) ∪ {(w,i)}
10:         end for
11:     end for
12:     return CLIQUE-REF(t) forall t ∈ TRACE-SET
13: end procedure
```

---

[2] Throughout this discussion, we adopt a notation whereas for the clique $V_C \subseteq V$, $C \subseteq E$ represents its (possibly empty) corresponding set of connecting edges.

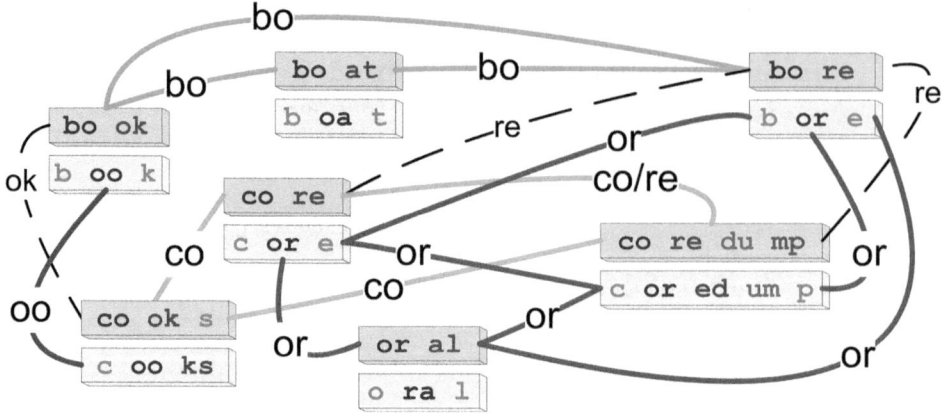

Fig. 1. Bouma2 graph for Example 4, highlighting the $M_{L|c1}$ clique partition

Theorem 4. *All maximal cliques in a Bouma2 graph are found in linear time.*

Proof. Algorithm 1 finds all the sets CLIQUE-REF *in* $O(sz(L))$ *time. Each such set corresponds to a trace and therefore to a maximal clique.* □

The original version of our benchmark (Section 6) employed a "greedy" heuristic using the CLIQUE-REF sets together with a cost function (see Definition 7). The heuristic iteratively determined the maximum clique, added it to the motif-set and pruned it from the graph.

5   The Bouma2 Match Process

5.1   The Match Structures

Unlike DFA-based algorithms, Bouma2 keeps the words in original form, maintaining a set of offsets to word-fragments. The match structure is thus very compact and linearly dependent on the total length of the set of words.

Definition 9. *A Motif's Match Function,* $\Gamma_\mu : \Sigma^* \times \mathbb{N} \to L \cup \{\epsilon\}$, *satisfies:*

For $m = m_p w_p \mu w_s m_s$ such that $|m_p w_p| = l$, and a given H_{M_L} :
$$H_{M_L}(w_p \mu w_s, |w_p|\%2) = \mu \implies \Gamma_\mu(m, l) = w_p \mu w_s$$
$$\text{Otherwise}: \quad \Gamma_\mu(m, l) = \epsilon . \quad (14)$$

Definition 10. *The Bouma2 Match Structure is defined as:*

$$B_L := (L, M_L, \{\Gamma_\mu : \mu \in M_L\}). \quad (15)$$

Theorem 5. *The memory consumption of* B_L *is* $O(sz(L) + sz(\Sigma^2) + |L|)$.

Proof. We keep the original set of words, hence $O(sz(L))$. $M_L \subseteq \Sigma^2$, contributing $O(sz(\Sigma^2))$. Finally, the match functions require offset mapppings within the words, maximum 4 per word (1 prefix offset and 1 suffix offset for each word's even motif, and another such couple for the odd motif), giving $O(|L|)$.   □

5.2 Match-Time Algorithm

Given an arbitrary input message $m \in \Sigma^*$, Algorithm 4 follows a 2-stage Fast-Path/Slow-Path discipline: the fast-path phase (Algorithm 2) first advances efficiently in 2-symbol strides, "harvesting" motif occurrences within the input, over a block of predefined size. The order in which the 2-symbol checks are performed within the block is not important. If any motifs are found, the slow-path phase (Algorithm 3) analyzes each occurrence to verify a complete match.

Algorithm 2 BOUMA2-FAST-PATH($B_L = (L, M_L, \{\Gamma_\mu : \mu \in M_L\}), m \in \Sigma^*$)

1: **procedure** BOUMA2-FAST-PATH(B_L,m)
2:     HARVEST={}
3:     **for all** $i \in \{x : 0 \leq x < |m| \wedge x\%2 = 0\}$ **do**
4:         **if** $m = m_p t m_s : |m_p| = i \wedge t \in M_L$ **then**
5:             HARVEST = HARVEST $\cup$ $\{(i, t)\}$
6:         **end if**
7:     **end for**
8:     **return** HARVEST
9: **end procedure**

Algorithm 3 BOUMA2-SLOW-PATH($B_L = (L, M_L, \{\Gamma_\mu : \mu \in M_L\}), m \in \Sigma^*$)

1: **procedure** BOUMA2-SLOW-PATH(B_L,m,HARVEST)
2:     **for all** $(i, \mu) \in$ HARVEST **do**
3:         **if** $\Gamma_\mu(m, i) \neq \epsilon$ **then**
4:             MATCH = MATCH $\cup$ $\{(i, \Gamma_\mu(m, i))\}$
5:         **end if**
6:     **end for**
7:     **return** MATCH
8: **end procedure**

Algorithm 4 BOUMA2-MATCH-TIME($B_L = (L, M_L, \{\Gamma_\mu : \mu \in M_L\}), MSG \in \Sigma^*, l \in \mathbb{N}$)

1: **procedure** BOUMA2-MATCH-TIME(B_L, MSG, l)
2:     MATCH={}
3:     **for all** $m \in \{x : MSG = m_p x m_s \wedge |m_p|\%l = 0\}$ **do**
4:         HARVEST = BOUMA2-FAST-PATH(B_L,m)
5:         MATCH = MATCH $\cup$ BOUMA2-SLOW-PATH(B_L,m,HARVEST)
6:     **end for**
7:     **return** MATCH
8: **end procedure**

5.3 The Resolve Process

In the slow-path phase, the match function Γ_μ first has to attempt to rule-out the words that H_{M_L} maps to μ (Definition 11 formalizes this requirement). This may be accomplished by looking for distinct symbols at offsets around the motif match. Proper motif selection in compile-time may be used to reduce the number of resolve-points per motif (see Section 5.4).

Definition 11. *The Bouma2 Resolving Problem is defined as follows: Given a resolve-set R_μ and an input message $m = m_p \mu m_s$, find an algorithm that would yield $R'_\mu \subseteq R_\mu$ in the smallest number of steps, whereas:*

$$\forall w_{p'} \mu w_{s'} \in R'_\mu \; : \; m_p = m_{pp} w_{p'} \; \wedge \; m_s = w_{s'} m_{ss} \,. \tag{16}$$

If not all the words could be ruled-out, the second task is to match the remaining words' prefix and suffix (before and after the motif occurrence, respectively) against the input. The prefix and suffix matches cause the input to be traversed multiple times. Nevertheless, these procedures can be considereably optimized (e.g. by 4-symbol strides), since they involve a simple comparison.

5.4   The Resolve Dimension

We loosely define the *Resolve Dimension*, $D(L'_t)$, as follows: given a set of words $L'_t \subseteq L$, for which $\forall w \in L'_t : assoc_0(w, t) = 1 \; \vee \; assoc_1(w, t) = 1$, $D(L'_t)$ is the minimum number of symbols within an input that need to be checked for ruling out at least $|L'_t| - 1$ words[3]. Each symbol check at an offset relative to the motif match is named a *Resolve Point*. Alternatively, we declare the notion of a *Mangled Trie* (see Example 5), which is an optimal offset-based decision-tree, and state that if d is the depth of a mangled-trie that completely resolves L'_t, then $D(L'_t) = d - 1$.

Example 5 (Resolve Dimension). Consider $H_{M_{L|c1}}$ in Example 4, illustrated in Figure 2. We specify the motif occurrence position as offset 0. $D(R_{ra})$ and $D(R_{oa})$ are both 0, since the resolve-sets contain a single word each. R_{oo} can be resolved at offset -1, and R_{bo} at offset 2 or 3, hence $D(R_{oo})$ and $D(R_{bo})$ are both 1. R_{co} cannot be resolved without examining at least two of offsets 2, 3 and 4. Thus, $D(R_{co})$ is 2. $D(R_{or}) = 2$ since both offsets -1 and 2 need to be examined.

6   Experimental Results

All tests were done on a Dell^{TM} computer with Intel®Core^{TM}2 Duo CPU 2.53 GHz with 1.95 GB RAM, running Windows XP SP3. The Bouma2 compiler and matcher were written in C++ using Microsoft®Visual Studio®2010 Premium. The Aho-Corasick benchmark was taken from [15] and adapted for Microsoft® Windows. The benchmark included all the algorithms under test running together, repeatedly scanning an input file for matches. The match results and

---

[3] For simplicity, this discussion ignores some special cases; for example, special treatment is required for multiple words that partially overlap by 4 symbols or more, such as in $R_{en} = \{Developm\underline{ent}, m\underline{ent}ality\}$.

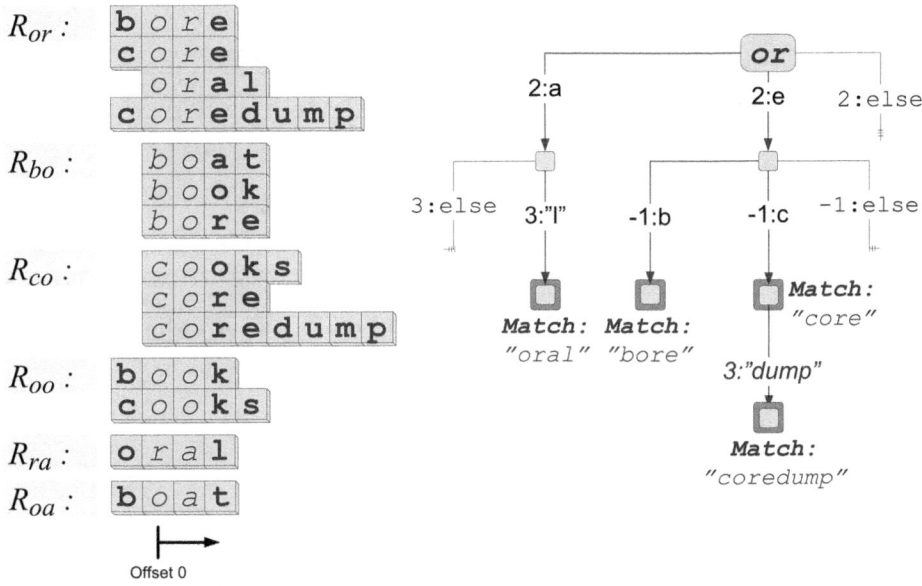

Fig. 2. Resolve-sets for $M_{L|c1}$ in Example 4, and a *Mangled Trie* for resolving R_{or}

Table 2. Experimental Results (Clique-Partition Heuristic)

Test #	Motif Set Size	Word Count	Total Word Length	Input Length	Match Count	Test Iterations
1	524,296	22	394	205,748	70	40,960

Matcher:	AC	Bouma2:Opt. Cmpl-Time	Bouma2:Min. Motif FPs
Profiler Samples	19,008	9,839	7,857
Match-Time (% of AC)	100	51.76241582	41.33522727
Total Memory (bytes)	54,944	528,204	526,884
Resolve-Sets Size	N/A	3,908	2,588
Motifs	N/A	15	33
Motif False-Positives	N/A	13,021	6,939

Test #	Motif Set Size	Word Count	Total Word Length	Input Length	Match Count	Test Iterations
2	524,296	100	3,994	22,445,535	99	40,960

Matcher:	AC	Bouma2:Opt. Cmpl-Time	Bouma2:Min. Motif FPs
Profiler Samples	56,361	55,259	33,119
Match-Time (% of AC)	100	98.04474725	58.76226469
Total Memory (bytes)	348,524	541,852	537,096
Resolve-Sets Size	N/A	17,556	12,800
Motifs	N/A	56	151
Motif False-Positives	N/A	2,671,687	2,791

Test #	Motif Set Size	Word Count	Total Word Length	Input Length	Match Count	Test Iterations
3	524,296	1,334	35,838	7,538,346	3,101	512

Matcher:	AC	Bouma2:Opt. Cmpl-Time	Bouma2:Min. Motif FPs
Profiler Samples	40,770	12,992	8,987
Match-Time (% of AC)	100	31.86656856	22.043169
Total Memory (bytes)	2,415,212	717,968	761,088
Resolve-Sets Size	N/A	193,672	236,792
Motifs	N/A	492	1,035
Motif False-Positives	N/A	2,430,868	1,574,338

Table 3. Experimental Results (Branch-and-Cut)

Test #	Motif Set Size	Word Count	Total Word Length	Input Length	Match Count	Test Iterations
4	524,572	3,093	24,301	355,228	14,774	8,192

Matcher:	AC	Bouma2:No Motif Priority	Bouma2:Min. Motif FPs
Profiler Samples	17,314	10,146	8,800
Match-Time (% of AC)	100	58.5999769	50.82592122
Total Memory (bytes)	1,260,092	850,665	869,597
Resolve-Sets Size	N/A	326,093	345,025
Motifs	N/A	232	293
Motif False-Positives	N/A	70,383	56,565

Test #	Motif Set Size	Word Count	Total Word Length	Input Length	Match Count	Test Iterations
5	524,572	3,408	26,976	355,228	16,403	8,192

Matcher:	AC	Bouma2:No Motif Priority	Bouma2:Min. Motif FPs
Profiler Samples	17,554	10,275	8,941
Match-Time (% of AC)	100	58.53366754	50.93426
Total Memory (bytes)	1,340,372	881,924	886,708
Resolve-Sets Size	N/A	357,352	362,136
Motifs	N/A	234	281
Motif False-Positives	N/A	68,775	57,180

positions of the competing algorithms were verified to be identical. The match procedure for each algorithm was instantiated within the same loop, which was repeated a large number of times. The comparison was made with the aid of the built-in sampling profiler in Microsoft®Visual Studio®. We compared the relative running-times of each competing algorithm, as expressed in aggregated profiler samples (samples were taken once every 10,000,000 clock cycles).

The compiler for the original benchmark was based on the heuristic described in Section 4.2. Table 2 shows the results of this benchmark with Aho-Corasick vs. Bouma2 (optimized compile-time) vs. Bouma2 (minimum motif false-positives). These tests were performed over binary texts and search-strings (virus signatures taken from [20]).

Since the original tests, the Bouma2 compiler was completely rewritten based on the Branch-and-Cut implementation within the COIN-OR[17] BCP[18] package, for solving the ILP described in Section 4.1. This allowed support of considerably larger languages and improved the quality of the resulting motif-sets (through finding a global maximum instead of a local maximum), at a fraction of the original compilation-time. Table 3 shows results of the new benchmark with Aho-Corasick vs. Bouma2 (no motif prioritization) vs. Bouma2 (minimum motif false-positives). These tests consisted of searching the book "Cyrus the Great" [21] for randomly extracted words.

One observation from these tests is that properly applying statistics to improve the quality of the motif-set indeed has a direct impact on the Bouma2 performance. The fast-path overhead is deterministic and allows little room for optimization in S/W. Nevertheless, the slow-path overhead can be minimized both by optimizing the resolving procedure and by accessing it fewer times through the use of a better motif-set.

7   Extensions and Future Work

Compile-Time: Methods to further improve compilation time and the quality of the solution motif-set are being researched. Furthermore, if for a certain word the motif-set includes more than one odd motif or more than one even motif, the compiler has to decide on a single motif which will represent this word. This duplicates-removal process currently considers only the number of words per motif, and can probably be optimized through the use of occurrence statistics. Finally, the design of an adaptive system that extracts statistics while analyzing the input and improves the resolve-sets in the background may be considered.

Match-Time: Optimizing the resolving algorithm and the structure of the Mangled Trie should improve performance considerably: currently, the resolving process may sometimes amount to more than 70% of the Bouma2 match procedure. For this aim, we should find a complete solution for the problem stated in Definition 11.

H/W implementation: The fast-path phase can be highly optimized for H/W: it is completely stateless, such that separate 2-symbol sequences can be checked in no particular order. E.g. for 8-bit words we can have several copies of a 64K-entry direct-access table, and use each copy for motif searches at different positions. The ability to pipeline fast-path and slow-path procedures, the small footprint, and the inherent cache-sensitivity also make Bouma2 suitable for H/W.

Match Scenarios: Currently unsupported 2-symbol words can be expanded to $|\Sigma|$ 3-symbol words with a 1-symbol prefix (actually, when matching e.g. `"\r\n"` in HTTP, there is a specific character-subset to which the preceding character must belong, reducing the expansion to less than $|\Sigma|$); For ASCII characters the match structure may hold bitmasks for normalizing the input's case when performing a case-insensitive match (also requiring at compile-time to assign the complete set of possible motifs - up to 8 - to the same word); Regular expressions (wildcards, character-sets, etc.) can be translated to checks that accompany the exact match process (thus a Bouma2-powered variant of the PCRE [19] package can be considered); The issue of pattern-match across input fragments (e.g. fragmented IP packets), which is handled well by Aho-Corasick (the match can be paused and the state stored at any point), requires special treatment with Bouma2, which may need to examine the input more than once (we may consider e.g. calculating at compile-time the maximum required first-fragment suffix that we will need to store, as a function of the values at the end of the fragment); Finally, we may consider the relevance of the Bouma2 hashing scheme to other applications besides pattern-match (e.g. text indexing).

References

1. Aho, A., Corasick, M.: Efficient string matching: An aid to bibliographic search. Comm. ACM 18, 333–340 (1975)
2. Commentz-Walter, B.: A string matching algorithm fast on the average. In: Maurer, H.A. (ed.) ICALP 1979. LNCS, vol. 71, pp. 118–132. Springer, Heidelberg (1979)

3. Wu, S., Manber, U.: A fast algorithm for multi-pattern searching., Technical Report TR-94-17, Department of Computer Science, University of Arizona (1994)
4. Becchi, M., Crowley, P.: Efficient regular expression evaluation: theory to practice. In: Proceedings of (ANCS 2008), pp. 50–59. ACM Press, New York (2008)
5. Karp, R., Rabin, M.: Efficient randomized pattern-matching algorithms. IBM Journal of Research and Development 31(2), 249–260 (1987)
6. Heger, A., Lappe, M., Holm, L.: Accurate detection of very sparse sequence motifs. In: Proceedings of the Seventh Annual International Conference on Computational Molecular Biology, pp. 139–147. ACM Press, Berlin (2003)
7. Altschul, S., Gish, W., Miller, W., Myers, E., Lipman, D.: A basic local alignment search tool. J. Mol. Biol. 215, 403–410 (1990)
8. Thompson, J., Higgins, D., Gibson, T.: ClustalW: improving the sensitivity of progressive multiple sequence alignment through sequence weighting, position-specific gap penalties and weight matrix choice. Nucl. Acids Res. 22, 4673–4690 (1994)
9. Liang, S.: cWINNOWER algorithm for finding fuzzy DNA motifs. In: IEEE Computer Society Bioinformatics Conference, pp. 260–265 (2003)
10. Salmela, L., Tarhio, J., Kytojoki, J.: Multi-pattern string matching with q-grams. ACM Journal of Experimental Algorithmics 11 (2006)
11. Kumar, S., Dharmapurikar, S., Yu, F., Crowley, P., Turner, J.: Algorithms to accelerate multiple regular expressions matching for deep packet inspection. In: Proc. of SIGCOMM 2006, pp. 339–350. ACM Press, New York (2006)
12. Ficara, D., Giordano, S., Procissi, G., Vitucci, F., Antichi, G., Pietro, A.D.: An improved dfa for fast regular expression matching. SIGCOMM Comput. Commun. Rev. 38(5), 29–40 (2008)
13. Smith, H.O., Annau, T.M., Chandrasegaran, S.: Finding sequence motifs in groups of functionally related proteins. Proc. Natl. Acad. Sci. 87, 826–883 (1990)
14. Hopcroft, J.E., Motwani, R., Ullman, J.D.: Introduction to Automata Theory, Languages and Computation. Addison Wesley, Reading (2000)
15. Bos, H., de Bruijn, W., Cristea, M., Nguyen, T., Portokalidis, G.: FFPF: Fairly Fast Packet Filters. In: Proceedings of OSDI 2004, San Francisco, CA (December 2004)
16. Nemhauser, G.L., Wolsey, L.A.: Integer and Combinatorial Optimization. John Wiley & Sons, New York (1988)
17. Lougee-Heimer, R.: The Common Optimization INterface for Operations Research. IBM Journal of Research and Development 47(1), 57–66 (2003), http://www.coin-or.org/
18. Branch-Cut-Price Framework, https://projects.coin-or.org/Bcp/
19. Perl Compatible Regular Expressions, http://www.pcre.org/
20. Clam Antivirus, http://www.clamav.net/lang/en/
21. The Project Gutenberg EBook of Cyrus the Great, by Jacob Abbott, http://www.gutenberg.org/cache/epub/30707/pg30707.txt
22. Saenger, P.: Space Between Words: The Origins of Silent Reading. Stanford University Press, Stanford (2000) ISBN 0-8047-4016-X

Random Generation of Deterministic Acyclic Automata Using Markov Chains

Vincent Carnino and Sven De Felice*

LIGM, UMR 8049, Université Paris-Est et CNRS,
77454 Marne-la-Vallée, France
vcarnino@etudiant.univ-mlv.fr, defelic@univ-mlv.fr

Abstract. In this article we propose an algorithm, based on Markov chain techniques, to generate random automata that are deterministic, accessible and acyclic. The distribution of the output approaches the uniform distribution on n-state such automata. We then show how to adapt this algorithm in order to generate minimal acyclic automata with n states almost uniformly.

1 Introduction

In language theory, acyclic automata are exactly the automata that recognize finite languages. For this reason, they play an important role in some specific fields of applications, such as the treatment of natural language. From an algorithmic point of view, they often enjoy more efficient solutions than general automata; a famous example is the linear minimization algorithm proposed by Revuz for deterministic acyclic automata [15]. They also appear as first steps in some algorithms, two examples of which are related to Glushkov construction [3,4,5] and some extension of Aho-Corasick automaton [14].

In the design and analysis of algorithms it is of great use to have access to exhaustive and random generators for the inputs of the algorithm one wants to study: the exhaustive generator is used to analyze the behavior of the algorithm for small inputs, but cannot be used for large inputs since there are too many of them; typically the number of size-n inputs often grows at least exponentially in n. Those generators can be used either to test the correctness and the efficiency of an implementation, or to help the researcher while establishing theoretical results about the average case analysis of the algorithm.

An exhaustive generator for minimal deterministic acyclic automata has been given by Almeida, Moreira and Reis [1], and in this paper we propose an algorithm to generate at random deterministic, accessible and acyclic automata, with a distribution that is almost uniform, using Markov chain techniques. With just a few changes, this algorithm can be turned into a generator for minimal acyclic automata. The idea is to start with a n-state acyclic automaton, then to perform a certain amount T of mutations of this automaton, a mutation being a

---

* The second author was supported by ANR MAGNUM - project ANR-2010-BLAN-0204.

B. Bouchou-Markhoff et al. (Eds.): CIAA 2011, LNCS 6807, pp. 65–75, 2011.
© Springer-Verlag Berlin Heidelberg 2011

small local transformation that preserves the required properties (deterministic, accessible and acyclic with the same number of states). Since each mutation is performed in time $O(n)$, the complexity of our algorithm is $O(nT)$. The bigger T is, the more the output distribution approaches the uniform distribution. For a given distance to uniformity, it is a general difficult problem to give a good estimation of a corresponding value of T; this is directly related to the *mixing time* of the Markov chain, which is generally a difficult problem [10]. Nonetheless, the diameter of the Markov chain and the simulations we performed seems to indicate that a choice of T in $\Theta(n)$ already gives a correct random generator, at least for most applications, of complexity $O(n^2)$.

Note that the other generic methods to generate combinatorial structures uniformly at random seem to fail here. For instance, recursive methods [8] or Boltzmann samplers [7], which have been used for deterministic automata [6,2,9], rely on a good recursive description of the input, which is not known for acyclic automata. To our knowledge, the only combinatorial result on acyclic automata is due to Liskovets [11], who gave a close formula for the number of acyclic automata, but which cannot be directly translate into a good recursive description.

Related work: as mentioned above, our algorithm is a complement of the exhaustive generator of Almeida, Moreira and Reis [1] for testing conjectures and algorithms based on deterministic acyclic automata. The idea of using Markov chain for that kind of objects starts with works on acyclic graphs, which has been done for graph visualization purposes [12,13]. Though using the same general idea, deterministic acyclic automata do not resemble acyclic graphs that much, mainly because they only have a linear number of edges (transitions). In particular, the diameter of the Markov chain, which is a lower bound for the mixing time, is quadratic for acyclic graphs but linear in our case. Moreover, automata considered in this article must be accessible, which is not a natural condition for graphs (there is no notion of distinguished initial vertex); Melançon and Philippe considered simply connected acyclic graphs in [13], but this is not the same as accessibility. For instance, they use a nice optimization based on reversing an edge, which preserves connectedness but not accessibility; hence it cannot be reused here.

The paper is organized as follows. In Section 2, we recall basic notations about automata; and in Section 3 classical Markov chain concepts are detailed. The algorithm is described in Section 4, and its correctness is given in Section 5. We present a generator for minimal acyclic automata in Section 6. Finally, in Section 7 we perform some experimentations.

2   Notations

Throughout this paper, a *deterministic finite automaton* is a tuple $\mathcal{A} = (Q, A, \delta, i_0, F)$, where Q is a finite set of *states*, A is a finite set of *letters* called the *alphabet*, $\delta : Q \times A \rightarrow Q$ is the (partial) *transition function*, $i_0 \in Q$ is the *initial state* and $F \subseteq Q$ is the set of *final states*. In the sequel we always suppose that $|A| > 1$. For any state $q \in Q$, the transition function $\delta(q, \cdot)$ is inductively extended to the set A^* of all finite words over A: $\delta(q, \varepsilon) = q$, where ε is the *empty word*, and for all $w \in A^*$ such that $w = w_1 w_2 \ldots w_n$, then $\delta(q, w)$ $:= \delta(\delta(\ldots \delta(\delta(q, w_1), w_2) \ldots), w_n)$.

In this paper, we represent a transition $\delta(p, a) = q$, with $(p, q) \in Q^2$ and $a \in A$, by $p \xrightarrow{a} q$. The notation $\mathcal{A} \oplus p \xrightarrow{a} q$ represents the automaton $\mathcal{A}$ with the additional transition $p \xrightarrow{a} q$. Similarly, the notation $\mathcal{A} \ominus p \xrightarrow{a} q$ represents the automaton $\mathcal{A}$ where the transition $p \xrightarrow{a} q$ has been removed, if it exists.

A state $q \in Q$ is *accessible* (resp. *co-accessible*) when there exists $w \in A^*$ such that $\delta(i_0, w) = q$ (resp. $\delta(q, w) \in F$). An automaton is *accessible* (resp. *co-accessible*) when all its states are accessible (resp. *co-accessible*).

A state $q \in Q$ is *transient* if for all $w \in A^+$, $\delta(q, w) \neq q$. A state that is not transient is called *recurrent*. An automaton is *acyclic* when every state is transient. Another definition of acyclic automata is that the underlying directed graph is an acyclic graph. Remark that it is impossible for a complete automaton to be acyclic.

In the sequel, without loss of generality, the set of states Q of an n-state deterministic automaton will always be $\{1, \ldots, n\}$ and 1 will always be the initial state. The *size* of an automaton is its number of states, and we furthermore assume from now on that $n \geq 2$. Moreover, since we always consider deterministic, accessible and acyclic automata in this article, we shall just denote them by "acyclic automata" for short. The set of all n-state acyclic automata is denoted by $\mathbb{A}_n$.

Also, except in Section 6, we are not considering the set of final states in our random generator. We assume that final states are chosen independently once the underlying graph of the automaton is generated.

3   Markov Chains and Random Generation

In this section we describe our algorithm to generate an acyclic automaton $\mathcal{A}$ of size n over the alphabet A, with the uniform probability on $\mathbb{A}_n$. The input of algorithm is two positive integers: n, the number of states, and T, the number of iterations.

The algorithm relies on a *Markov chain* process: it randomly moves in the set $\mathbb{A}_n$ and returns the last automaton reached after T steps. The Markov chain of the algorithm can be seen as a directed graph whose vertices are elements of $\mathbb{A}_n$. An edge from an automaton $\mathcal{A}$ to another automaton $\mathcal{B}$ is labelled by a real $r \in [0, 1]$, which represents the probability to move from automaton $\mathcal{A}$ to automaton $\mathcal{B}$ in one step. For two automata $\mathcal{A}, \mathcal{B} \in \mathbb{A}_n$ we denote by $\mathcal{P}_{\mathcal{A}, \mathcal{B}}$ the label of the edge from $\mathcal{A}$ to $\mathcal{B}$, if it exists, otherwise we set $\mathcal{P}_{\mathcal{A}, \mathcal{B}} = 0$. Since it is a probability, we have:

$$\forall \mathcal{A} \in \mathbb{A}_n, \sum_{\mathcal{B} \in \mathbb{A}_n} \mathcal{P}_{\mathcal{A}, \mathcal{B}} = 1.$$

A distribution on $\mathbb{A}_n$ is a mapping p from $\mathbb{A}_n$ to $[0, 1]$ such that $\sum_{\mathcal{A} \in \mathbb{A}_n} p(\mathcal{A}) = 1$. A *stationary distribution of a Markov chain* π is a distribution that remains globally unchanged after each random move, that is,

$$\forall \mathcal{B} \in \mathbb{A}_n, \pi(\mathcal{B}) = \sum_{\mathcal{A} \in \mathbb{A}_n} \pi(\mathcal{A}) \times \mathcal{P}_{\mathcal{A}, \mathcal{B}}.$$

A Markov Chain is called *irreducible* when its graph is strongly connected. For $i \in \mathbb{N}$, let $\mathcal{P}_{\mathcal{A}, \mathcal{B}}^{(i)}$ be the probability to move from $\mathcal{A}$ to $\mathcal{B}$ in i steps of the algorithm. We define the *period* of a vertex $\mathcal{A}$ as the gcd of the lengths of all circuits on $\mathcal{A}$: $gcd(\{i \in \mathbb{N} \mid \mathcal{P}_{\mathcal{A}, \mathcal{A}}^{(i)} > 0\})$. If there is a loop of length 1 on $\mathcal{A}$, the period of $\mathcal{A}$ is 1 by definition. A vertex is *aperiodic* if its period is 1. A Markov chain is *aperiodic* when all its states are aperiodic. A Markov chain is *ergodic* when it is both irreducible and aperiodic.

A famous property of ergodic Markov chains with a finite number of vertices is that they have a unique stationary distribution and that starting at any vertex the distribution obtained after T steps tends to this stationary distribution as T tends to infinity [10]. This gives a general framework to build a random generator on a non-empty finite set E: design an ergodic Markov chain whose set of vertex is E and such that the stationary distribution is the uniform distribution. Start from any vertex, then move randomly for a long enough time to obtain a random element of E almost uniformly.

This is exactly what we do in this article. A part of the Markov chain that is behind our algorithm is depicted in Figure 1. Each step consists either in doing nothing or in changing a transition. The complete description of the algorithm is done in Section 4. Our main result, which is proved in Section 5 is the following:

Theorem 1. *The Markov chain of the algorithm is ergodic and its stationary distribution is the uniform distribution.*

Since 1 is always the initial state and since there are $(n - 1)!$ different way to label the other states there are exactly $(n - 1)!$ automatata isomorphic to any element of $\mathbb{A}_n$. Consequently, our uniform random generator on $\mathbb{A}_n$ yields a generator on isomorphic classes of automata which is also uniform. Note that the number of iterations T must be large enough in order to approach closely the uniform distribution. The choice of T is a difficult problem [10] and it is not entirely cover in this paper. The diameter of the Markov chain's graph is a lower bound for T, and we will show in Section 5 that this diameter is linear in our case. In Section 7, we will see that the uniform distribution seems to be well approximated using a linear number of iterations, at least well enough for most simulation purposes.

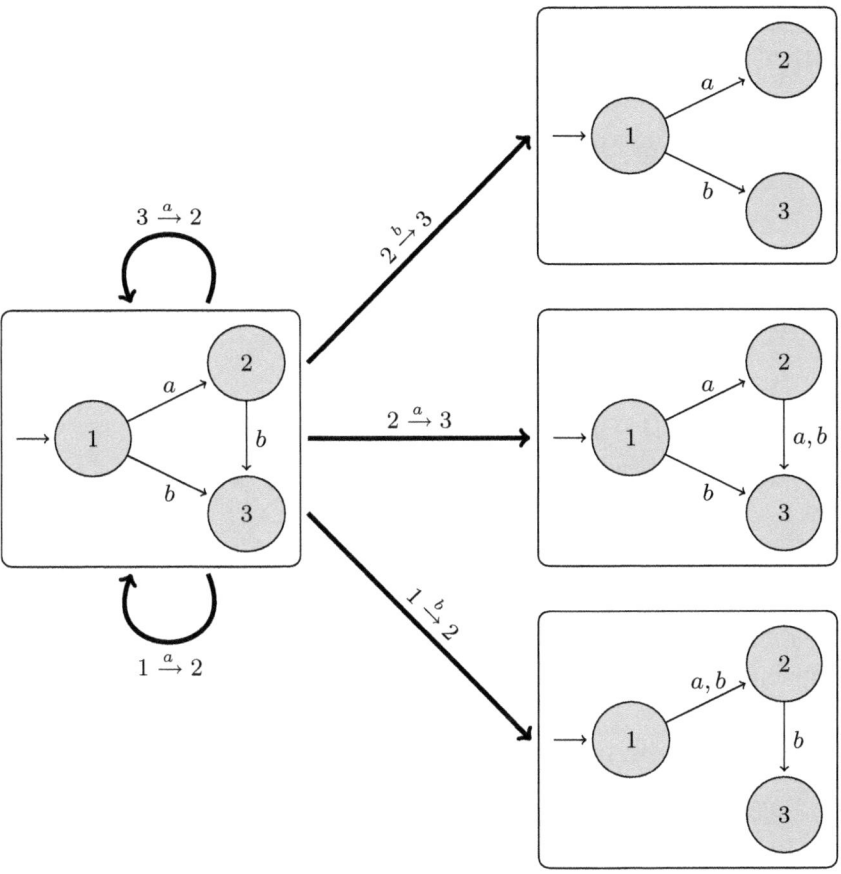

Fig. 1. Part of the Markov chain: at each iteration an element $p \xrightarrow{a} q$ of $Q \times A \times Q$ is chosen randomly. If it corresponds to a transition of the automaton, as $2 \xrightarrow{b} 3$, then it is removed. If there is no transition labelled by a and starting at p it is added; this is the case for $2 \xrightarrow{a} 3$. When there already is a transition labelled by a and starting at p, it is redirected to q; this is the case for $1 \xrightarrow{b} 2$. The mutation is not done if the automaton is not acyclic anymore ($3 \xrightarrow{a} 2$) or if it is not accessible anymore ($1 \xrightarrow{a} 2$).

4   Algorithm

The algorithm has two arguments: the number n of states and a value T which indicates the desired number of iterations (it is quite difficult to know when the uniform distribution is reached so it is convenient to specify it). After choosing any acyclic automaton $\mathcal{A} \in \mathbb{A}_n$ to start with, the algorithm repeats the following steps T times: choose uniformly a labelled edge $p \xrightarrow{a} q$ with $p \neq q$ ($p = q$ is not interesting since we are considering acyclic automata). Then there are three possible cases:

```
AcyclicAutomatonGeneration(n, T)
```

1 $\mathcal{A} \leftarrow$ any deterministic, accessible and acyclic automaton with n states
2 $i \leftarrow 0$
3 **while** $i < T$ **do**
4 $\quad$ $p \leftarrow Uniform(Q),\ a \leftarrow Uniform(A),\ q \leftarrow Uniform(Q \setminus \{p\})$
5 $\quad$ **if** $\delta(p,a)$ *is undefined* **then**
6 $\quad\quad$ **if** *IsAcyclic*$(\mathcal{A} \oplus p \xrightarrow{a} q)$ **then** $\mathcal{A} = \mathcal{A} \oplus p \xrightarrow{a} q$
7 $\quad$ **else if** $\delta(p,a) = q$ **then**
8 $\quad\quad$ **if** *IsAccessible*$(\mathcal{A} \ominus p \xrightarrow{a} q)$ **then** $\mathcal{A} = \mathcal{A} \ominus p \xrightarrow{a} q$
9 $\quad$ **else**
10 $\quad\quad$ $r \leftarrow \delta(p,a)$
11 $\quad\quad$ **if** *IsAccessible*$(\mathcal{A} \ominus p \xrightarrow{a} r)$ **then**
12 $\quad\quad\quad$ $\mathcal{A} = \mathcal{A} \ominus p \xrightarrow{a} r$
13 $\quad\quad\quad$ **if** *IsAcyclic*$(\mathcal{A} \oplus p \xrightarrow{a} q)$ **then**
14 $\quad\quad\quad\quad$ $\mathcal{A} = \mathcal{A} \oplus p \xrightarrow{a} q$
15 $\quad\quad\quad$ **else**
16 $\quad\quad\quad\quad$ $\mathcal{A} = \mathcal{A} \oplus p \xrightarrow{a} r$

17 $\quad$ $i \leftarrow i + 1$
18 Randomly choose the set of final states of $\mathcal{A}$
19 **return** $\mathcal{A}$

- There is no transition starting from p and labelled with a. In such a case, we try to add $p \xrightarrow{a} q$ to $\mathcal{A}$ and test if it is still acyclic. The transition is added only if it is.
- There already is a transition $p \xrightarrow{a} q$ in $\mathcal{A}$. In that case, we test if $\mathcal{A}$ is still accessible if we remove it. If it is, the transition is removed, else $\mathcal{A}$ remains unchanged.
- There is a transition starting from p, labelled with a and reaching a state r, with $r \neq q$. In this last case, we first test whether $\mathcal{A}$ is still accessible if we redirect $\delta(p,a)$ to q. If it is, we do the redirection, otherwise $\mathcal{A}$ remains unchanged.

In this process, we need to check regularly the accessibility and the acyclicity of $\mathcal{A}$.

The accessibility test is implemented the following way. We keep up-to-date, for each state q, a counter that indicates the total amount of transitions ending in q. Each time we add or remove such a transition, this counter is increased or decreased. Thus, to test the accessibility, we just have to check, after the transition has been removed, whether the counter on the state that ends the transition reaches 0 or not; this is a consequence of Lemma 1 (see Section 5). It clearly has a $O(1)$ time complexity.

The acyclicity is tested by the classical algorithm, using a depth-first-search algorithm which runs in time $O(n)$, since the number of transitions is linear in a deterministic automaton.

We therefore get the following result.

Proposition 1. *Each iteration of the algorithm is performed in time $O(n)$. The worst case time complexity of the algorithm is $O(Tn)$ and its space complexity is $O(n)$.*

The experimental results of Section 7 suggest that choosing $T \in \Theta(n)$ should be good enough; with this choice, the complexity of our algorithm would be quadratic.

5   Proofs

In this section, we prove the main facts that are used for our algorithm to correctly generate an acyclic automaton with almost uniform distribution, and with the announced complexity.

An operation which consists in removing, adding or changing a transition is called an *elementary operation*.

Lemma 1. *Let $\mathcal{A}$ be an acyclic automaton of size n and $\mathcal{B} = \mathcal{A} \ominus p \xrightarrow{a} q$, where $p \xrightarrow{a} q$ is any transition of $\mathcal{A}$. The automaton $\mathcal{B}$ is acyclic, and it is accessible if and only if there is at least one transition that ends in q in the automaton $\mathcal{B}$.*

Proof. First note that $q \neq 1$, since 1 is the initial state of $\mathcal{A}$, which is an acyclic and accessible automaton with at least two states.

Suppose that there is no transition that ends in q in $\mathcal{B}$. Since $q \neq 1$, q is not accessible and neither is $\mathcal{B}$.

Suppose now that $\mathcal{B}$ has a transition $r \xrightarrow{b} q$, for some state r and some letter b. The state r is accessible in $\mathcal{A}$, and $r \neq q$. Since $\mathcal{A}$ and $\mathcal{B}$ only differ by a transition that ends in q, r is still accessible in $\mathcal{B}$. Therefore, q is accessible in $\mathcal{B}$ because one can follow a path from 1 to r, then use the transition $r \xrightarrow{b} q$. Since all other states are accessible for the same reason as r is, $\mathcal{B}$ is accessible.   □

Note that the result of Lemma 1 does not hold for automata that are not acyclic.

Lemma 2. *The Markov chain of the algorithm is symmetric, that is, for all $\mathcal{A}, \mathcal{B} \in \mathbb{A}_n$, $\mathcal{P}_{\mathcal{A},\mathcal{B}} = \mathcal{P}_{\mathcal{B},\mathcal{A}}$.*

Proof. Recall that the probability to draw a given triplet (p, a, q) with $p \in Q$, $q \in Q \backslash \{p\}$, and $a \in A$ is $\frac{1}{n(n-1)|A|}$. Let $\mathcal{A}, \mathcal{B}$ be in $\mathbb{A}_n$ such that $\mathcal{P}_{\mathcal{A},\mathcal{B}} > 0$. Then there exists an elementary operation that transforms $\mathcal{A}$ into $\mathcal{B}$. Suppose $\mathcal{B} = \mathcal{A} \oplus p \xrightarrow{a} q$. The probability to draw the triplet (p, a, q) is $\frac{1}{n(n-1)|A|}$. Now from $\mathcal{B}$ the only possible elementary operation to reach $\mathcal{A}$ is to remove the transition $p \xrightarrow{a} q$. Thus, we need to draw the triplet (p, a, q) and the probability of this event is $\frac{1}{n(n-1)|A|}$ too. If $\mathcal{B} = \mathcal{A} \ominus p \xrightarrow{a} q$ then $\mathcal{A} = \mathcal{B} \oplus p \xrightarrow{a} q$ thus we are in the same case as above and $\mathcal{P}_{\mathcal{A},\mathcal{B}} = \mathcal{P}_{\mathcal{B},\mathcal{A}}$.

Suppose the elementary operation that transforms $\mathcal{A}$ to $\mathcal{B}$ is to redirect the transition $p \xrightarrow{a} q$ of $\mathcal{A}$ to obtain $p \xrightarrow{a} s$ in $\mathcal{B}$. To get this, we need to draw the triplet (p, a, s) and the probability of this event is $\frac{1}{n(n-1)|A|} = \mathcal{P}_{\mathcal{A},\mathcal{B}}$. The only possible elementary operation to reach $\mathcal{A}$ from $\mathcal{B}$ is to redirect the new transition $p \xrightarrow{a} s$ to $p \xrightarrow{a} q$ which has the same probability, for the same reasons. Hence $\mathcal{P}_{\mathcal{A},\mathcal{B}} = \mathcal{P}_{\mathcal{B},\mathcal{A}}$ in this case too. □

Lemma 3. *The Markov chain of the algorithm is ergodic.*

Proof. We need to prove that it is both irreducible and aperiodic.

To prove the irreducibility, we show that, in the Markov chain, there is a path from any acyclic automaton $A \in \mathbb{A}_n$ to an automaton $S_n \in \mathbb{A}_n$, where S_n is the acyclic automaton whose only transitions are $i \xrightarrow{a} i + 1$, for $i \in \{1, \ldots, n-1\}$:

Let $\mathcal{A}$ be any acyclic automaton and let a be a letter in A. Let E be the set of states that are accessible from the initial state by reading only a's. E is not empty since it contains at least the initial state 1. Repeatedly remove every transition $p \xrightarrow{\alpha} q$ where $q \in E$ and $p \notin E$. Then repeatedly remove every remaining transition $p \xrightarrow{\alpha} q$ where $p, q \in E$ and $\alpha \neq a$. This actions are valid moves in the Markov chain by Lemma 1 since we always keep the transitions $p \xrightarrow{a} q$ with $p, q \in E$. Let ℓ be the only state in E with no outgoing transition labelled with a.

If $|E| < n$, choose a state s of $\mathcal{A}$ that is not in E and add a transition $\ell \xrightarrow{a} s$. Because there is no path between s and a state of E, this operation cannot create a cycle. Repeatedly remove all transitions directed toward s except $\ell \xrightarrow{a} s$. Add s to E, the set E is one state bigger. The size of E being finite, this operations can be repeated until E contains all states of $\mathcal{A}$.

Hence, at some point $|E| = n$ and $\mathcal{A}$ is isomorph to S_n, since every state but the initial one has exactly one incoming transition, which is labelled by a. The only difference with S_n is that the states are not necessarily in the correct order. We now explain how they can be re-ordered.

Let $b \in A$, $b \neq a$ for each transition $p \xrightarrow{a} q$ of $\mathcal{A}$, we add to $\mathcal{A}$ the transition $p \xrightarrow{b} q$ by elementary operations, which do not create any cycle. Now we remove all transitions labelled by a, $\mathcal{A}$ remains accessible because of the transitions labelled by b. We are in the case $|E| < n$ above, where the set E contains the state 1 only. To reach the automaton S_n, it is sufficient to choose the new states added to E in the order of their label. After removing all transitions labelled by b, we finally obtain the automaton S_n.

Hence for every $\mathcal{A} \in \mathbb{A}_n$, there exists a path from $\mathcal{A}$ to S_n in the Markov chain. By Lemma 2 there also exists a path from S_n to $\mathcal{A}$: the Markov chain is therefore irreducible. For every automaton $\mathcal{A} \in \mathbb{A}_n$ and any state $p \neq 1$ and any letter $a \in A$, if the edge chosen by the algorithm is $(p, a, 1)$ then $\mathcal{A}$ remains the same: adding the transitions would make $\mathcal{A}$ cyclic. Hence every vertex has a loop of length 1 in the Markov chain, it is therefore aperiodic.   □

Lemma 4. *The diameter of the Markov chain is in $\Theta(n)$.*

Proof. Using the construction proposed in the proof of Lemma 3, every $\mathcal{A} \in \mathbb{A}_n$ is at distance at most $(|A| + 5)n$ of S_n. The diameter of the Markov chain is thus at most $2(|A| + 5)n$, which is $O(n)$. The lower bound in $\Omega(n)$ is obtained by considering the distance from S_n to an acyclic automaton whose edges are all labelled by a letter $b \neq a$.   □

Theorem 1 is a consequence of the lemmas above: By Lemma 3 the Markov chain of the algorithm is ergodic and by Lemma 2 it is symmetric. According to a classical result in Markov chain theory [10], its stationary distribution is the uniform distribution on $\mathbb{A}_n$.

6   Minimal Acyclic Automata

In this section we briefly describe how to adapt our algorithm in order to generate minimal acyclic automata. Due to the lack of place, we do not give all the details here, but the adaptation is quite straightforward.

An acyclic automaton $\mathcal{A}$ of $\mathbb{A}_n$ is a *hammock acyclic automaton* (or *hammock automaton* for short) if $\mathcal{A}$ has only one state with no outgoing transition. This state is called the *target* state of the hammock automaton. We denote by $\mathbb{H}_n \subset \mathbb{A}_n$ the set of size-n hammock automaton whose target state is n.

Our random generator can readily be adapted to generate elements of $\mathbb{H}_n$: never choose $p = n$, in order to keep n without outgoing transition, and do not perform a deletion of $p \xrightarrow{a} q$ if it is the only outgoing transition of p.

Adapting the proof of Lemma 3 to hammock automata, we can prove that the Markov chain is still ergodic and symmetric. Its stationary distribution is therefore the uniform distribution on $\mathbb{H}_n$. The diameter is also in $\Theta(n)$ for this new chain.

Let $\mathbb{M}_n$ denote the set of minimal acyclic automata with n states. One can verify that such an automaton is necessarily an hammock automaton whose target state is final. This is of course not a sufficient condition. However, we can use this property to generate elements of $\mathbb{M}_n$ using a rejection algorithm: repeatedly draw a random hammock automaton (whose target state is final) until the automaton is minimal. This pseudo-algorithm may never halt, but if the proportion of minimal automata is large enough, the average number of rejections is polynomial or even bounded above by a constant. The important point is that no bias is introduced by this method: if hammock automata are generated uniformly at random, the induced probability on the output is the uniform distribution on $\mathbb{M}_n$.

We have no asymptotic result yet about the proportion of minimal automata amongst hammock automata. This may be a difficult problem, since it is still open for general deterministic automata. But experiments indicate that this proportion should be non-negligible: amongst 1000 random hammock automata of size 100, on a two-letter alphabet, we found 758 minimal automata. If we accept the conjecture that the proportion of minimal automata is at least $c > 0$, this yields a random generator for minimal acyclic automata with no increase of complexity, in average: the average number of rejections is bounded from above by a constant, and the test of minimality is linear, using Revuz algorithm [15].

7 Experiments

In this section, we present some experiments we did in order to evaluate the rate of convergence of our algorithm as T grows. For this purpose we use the Kolmogorov-Smirnov statistic test, which, roughly speaking, computes a value that measures the distance to the uniform distribution. This testing protocol is limited to small values of n: we need to store, for each isomorphism class of $\mathbb{A}_n$ the number of times it has been generated when performing a large number N of random generations. For the test to be meaningful, all isomorphism classes of $\mathbb{A}_n$ must have been generated, and there are many of them, even for small values of n [11,1].

We generated a large number of acyclic automata with our generator and reported the value of the Kolmogorov-Smirnov statistic test. The results are given in Figure 2 below.

n	3	4	5	6
$\|(\mathbb{A}_\sim)_n\|$	16	127	13183	18628
$T = 2n$	0.2	0.3	0.077	0.05
$T = 8n$	**0.026**	0.02	0.013	0.003
$T = 16n$	**0.016**	**0.0070**	**0.0015**	**0.00068**
$T = 24n$	**0.02**	**0.0074**	**0.0014**	**0.00044**

Fig. 2. The values of the uniform Kolmogorov-Smirnov statistic test depending on n and of the number T of iterations in the algorithm. The tests are performed on a population of $100|(\mathbb{A}_\sim)_n|$ automata generated by the algorithm, where $(\mathbb{A}_\sim)_n$ is the set of isomorphism classes of $\mathbb{A}_n$. We indicated in bold when the test of uniformity is successful.

8 Conclusion

Our random generators are already usable in practice, and easy to implement. Two questions remain to justify fully their good behavior, which are ongoing works:

- The complete analysis of the main algorithm requires a good estimation of the mixing time of the underlying Markov chain.
- The efficiency of our algorithm that generates minimal acyclic automata relies on an estimation of the proportion of minimal automata amongst hammock automata.

Acknowledgement. we would like to thanks Cyril Nicaud for his precious help in most stages of this work.

References

1. Almeida, M., Moreira, N., Reis, R.: Exact generation of minimal acyclic deterministic finite automata. Int. J. Found. Comput. Sci. 19(4), 751–765 (2008)
2. Bassino, F., Nicaud, C.: Enumeration and random generation of accessible automata. Theor. Comput. Sci. 381(1-3), 86–104 (2007)
3. Caron, P., Champarnaud, J.-M., Mignot, L.: Small Extended Expressions for Acyclic Automata. In: Maneth, S. (ed.) CIAA 2009. LNCS, vol. 5642, pp. 198–207. Springer, Heidelberg (2009)
4. Caron, P., Champarnaud, J.-M., Mignot, L.: Acyclic automata and small expressions using multi-tilde-bar operators. Theor. Comput. Sci. 411(38-39), 3423–3435 (2010)
5. Caron, P., Ziadi, D.: Characterization of glushkov automata. Theor. Comput. Sci. 233(1-2), 75–90 (2000)
6. Champarnaud, J.-M., Paranthoën, T.: Random generation of DFAs. Theor. Comput. Sci. 330(2), 221–235 (2005)
7. Duchon, P., Flajolet, P., Louchard, G., Schaeffer, G.: Boltzmann samplers for the random generation of combinatorial structures. Combinatorics, Probability & Computing 13(4-5), 577–625 (2004)
8. Flajolet, P., Zimmermann, P., Van Cutsem, B.: A calculus for the random generation of labelled combinatorial structures. Theor. Comput. Sci. 132(2), 1–35 (1994)
9. Héam, P.-C., Nicaud, C., Schmitz, S.: Parametric random generation of deterministic tree automata. Theor. Comput. Sci. 411(38-39), 3469–3480 (2010)
10. David, A., Peres, Y., Wilmer, E.L.: Markov Chains and Mixing Times. AMS, Providence (2008)
11. Liskovets, V.A.: Exact enumeration of acyclic deterministic automata. Discrete Applied Mathematics 154(3), 537–551 (2006)
12. Melançon, G., Dutour, I., Bousquet-Mélou, M.: Random generation of directed acyclic graphs. Electronic Notes in Discrete Mathematics 10, 202–207 (2001)
13. Melançon, G., Philippe, F.: Generating connected acyclic digraphs uniformly at random. Inf. Process. Lett. 90(4), 209–213 (2004)
14. Mohri, M.: String-matching with automata. Nord. J. Comput. 4(2), 217–231 (1997)
15. Revuz, D.: Minimisation of acyclic deterministic automata in linear time. Theor. Comput. Sci. 92(1), 181–189 (1992)

Variable and Clause Ordering in an FSA Approach to Propositional Satisfiability

José M. Castaño and Rodrigo Castaño

Depto. de Computación, FCEyN, UBA, Argentina
{jcastano,rcastano}@dc.uba.ar

Abstract. We use a finite state (FSA) construction approach to address the problem of propositional satisfiability (SAT). We use a very simple translation from formulas in conjunctive normal form (CNF) to regular expressions and use regular expressions to construct an FSA. As a consequence of the FSA construction, we obtain an ALL-SAT solver and model counter. We compare how several variable ordering (state ordering) heuristics affect the running time of the FSA construction. We also present a strategy for clause ordering (automata composition). We compare the running time of state-of-the-art model counters, BDD based sat solvers and we show that this FSA approach obtains state-of-the-art performance on some hard unsatisfiable benchmarks. This work brings up many questions on the possible use of automata to address SAT.

Keywords: ALL-SAT, model counting, FSA intersection, regular expression compilation.

1   Introduction

There is a long tradition that analyzed transformations of logic formulas and automata formally [7,10,24,23]. Propositional satisfiability (SAT) solving has many practical applications ranging from artificial intelligence to software verification. Search-based techniques in SAT solving have been enormously successful. State-of-the-art SAT solvers are based on the DPLL (Davis-Putnam-Logemann-Loveland) algorithm, augmented with a number of features. Much of current research in this area involves refinements and extensions of the DPLL technique. Little effort has gone into investigating alternative techniques.

There are applications that require not only a boolean answer but also the number of models for a propositional formula, or to know which are those models (ALL-SAT), or testing for functional equivalence. These tasks are performed using knowledge compilation. In knowledge compilation, a representation in a source language is compiled into a target language in order to perform reasoning tasks in polynomial time. Popular target languages are binary decision diagrams (BDD) and decomposable negation normal form (d-NNF).

Model counters [12] haven't progressed as much as SAT solvers because SAT heuristics designed to reduce the search space are, in many cases, not applicable, or their effectiveness is heavily reduced. BDD based solving has been an active

B. Bouchou-Markhoff et al. (Eds.): CIAA 2011, LNCS 6807, pp. 76–86, 2011.

research topic and there are efficient BDD based SAT solvers and model counters available, since the model count of a formula can be obtained from a BDD encoding.

> Surprisingly, model counting is the canonical #P-complete problem. Model counting is hard even for some polynomial-time solvable cases like 2-SAT and Horn-SAT. Efficient algorithms for this problem will have a significant impact on many application areas that are inherently beyond SAT, perhaps most importantly, general probabilistic inference. [12]

This work focuses on finite state techniques for SAT solving, an almost totally ignored approach. Approaching SAT as an FSA construction problem offers knowledge compilation capabilities. We can obtain model counting, equivalence testing and ALL-SAT answers from the constructed FSA. An FSA approach offers the advantages of a vast body of research and very simple and thoroughly studied algorithms.

Given the similarities with BDDs [13], it is rather surprising that this approach has not been explored more deeply in the context of propositional satisfiability, at least to our knowledge.[1] We found only one reference [25] that uses an FSA approach in the context of constraint satisfaction programming.

In order to use FSAs in the context of SAT, every valuation satisfying a propositional formula with variables v_i with $i \in [1, n]$ can be represented by a string in e^n, where e is either 1 or 0. If the i-th character of the string is 1 then v_i is *True* in that valuation, otherwise, v_i is *False* (cf. [25] and Theorem 7.3.8 and its corollary in [16][2]). Two crucial aspects are important for this approach to have any reasonable performance: variable ordering (same as in BDD), and clause ordering, if the formula is in conjunctive normal form (CNF). This approach was briefly described in [8]. In the present work we report experiments on a number of benchmarks and show how variable ordering and clause ordering affect considerably the performance. An FSA can be constructed in competitive time compared to state-of-the-art sat solvers.

The remainder of this paper is organized as follows. In Section 2 we provide some basic definitions. Section 3 describes the approach presented in [8] to construct an automaton that defines the language of possible valuations of a propositional formula. Section 4 describes the variable ordering and clause ordering heuristics. In Section 5 we describe the experiments we performed that show the possibilities of an FSA approach to SAT. Section 6 presents conclusions, and questions for future work.

2   Definitions

Most of SAT related definitions and notation follow the ones given in [17]. **L(A)** denotes the language generated by an Automaton or Grammar, A.

---

[1] There is no reference to an FSA approach in the recently published [6].

[2] Thanks to an anonymous reviewer for pointing out this reference.

A clause is a propositional formula of the form $l_1 \vee \ldots l_n$, where each l_i is a *literal* a positive or negated propositional variable.

A term is a propositional formula of the form $l_1 \wedge \ldots l_n$.

Valuations, are defined as functions v on a set of variables Var and with values in $\{0,1\}$. Valuations assign a truth value from $\{0,1\}$ to each propositional variable $p \in Var$. We denote the set of literals (positive or negated variables) determined by the set of variables Var by Lit. Then if $|Var| = n$, $|Lit| = 2n$. We say that a valuation v *satisfies* a formula ϕ or $v \models \phi$.

Complete set of literals: A complete set of literals is a set $S \subseteq Lit$ such that for every $p \in Var$ exactly one of $p, \neg p$ belongs to S. There is a bijective correspondence between valuations and complete sets of literals. One such mapping associates positive literals with 1 and negative literals with 0. An alternative mapping associates positive literals with 0, and negative literals with 1. It follows that if $|Var| = n$, then there are 2^n complete sets of literals over the set Var.

Valuations as strings: There is a correspondence between valuations and strings in $\{0,1\}^n$, therefore valuations can be ordered (anti-)lexicographically. We will say that a word w satisfies a formula ϕ ($w \models \phi$) iff w is the string representation of an element $v \in Val$ and $v \models \phi$.

3   Satisfiability as FSA Construction

Barton [4] uses a finite state machine (FSM) to solve propositional SAT in order to show that descriptive and generative power of PC-Kimmo and Two Level Morphology (TLM) as a grammar device are NP-complete. TLM and PC-Kimmo aimed to the description of morphological properties in a computational linguistics frame. The approach presented in [25] is in the more general framework of constraint satisfaction and introduces a representation of valuations as tuples. However [25] focuses on the construction of the minimized finite state automata (MDFA), an issue that we will ignore. We believe that the prohibitive cost of a direct translation is the reason why such an approach was not further explored. In order to construct the MDFA we assume we have a library that takes as input a regular expression and builds the MDFA. There will be issues of efficiency that will be idiosyncratic and dependant for each implementation of well known algorithms.

We describe how to construct an FSA automaton A for each formula ϕ in CNF,[3] such that the formula is satisfiable iff the language of A is not empty and for every word w in the language of A, $w \models \phi$, i.e. $L(A) = \{w \in \{0,1\}^n | w \models \phi, n = |V_\phi|\}$. This means that the language of the automaton is the string representation of the set of valuations v such that $v \models \phi$.

The construction is based on the mapping between clauses and the dual terms. It is also based on the direct translation between boolean formulas and regular

---

[3] The translation can be easily extended to formulas not in CNF. We performed experiments on formulas not in CNF, in edimacs and iscas format, but that topic exceeds the scope of this paper.

expressions, given the direct correspondence between $\vee, \wedge, \neg$ and $|, \&, \tilde{}$, respectively and the closure properties of finite state automata.

Each clause in a CNF formula will be interpreted as a regular expression that describes the automaton representing the set of valuations that satisfy that clause. For a formula with m clauses the automaton to be constructed will be equal to the intersection of the corresponding m sub-automata. Therefore the asymptotic complexity of this construction is $O(|Var|^m)$. This is probably another reason why this approach was not pursued.

For instance, a clause such as $v_1 \vee v_2 \vee v_3$, from a formula in CNF, with $|Var| = 10$, will be translated as the regular expression $\tilde{}[0\ 0\ 0\ ?\ ?\ ?\ ?\ ?\ ?\ ?]$. We use the notation used in XFST (Xerox Finite State Tool [5]) which we used to do the experimentation. This is equivalent to $\hat{}$'000.......' in languages like python, awk or perl, with an extended use of the complement operator ($\hat{}$), which is used in these languages as a single character complement. Thus $\tilde{}[0\ 0\ 0\ ?\ ?\ ?\ ?\ ?\ ?\ ?]$ matches any string in $(0|1)^{10}$, that does not start with **000**.[4]

In Table 1 we show how a formula with ten variables is translated into a regular expression (second column) and a string (third column). Each clause and the corresponding regular expression are matched in a line. The first line in the regular expression column specifies the valuation space.

Table 1. A propositional formula translated into a regular expression

Formula	Regular Expression	String in $\{a, b, \_\}^n$
	$[1\|0]^{10}$ &	
$(\neg v_1 \vee \neg v_3 \vee \neg v_5) \wedge$	$\tilde{}[1\ ?\ 1\ ?\ 1\ ?\ ?\ ?\ ?\ ?]$ &	a_a_a_____
$(\neg v_1 \vee v_3 \vee v_6) \wedge$	$\tilde{}[1\ ?\ 0\ ?\ ?\ 0\ ?\ ?\ ?\ ?]$ &	a_b__b____
$(v_1 \vee v_4 \vee v_6) \wedge$	$\tilde{}[0\ ?\ ?\ 0\ ?\ 0\ ?\ ?\ ?\ ?]$ &	b__b_b____
$(\neg v_3 \vee v_5 \vee v_8) \wedge$	$\tilde{}[?\ ?\ 1\ ?\ 0\ ?\ ?\ 0\ ?\ ?]$ &	__a_b__b__
$(\neg v_2 \vee v_5 \vee \neg v_8) \wedge$	$\tilde{}[?\ 1\ ?\ ?\ 0\ ?\ ?\ 1\ ?\ ?]$ &	_a__b__a__
$(v_2 \vee v_7 \vee v_9) \wedge$	$\tilde{}[?\ 0\ ?\ ?\ ?\ ?\ 0\ ?\ 0\ ?]$ &	_b____b_b_
$(\neg v_2 \vee v_7 \vee v_9) \wedge$	$\tilde{}[?\ 1\ ?\ ?\ ?\ ?\ 0\ ?\ 0\ ?]$ &	_a____b_b_
$(v_2 \vee \neg v_7 \vee v_9) \wedge$	$\tilde{}[?\ 0\ ?\ ?\ ?\ ?\ 1\ ?\ 0\ ?]$ &	_b____a_b_
$(\neg v_2 \vee \neg v_7 \vee v_9) \wedge$	$\tilde{}[?\ 1\ ?\ ?\ ?\ ?\ 1\ ?\ 0\ ?]$ &	_a____a_b_
$(v_2 \vee \neg v_7 \vee \neg v_9) \wedge$	$\tilde{}[?\ 0\ ?\ ?\ ?\ ?\ 1\ ?\ 1\ ?]$ &	_b____a_a_
$(\neg v_2 \vee \neg v_7 \vee \neg v_9) \wedge$	$\tilde{}[?\ 1\ ?\ ?\ ?\ ?\ 1\ ?\ 1\ ?]$ &	_a____a_a_
$(v_4 \vee \neg v_6 \vee v_{10}) \wedge$	$\tilde{}[?\ ?\ ?\ 0\ ?\ 1\ ?\ ?\ ?\ 0]$ &	___b_a___b
$(\neg v_4 \vee \neg v_9 \vee v_{10}) \wedge$	$\tilde{}[?\ ?\ ?\ 1\ ?\ ?\ ?\ ?\ 1\ 0]$ &	___a____ab
$(v_7 \vee \neg v_9 \vee \neg v_{10})$	$\tilde{}[?\ ?\ ?\ ?\ ?\ ?\ 0\ ?\ 1\ 1]$	_____b_aa

An automaton constructed this way may be used to check which are the strings generated. The translated regular expression is used directly by XFST to compute the automaton. In this case the string generated by the automaton, representing the satisfying valuation of ϕ, will be '1010000110' (or 'ababbbbaab' using the third column representation).

---

[4] We used a for 1 and b for 0, due to restrictions on XFST use of 0.

4   Variable and Clause Ordering Heuristics

The variable ordering heuristics we decided to test ranged from very simple ones to some very elaborate.

Freq. This is the simplest variable ordering heuristic. We sort variables according to the number of clauses they participate in, placing the most frequent first. In this way frequent variables will correspond to states located at the beginning of the FSA. The increased probability of obtaining a single path or no path at the early states, will reduce the size of the constructed automata. If there are ties there is no preference strategy.

Max and Min. These two heuristics are extensions of the previous heuristic (Freq). In Freq, the order of a variable was not affected by its frequency as a positive or negated literal. **Max** gives preference in the ordering to variables that appear most of the times either negated or positive. **Min** will order first those variables that appear a similar number of times as a positive and negated literal.

Johnson. This heuristic is based on the heuristic for the maximum satisfiability (Max-SAT) problem proposed by Johnson [14]. Johnson's heuristic will iteratively satisfy the most frequent literal. Clauses that contain this literal are removed, and the dual literal is removed from the clauses that contain it. The proposed order is the one in which the variables are set. It is another variation of **Freq** heuristics.

Force. Force is a variable ordering heuristic intended to be used with BDDs and SAT solvers [2]. Force is particularly suitable for problems that possess a structure. This iterative algorithm like MINCE (Min-cut vertex/variable reordering), tries to get the minimal cut value for variables (vertex or state). The cut value of a variable with index i is the number of clauses that contain variables with indices both $> i + 0.5$ and $\leq i + 0.5$. This also reduces the average clause span.

Anti-Lexicographic Clause Reordering. Anti-Lexicographic ordering was used in [8] in order to decide satisfiability of formulas in CCNF (i.e., where each clause has the full set of variables) in polynomial time $O(n^6)$.

Anti-lexicographic ordering is equivalent to reverse lexicographic ordering, i.e. ordering starting from the right. In the translation of CNF formulas to XFST regular expressions the following orders were used: ? $< a < b$ (a was used instead of 1 and b instead of 0, as we mentioned above). The order of second and third columns in Table 1 follows the anti-lexicographic ordering.For instance if we have three variables, and each clause has exactly two literals, the anti-lexicographic order in XFST regex will be 1) [a a ?], 2) [b a ?], 3) [a b ?],... 27) [? b b].

This heuristic combined with a variable ordering heuristics, has the effect of computing first the intersection of clauses with variables that have higher priority order and postpone the computation of intersection in clauses with variables that have less priority. Also due to the lexicographic ordering, clauses with smaller *span* (less difference between smallest and largest variable) will be given priority over clauses with bigger span. A third consequence is that clauses that share variables will be placed together in the ordering. All these facts are exemplified in Table 1, third column, above.

5   Experimentation

5.1   Tool and Setting

In order to test the possibilities of using the FSA construction approach as an ALL-SAT and model counter, we translated CNF encoded formulas into regular expressions as explained above. Then we used XFST to build the automata. XFST has the advantage of having a team with a sound experience on FSA tools. These tools have been developed with other purposes in mind (natural language processing, NLP). There are many open source tools that can be more attractive due to the possibility to modify them to try optimizations or profiling. Given this is a first approach in order to build a proof of concept, we considered that it would be a better choice to use a heavily tested and widely used tool. At the same time, the goal of these experiments were not about the FSA implementation of well known algorithms, but testing on differences in the running time due to variable (state) ordering and clause (sub-automata) intersection. XFST documentation is described extensively in [5]. We used XFST PARC version 2.15.2 available online. Variable ordering heuristics as well as clause reordering were developed in C++ and Python. The applications were attached together with Python and bash scripts. Both variable ordering and clause ordering heuristics were done without prioritizing performance. The goal was to compare as many heuristics as possible. Running time seems negligible. Of course if we were considering the strict performance of solvers, all these details have to be measured. All these algorithms have limited time and space complexity.

The sequence is as follows: a) Compute a variable order b) compute the anti-lexicographic order of clauses c) translate into a regular expression d) run XFST on the regular expression to build the automaton. The output of XFST shows the properties of the automaton built with the number of solutions.

The running time of each application was observed using the Python time-it module. Each running time corresponds to a single run of the heuristics followed by XFST, except in the case of Force. Since Force has stochastic behaviour, we decided to run the heuristic with XFST 5 times for each test case, computing the average running time. Experiments were run in a Linux machine with processor Intel Xeon X3430, 2.40GHZ with 8GB of memory.

5.2   Initial Experiments

Previously we had run some experiments in a slower machine. It was observed that the direct translation of the formula to a regular expression had a pro-hibitive processing time, using the XFST regular expression compilation. Those experiments were run on some randomly chosen formulas. For example, processing the translated formula uf50-03.cnf from SATLIB (50 variables/218 clauses), took 864 seconds by XFST. However, after reordering variables by frequency and then reordering the formula in anti-lexicographic ordering took only 0.533 seconds. Processing time reflects the size of intermediate automata in the construction. The final size of the constructed automaton is not representative of the complexity of intermediate steps. Reordering the same formula with the well known static variable reordering algorithm, Force[2], did not improve as much the running time. For instance on the same formula uf50-03, after reordering variables with Force processing time of XFST took 527 sec.

Later, we ran all the heuristics on the 1000 formulas of the uf50 SATLIB benchmark. Most of the heuristics exceeded the time limit of 30 sec. that we had set. The rest had the following average timings: a) Freq-AL: 1.62 sec. b) Force-AL: 2.14 sec. and Johnson-AL: 4.20 sec.

The first part of the heuristic names refers to the variable ordering heuristics described above in Section 4. The second part of the names, AL, refers to anti-lexicographic ordering. In the next subsection, we use NR, denoting no reordering of the original clause ordering. The anti-lexicographic ordering has shown to be a consistent strategy to limit the explosion of state size in the computation of automata intersection.

5.3   Hard Benchmarks Experiments

Initially we performed some tests using some of the ebddres benchmarks used in [19].[5] We chose them in order to compare with Ebddres, given it is a BDD based solver. These are hard unsatisfiable problems. Many of these problems have been looked at even with local search solutions [3,1]. **ph** files are instances of the pigeon hole problems. **Chnl** are unsatisfiable instances that model the routing of X wires in N channels [1]. **Urq** files are unsatisfiable randomized instances based on expander graphs [22]. **Fpga** are some satisfiable and unsatisfiable instances from FPGA routing. **Mutcb** instances correspond to the mutilated checker board. Then we added some other classes known to be hard, instances from the Beijing and Hanoi set (**2bit, hanoi**) from SATLIB. We also added some of the BMC-dimacs benchmark (**barrel,queueinv,longmult**). The details and properties of these benchmarks can be found in Table 6.

The results were very good considering we were just implementing. However, they were rather disparate on some benchmarks.

Table 2 summarizes Min, Max, and Force variable reordering heuristics, with NR (no clause reordering). Null-NR, corresponds to the direct translation of the formula into a regular expression (no variable nor clause reordering). As it can be seen Force-NR is the best in this set, although Null-NR (plain translation) is pretty close. The good performance of the *NR* class follows from the fact that these instances were constructed with some sort of anti-lexicographic ordering. Also in some cases the ordering of the variables seems to be close to the ordering computed by some heuristics.

---

[5] They are available at the Ebddres web page.

Table 2. FSA Variable Ordering Heuristics with No Clause Reordering

File	Min-NR	Force-NR	Null-NR	Max-NR
Total sec	20618	16455	18384	21771
# Not solved	29	24	25	33

Table 3 summarizes Freq, Min, Max, and Force, Johnson variable reordering heuristics combined with anti-lexicographic ordering (AL). Null-AL, corresponds to no variable re-ordering. Given the performance on the uf50 benchmark, the heuristics combined with AL were expected to have better performance. Force-AL was the best performing in both classes (NR and AL).

Table 3. FSA with variable and anti-lexicographic clause reordering heuristics

File	Min-AL	Force-AL	Johnson-AL	Null-AL	Freq-AL	Max-AL
Total sec	18052	10847	25297	19710	21017	19129
# Not Solved	26	17	40	28	30	27

In order to have an approximate comparison with alternative approaches with model counting or ALL-SAT capabilities, we ran experiments with the following solvers: clasp, Ebddres, sbsat, sharpSAT, c2d and relsat. Model counting and ALL-SAT, is not relevant for unsatisfiable instances anyway (most of them, given the number of models is zero).[6]

Clasp obtained the gold medal for SAT/UNSAT crafted problems, in the last (2009) competition. It contains many advanced features, and also the capability to obtain a model counter or all-sat.

Ebddres [19] (version 1.0), is a BDD based SAT solver that can generate extended resolution proof traces.

Sbsat [11] is a state-based, BDD-based satisfiability solver. We used version sbsat-2.7b.

Relsat is a model counter that was developed a few years ago.

Sharpsat [21], is a #SAT solver that is based on DPLL algorithm. It is supposed to have a good performance on large structure problems.

C2d [9], compiles CNF into d-NNF (decomposable negation normal form) a generalization of BDD. SharpSAT, was also used to compile CNF into d-NNF[18].

---

[6] These are the parameters we used to run each solver when we did not use the default values:
- clingo –clasp -n 0 -q (clasp mode, enumerate all models, quite mode)
- NetPlacer -c 6 (affects the output variable order, this is one of Force executables)
- relsat -♯count -t600 (Count models, time limit)
- sbsat -All 0 -In 0 –max-solutions 0 -t –debug 0 (disable preprocessing options, disable inferences, find all solutions, start a stripped down version of the SMURF solver, disable debug).

In the following Table 4 we compare Force-AL, the best performance of the heuristics we tried, against the above mentioned solvers. As it can be seen its overall performance is very good. It is almost tied with ebddres in total time used but Force-AL solved more problems. However the average time used by Force-AL is lower than the one used by ebddres and sharpsat.

Table 4. FSA with Force and AL clause ordering vs other solvers

Heuristics	Force-AL	sbsat	ebddres	c2d	relsat	sharpsat	clasp
Total time	10847	18051	10764	16969	21638	18321	15471
Not Solved	15	28	17	27	29	29	21
Solving Time	1847	4250	564	2569	4237	920	2871
Solved	36	23	34	24	22	22	30
Average	51,39	184.78	16.58	107.04	192.58	41.81	95.7

In Table 5, we present the first four solvers or FSA construction strategy that had the best timings for each subset of problems. It can be seen that, for a number of problems (ph,Urq,chnll,fpga), the first positions are dominated by the FSA construction strategies, Force-AL, being the most predominant. However for other subsets, current solvers perform much better.

Table 5. Best timings for problem subsets

Problem set	First	Time	Second	Time	Third	Time	Fourth	Time
ph	Force-NR	844	Force-AL	1847	Max-AL	4060	sbsat	4068
mutcb	ebddres	7	c2d	264	clasp		844 Null-NR	906
Urq	ebddres	629	Force-AL	700	Force-NR	825	Max-AL	844
chnll	Force-AL	33	ebddres	49	Null-NR	999	Max-NR	1211
fpga	Force-AL	908	ebddres	1266	Null-NR	2644	*	*
sat-grid	sbsat	3	ebddres	7	Max-AL	7.15	c2d	92
barrel	sbsat	3	relsat	3	clasp	3	c2d	3
queueinv	sbsat	2	relsat	2	clasp	2	c2d	5
hanoi	sbsat	2	clasp	9	relsat	605	ebddres	609
2bit	clasp	1213	ebddres	1242	c2d	1242	sbsat	1968

If we analyze the data from these set of instances, we can observe significant differences between them. It looks like problems like ph,unsat-fpga, cannot be solved with usual features in most solvers, due to a similar distribution of variables in clauses (e.g. each variable occurs the same number of times).

6 Conclusions and Future Work

SAT solving and model counting is now present in many practical applications but these are NP and #NP complete problems. This paper evaluates the performance impact of several variable ordering heuristics on an FSA based SAT approach with ALL-SAT and model counting capabilities. Variable ordering and clause ordering (automata intersection) are problems known to be NP-complete. Force heuristic generated far better variable orderings in problems with a particular structure. That was not the case in random class problems (uf50), where Freq is better than Force. The anti-lexicographic ordering proved to be a consistent strategy, with still room for improvement. Most importantly the results of these experiments show that the FSA approach is very competitive versus the traditional DPLL approach in some hard problems (problems where most of the features added to the basic DPLL algorithm don't help). This should not be interpreted as saying that the FSA approach is better than the DPLL or BDD/NNF approaches. There are decades of research and experience that cannot be surpassed with this initial proposal. We believe that the most added value of these results are the questions they bring up front. Those questions should guide future research. Some of them are related to the knowledge built upon the DPLL and BDD tradition and concern mainly on how much of that experience can be used in an FSA approach. Other kind of questions are related to the FSA community and concern mainly on what can be the optimal way of constructing a MDFA for this class of languages. A second important question that should be elucidated is whether an approach to a fixed large window of k variables and m clauses can solve practical problems in a different way.[7]

Acknowledgements

This research was funded by UBACYT X-415 and ANPCyT PICT-00263-2008. Many thanks also to the anonymous reviewers who provided detailed and helpful comments to improve this version.

References

1. Aloul, F., Lynce, I., Prestwich, S.: Symmetry Breaking in Local Search for Unsatisfiability. In: 7th International Workshop on Symmetry and Constraint Satisfaction Problems, Providence, RI (2007)
2. Aloul, F.A., Markov, I.L., Sakallah, K.A.: FORCE: a fast and easy-to-implement variable-ordering heuristic. In: ACM Great Lakes Symposium on VLSI, pp. 116–119. ACM Press, New York (2003)
3. Aloul, F.A., Ramani, A., Markov, I.L., Sakallah, K.A.: Solving difficult SAT instances in the presence of symmetry. In: DAC, pp. 731–736. ACM Press, New York (2002)
4. Barton, G.E.: Computational complexity in two-level morphology. In: Proc. of the 24th ACL, New York, pp. 53–59 (1986)
5. Beesley, K., Karttunen, L.: Finite State Morphology. CSLI Publications (2003)
6. Biere, A., Heule, M., van Maaren, H., Walsh, T. (eds.): Handbook of Satisfiability. IOS Press, Amsterdam (2009)
7. Büchi, J.R.: Weak second-order arithmetic and finite automata. Zeit. Math. Logik. Grund. Math. 66–92 (1960)

---

[7] See [20,15] for intersection of a limited number of automata.

8. Castaño, J.: Two views on crossing dependencies, language, biology and satisfiability. In: 1st International Work-Conference on Linguistics, Biology and Computer Science: Interplays. IOS Press, Amsterdam (2011)
9. Darwiche, A.: New Advances in Compiling CNF into Decomposable Negation Normal Form. In: ECAI, pp. 328–332 (2004)
10. Elgot, C.C.: Decision problems of automata design and related arithmetics. Transactions of the American Mathematical Society (1961)
11. Franco, J., Kouril, M., Schlipf, J., Ward, J., Weaver, S., Dransfield, M.R., Vanfleet, W.M.: SBSAT: a state-based, BDD-based satisfiability solver. In: Giunchiglia, E., Tacchella, A. (eds.) SAT 2003. LNCS, vol. 2919, pp. 398–410. Springer, Heidelberg (2004)
12. Gomes, C.P., Sabharwal, A., Selman, B.: Model Counting. In: Handbook of Satisfiability. Frontiers in Artificial Intelligence and Applications, vol. 185, pp. 633–654. IOS Press, Amsterdam (2009)
13. Hadzic, T., Hansen, E.R., O'Sullivan, B.: On Automata. In: MDDs and BDDs in Constraint Satisfaction (2008)
14. Hansen, P., Jaumard, B.: Algorithms for the maximum satisfiability problem. Computing 44, 279–303 (1990)
15. Lange, K., Rossmanith, P.: The emptiness problem for intersections of regular languages. In: Havel, I.M., Koubek, V. (eds.) MFCS 1992. LNCS, vol. 629, pp. 346–354. Springer, Heidelberg (1992)
16. Lewis, H.R., Papadimitriou, C.H.: Elements of the Theory of Computation, 2nd edn. Prentice-Hall, Upper Saddle River (1997)
17. Marek, V.W.: Introduction to Mathematics of Satisfiability. Chapman and Hall/CRC (2010)
18. Muise, C., Beck, J.C., McIlraith, S.: Fast d-DNNF Compilation with sharpSAT (2010)
19. Sinz, C., Biere, A.: Extended resolution proofs for conjoining BDDs. In: Grigoriev, D., Harrison, J., Hirsch, E.A. (eds.) CSR 2006. LNCS, vol. 3967, pp. 600–611. Springer, Heidelberg (2006)
20. Tapanainen, P.: Applying a Finite-State Intersection Grammar. In: Roche, E., Schabes, Y. (eds.) Finite-State Language Processing, pp. 311–327. MIT Press, Cambridge (1997)
21. Thurley, M.: sharpSAT – counting models with advanced component caching and implicit BCP. In: Biere, A., Gomes, C.P. (eds.) SAT 2006. LNCS, vol. 4121, pp. 424–429. Springer, Heidelberg (2006)
22. Urquhart, A.: Hard examples for resolution. J. ACM 34(1), 209–219 (1987)
23. Vardi, M.: Logic and Automata: A Match Made in Heaven. In: Baeten, J.C.M., Lenstra, J.K., Parrow, J., Woeginger, G.J. (eds.) ICALP 2003. LNCS, vol. 2719, pp. 193–193. Springer, Heidelberg (2003)
24. Vardi, M.Y., Wolper, P.: Automata-Theoretic techniques for modal logics of programs. J. Comput. Syst. Sci. 32, 183–221 (1986)
25. Vempaty, N.R.: Solving Constraint Satisfaction Problems Using Finite State Automata. In: AAAI, pp. 453–458 (1992)

Table 6. Force-AL versus other solvers, time limit reached at 600s. indicated by *

File	#Var	#CL	Force-AL	sbsat	ebddres	c2d	relsat	SharpSAT	clasp
ph07	56	204	1.042	1.00	1.00	1.00	1.00	1.00	1.00
ph08	72	297	1.046	1.00	1.00	6.00	2.00	1.00	1.00
ph09	90	415	5.06	0.99	1.00	85.00	18.00	5.00	6.00
ph10	110	561	8.27	1.00	2.00	*	231.00	50.00	48.00
ph11	132	738	14.09	2.00	5.00	*	599.00	*	400.00
ph12	156	949	23.11	30.00	12.00	*	599.00	*	*
ph13	182	1197	42.34	432.00	39.00	*	*	*	*
ph14	210	1485	77.37	*	*	*	*	*	*
ph15	240	1816	151.21	*	*	*	*	*	*
ph16	272	2193	322.05	*	*	*	*	*	*
ph17	306	2619	*	*	*	*	*	*	*
mutcb8	121	344	1.048	1.00	1.0	2.00	1.00	1.00	1.00
mutcb9	155	451	3.05	1.00	1.0	2.00	2.00	1.00	1.00
mutcb10	193	572	6.059	3.00	1.0	3.00	12.00	5.00	1.00
mutcb11	235	707	22.075	*	1.0	5.00	*	98.99	5.00
mutcb12	281	856	70.08	*	1.0	13.00	*	441.00	32.00
mutcb13	331	1019	269.89	*	1.0	42.00	*	*	204.00
mutcb14	385	1196	*	*	2.0	197.00	*	*	*
Urq3_5.cnf	46	470	6.06	*	1.0	23.00	*	*	79.00
Urq4_5.cnf	74	694	94.07	*	28.0	234.00	599.99	*	*
Urq5_5.cnf	121	1210	*	*	*	*	*	*	*
chnl10_11.cnf	220	1122	2.09	*	2.0	*	*	49.00	32.00
chnl10_12.cnf	240	1344	3.10	*	3.0	*	599.99	53.00	36.00
chnl10_13.cnf	260	1586	3.11	*	3.0	*	599.99	55.00	42.00
chnl11_12.cnf	264	1476	4.11	*	6.0	*	*	*	319.00
chnl11_13.cnf	286	1742	5.12	*	7.0	*	*	*	469.00
chnl11_20.cnf	440	4220	15.27	*	28.0	*	*	*	566.00
fpga11_15_unsat	330	2340	12.16	*	*	*	*	*	*
fpga11_20_unsat	440	4220	20.48	*	*	*	*	*	*
fpga12_11_sat	198	968	173.08	*	14.99 r	*	*	*	*
fpga12_12_sat	216	1128	101.89	*	18.00 r	*	*	*	*
fpga13_9_sat	176	759	*	*	33.00 r	*	*	*	*
sat-grid-pbl-0010	110	191	1.06	1.00	1.00	2.00	*	1.00	1.00
sat-grid-pbl-0015	420	781	47.05	1.00	1.00	6.00	*	*	*
sat-grid-pbl-0020	930	1771	*	1.00	5.00	84	599.99	*	*
barrel2	50	159	1.03	1.00	1.00	1.00	1.00	1.00	1.00
barrel3	275	942	268.08	1.00	*	3.00	1.00	1.00	1.00
barrel4	578	2035	*	1.00	*	4.00	1.00	1.00	1.00
queueinv2	116	399	3.05	1.00	1.01	2.00	1.00	1.00	1.00
queueinv4	256	955	*	1.00	293.00	3.00	1.00	1.00	1.00
longmult0	437	1206	*	*	*	*	*	*	*
longmult1	791	2335	*	*	*	*	*	*	*
hanoi4	718	4934	63.38	1.00	*	9.00	5.00	2.00	1
hanoi5	1931	14468	*	1.00	*	*	*	*	8
2bitcomp_5	125	310	4.04	7.00	*	2.00	2.00	1.00	1
2bitmax_6	252	766	*	*	*	12.00	42.00	2.00	2
2bitadd_10	590	1422	*	161.00	*	28.00	322.00	149.00	10
2bitadd_11	649	1562	*	*	*	*	*	*	*
2bitadd_12	708	1702	*	*	*	*	*	*	*

Nondeterministic Moore Automata and Brzozowski's Algorithm

Giusi Castiglione, Antonio Restivo, and Marinella Sciortino

Dipartimento di Matematica e Informatica
Università di Palermo, via Archirafi, 34 - 90123 Palermo, Italy
{giusi,restivo,mari}@math.unipa.it

Abstract. Moore automata represent a model that has many applications. In this paper we define a notion of coherent nondeterministic Moore automaton (NMA) and show that such a model has the same computational power of the classical deterministic Moore automaton. We consider also the problem of constructing the minimal deterministic Moore automaton equivalent to a given NMA. In this paper we propose an algorithm that is a variant of Brzozowski's algorithm in the sense that it is essentially structured as reverse operation and subset construction performed twice.

1 Introduction

In this paper we consider finite-state automata with output, i.e. automata viewed as computers of functions, not as recognizers of languages. The simplest model of automata with output are Moore automata. A Moore automaton is a deterministic finite-state machine whose output values are determined by its current state. Moore automata are named for Edward Forrest Moore who first studied them in 1956 (cf. [15]). *Acceptors*, i.e. deterministic automata recognizing languages, can be considered as particular Moore automata having a binary output ${False, True}$. So, in acceptors we distinguish between accepting states (states associated to the output $True$) and rejecting states (states with output $False$).

The notion of nondeterministic acceptors was introduced by Rabin and Scott in [16]. A nondeterministic acceptor is a machine with many choices, in the sense that for a given input string, it may exhibit several different transition sequences (paths). An input string is accepted if at least one of the possible paths, defined by the input, leads to an accepting state (winning path). In the literature, there exist several notions of nondeterminism also for automata with output, and in particular for Moore automata, that have been introduced in specific areas and are often motivated by specific applications (see for instance [8,20,12,14,21]).

In this paper we are interested in a notion of nondeterministic Moore automaton (NMA) that takes into account its behavior as computer of functions. In particular, we introduce the model of NMA equipped with a property called coherency and we prove that such a model has the same computational power of the classical deterministic Moore automaton (DMA). In fact, by using an

B. Bouchou-Markhoff et al. (Eds.): CIAA 2011, LNCS 6807, pp. 88–99, 2011.

adaptation of the subset construction, we prove that to each coherent NMA corresponds an equivalent deterministic one (i.e. that computes the same function). In this sense, our nondeterministic model can be viewed as a succinct representation of a function, since it can be exponentially smaller than the equivalent deterministic model.

In this paper we face also with the problem of simulate a coherent NMA by the minimal equivalent DMA. In order to solve such a problem we define a minimization algorithm that is a variant of Brzozowski's algorithm (cf. [2]). This approach is not immediate since Brzozowski's algorithm has been introduced for nondeterministic acceptors in which there is an asymmetry on the outputs: the output $True$, corresponding to a winning path, is privileged with respect to the output $False$, corresponding to a non-winning path. In Moore automata we do not distinguish between winning paths and non-winning paths, so there is no privileged output symbol. As for Brzozowski's algorithm, the method we propose is essentially structured on the operations of reverse and subset construction performed twice but such operations in the context of Moore automata assume different meanings.

The paper is organized as follows. In the first section we give the definition of the nondeterministic Moore automaton and show that our model is computationally equivalent to the classical deterministic Moore automaton. The second section is devoted to the definition of the variant of Brzozowski's algorithm to construct the minimal deterministic Moore automaton equivalent to a given coherent nondeterministic Moore automaton. The last section contains some conclusions and new research directions on this topic.

2   Nondeterministic Moore Automata

A *Moore automaton* is a classical notion (cf. [15]) in the Theory of Automata. It is an automaton with output because an output is associated to each state and the system emits an output as a function of a given input. Because of its several applications in many areas, as for instance system modeling, natural languages processing, system verification, machine learning (cf. for instance [12,7,13]), it was useful to introduce some elements of nondeterminism in such a computational model.

In this paper we would highlight the computational aspect of a Moore automaton, and in particular its ability to compute functions. Therefore, in this section, we introduce a nondeterministic Moore automaton with a property related to this goal.

A *nondeterministic Moore automaton* (denoted by NMA) is a system $\mathcal{A} = (\Sigma, \Gamma, Q, I, \Delta, \lambda)$ where Σ is the set of *input* symbols, $\Gamma = \{\gamma_1, \gamma_2, \ldots, \gamma_k\}$ is the set of *output* symbols (also called *colors*), Q is the set of *states*, $I \subseteq Q$ is the set of *initial* states, $\Delta \subseteq Q \times \Sigma \times Q$ is the set of the *transitions* of $\mathcal{A}$. Finally, $\lambda : Q \mapsto \Gamma$ is a partial *output function* that assigns a color to some states of the automaton. Note that in such a model both the input symbols and the output symbols could not be defined for all the transition or all the states, respectively.

The set $Q \setminus dom(\lambda)$ contains the not colored states, i.e. the states that do not output any symbol. By the triple (p, σ, q) (with $\sigma \in \Sigma$) we denote the transition from the state p to the state q labeled by σ. A path π of $\mathcal{A}$ labeled by the word $v = v_1 v_2 \ldots v_n \in \Sigma^*$ is a sequence $\{(q_i, v_i, q_{i+1})\}_{i=1,\ldots,n}$ of consecutive transitions. If λ is defined for q_{n+1}, we say that $\lambda(q_{n+1})$ is an output produced by v and we say that π is *colored* and has $\lambda(q_{n+1})$ as color.

A word v is *applicable* for the state q if there exists at least a path π labeled by v starting from q. A word v is applicable for the automaton $\mathcal{A}$ if it is applicable for at least an initial state. To each applicable word v of $\mathcal{A}$ we can associate many paths labeled by v. We denote by $L(\mathcal{A})$ the language of all applicable words of $\mathcal{A}$. A nondeterministic Moore automaton is *complete* if the language $L(\mathcal{A})$ is equal to Σ^*.

The NMA $\mathcal{A} = (\Sigma, \Gamma, Q, I, \Delta, \lambda)$ is *coherent* if for each applicable word v of $\mathcal{A}$ there exists at least a colored path labeled by v and all the colored paths associated to v have the same color. One can deduce that in a coherent nondeterministic Moore automaton at least one initial state must be colored and all colored initial states must have the same color. From the definition it follows that a coherent nondeterministic Moore automaton implicitly defines a partial function $f_{\mathcal{A}}$ from Σ^* to Γ that to each applicable word v of $\mathcal{A}$ associates a color that is the color of an associated colored path. The domain of the function is the language $L(\mathcal{A})$. Equivalently, we can say that the coherent NMA $\mathcal{A}$ induces a partition of $L(\mathcal{A})$ into the languages $\{L_i\}_{1 \leq i \leq k}$ where $L_i(\mathcal{A}) = \{w \in L(\mathcal{A}) \mid f_{\mathcal{A}}(w) = \gamma_i\}$.

Recall that the classical definition of *deterministic Moore automaton* (DMA) can be obtained by a nondeterministic Moore automaton in which Δ is a function (not necessarily total and often denoted by δ) from $Q \times \Sigma$ to Q, $|I| = 1$ and λ is a total function. Note that the coherent NMA is a model that takes an intermediate place between NMA and DMA.

Example 1. In Fig. 1(a) a coherent NMA $\mathcal{A} = (\Sigma, \Gamma, Q, I, \Delta, \lambda)$ is depicted, where $Q = \{1, 2, 3, 4, 5, 6\}$, $I = \{1, 2\}$, $\Sigma = \{a, b, c\}$, $\Gamma = \{Red, Green, Yellow\}$, $\lambda(2) = \lambda(6) = Red$, $\lambda(4) = Yellow$, $\lambda(5) = Green$. Output symbols are denoted with the initial letter of the color. The language of applicable words is $L(\mathcal{A}) = (a + c)^*(b + bb)(a + c)^* + \epsilon$.

We say that two coherent NMA's $\mathcal{A}, \mathcal{B}$ are *equivalent* if they define the same functions $f_{\mathcal{A}}$ and $f_{\mathcal{B}}$, or equivalently $L(\mathcal{A}) = L(\mathcal{B})$ and the induced partition is the same (up to renaming the output symbols). A coherent NMA is *minimal* if it has minimal number of states among its equivalent ones. As in the case of nondeterministic acceptors (i.e recognizing regular languages), such a minimal nondeterministic model could be not unique.

Given an NMA $\mathcal{A} = (\Sigma, \Gamma, Q, I, \Delta, \lambda)$ one can pose the following problems: 1. to decide whether $\mathcal{A}$ is coherent; 2. if $\mathcal{A}$ is a coherent NMA, to find an equivalent DMA. An answer to both the problems is given in Proposition 1.

Firstly, we describe an operation that is an adaptation of the subset construction for NFA and it will be fundamental also in the next section.

We can associate to the NMA $\mathcal{A} = (\Sigma, \Gamma, Q, I, \Delta, \lambda)$ the labeled colored *state graph* $\mathcal{G} = (N_{\mathcal{G}}, E_{\mathcal{G}}, \lambda_{\mathcal{G}})$ that is obtained from $\mathcal{A}$ by neglecting the information

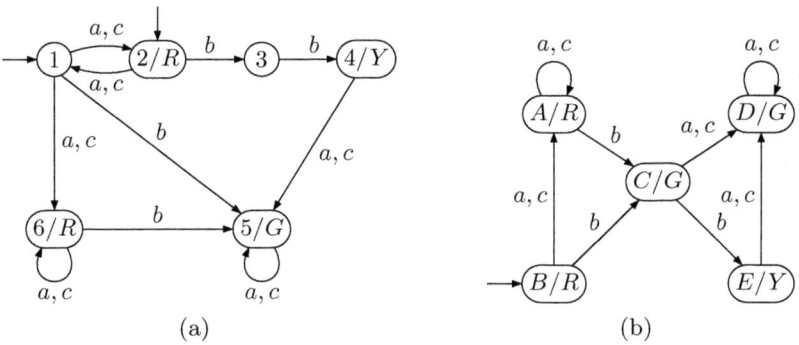

Fig. 1. A coherent nondeterministic Moore automaton $\mathcal{A}$ (a) and the equivalent DMA obtained by the subset construction on $\mathcal{A}$ and the set I of initial states (b)

Algorithm 1: Subset Construction on the pair $(\mathcal{G} = (N_\mathcal{G}, E_\mathcal{G}), P)$

1   $N_d = P \quad E_d = \emptyset \quad W = N_d$
2   **while** $W \neq \emptyset$ **do**
3   **extract** $\mathbb{p}$ **from** W
4   **for** $a \in \Sigma$ **do**
5          $\mathbb{q} = \{q \mid (p, a, q) \in E_\mathcal{G}, p \in \mathbb{p}\}$
6          $E_d = E_d \cup (\mathbb{p}, a, \mathbb{q})$
7   **if** $\mathbb{q} \notin N_d$ **then**
8              $W = W \cup \mathbb{q}$
9              $N_d = N_d \cup \mathbb{q}$

10   **return** $sub(\mathcal{G}) = (N_d, E_d)$

Fig. 2. Algorithm to compute the subset construction

about the initial states. The elements of $N_\mathcal{G}$ are called nodes or states of $\mathcal{G}$ and they are colored as in $\mathcal{A}$.

The *subset construction* takes as input a labeled graph $\mathcal{G}$ and a set P of subsets of $N_\mathcal{G}$. It produces a graph $sub(\mathcal{G}) = (N_d, E_d)$ in which the states are subsets of states of $\mathcal{G}$ accessible by the elements of P. Such an operation is described in Fig. 2.

Given an NMA $\mathcal{A} = (\Sigma, \Gamma, Q, I, \Delta, \lambda)$ and its state graph $\mathcal{G}$, for the graph returned by the subset construction of $(\mathcal{G}, P)$ we can define a coloring so that $sub(\mathcal{G})$ can be considered the state graph of an NMA. Therefore, we consider the *subset coloring function* λ_d defined as follows: $\lambda_d(\mathbb{p}) = \gamma_i$ if $\mathbb{p} \in N_d$ contains at least a state of Q colored by γ_i and it does not contain states of different color. Hence $sub(\mathcal{G})$ is the state graph of the NMA $sub_P(\mathcal{A}) = (\Sigma_d, \Gamma_d, Q_d, P, \Delta_d, \lambda_d)$, where $\Sigma_d = \Sigma$, $\Gamma_d = \Gamma$, $Q_d = N_d \subseteq \mathcal{P}(Q)$, $\Delta_d = E_d$. Note that, in $sub_P(\mathcal{A})$ the states are subsets of states of $\mathcal{A}$ and in particular Q_d is the set of all accessible states from the subsets (states) in P. Note also that, by construction, given

$\mathbb{p} \in Q_d$ and $a \in \Sigma$ there exists at most one $\mathbb{q}$ such that $(\mathbb{p}, a, \mathbb{q}) \in E_d$, so that Δ_d can be considered as a function from $Q_d \times \Sigma$ to Q_d.

Let us consider the subset construction applied to the state graph $\mathcal{G}$ of $\mathcal{A}$ and the set $P = \{I\}$. One can notice that such a construction works like the subset construction defined for the acceptors. By using the subset coloring function λ_d, we obtain the NMA $sub_{\{I\}}(\mathcal{A}) = (\Sigma, \Gamma, Q_d, \mathbb{q}_0, \Delta_d, \lambda_d)$ in which $\mathbb{q}_0 = I$ and Q_d are the states reachable from $\mathbb{q}_0$. Fig. 1(b) reports the automaton $sub_{\{I\}}(\mathcal{A})$ obtained by applying the subset construction to the NMA depicted in Fig. 1(a) and the set of its initial states. From next proposition, $sub_{\{I\}}(\mathcal{A})$ is a DMA.

Proposition 1. *A nondeterministic Moore automaton $\mathcal{A} = (\Sigma, \Gamma, Q, I, \Delta, \lambda)$ is coherent if and only if $sub_{\{I\}}(\mathcal{A}) = (\Sigma, \Gamma, Q_d, \mathbb{q}_0, \Delta_d, \lambda_d)$ is a deterministic Moore automaton. Moreover, $\mathcal{A}$ and $sub_{\{I\}}(\mathcal{A})$ are equivalent.*

Proof. It follows from the fact that $\mathcal{A}$ is a coherent NMA if and only if the subset coloring function λ_d is a total function. In fact each set of Q_d contains at least a colored state of Q and cannot contain states of Q of different color. Moreover, note that by construction Δ_d is a function. The equivalence follows from that fact that by construction the languages of applicable words in the coherent NMA $\mathcal{A}$ and in $sub_{\{I\}}(\mathcal{A})$ are the same as well as their induced partition. □

Let $\mathcal{A} = (\Sigma, \Gamma, Q, q_0, \delta, \lambda)$ be a DMA with initial state q_0. The function δ can be recursively extended to a partial function from $Q \times \Sigma^*$ to Q as follows. Let $q \in Q$, $w \in \Sigma^*$ and $a \in \Sigma$, we define $\delta(q, \epsilon) = q$ and $\delta(q, aw) = \delta(\delta(q, a), w)$, if $\delta(q, a)$ is defined. The notion of minimality of a DMA is connected to an equivalence relation among states of Q as follows (cf. [15]). Firstly, we say that two state $p, q \in Q$ are *distinguishable* if, either there exists $w \in \Sigma^*$ that is applicable for p or for q but not for both, or there exists $w \in \Sigma^*$ applicable for both and $\lambda(\delta(p, w)) \neq \lambda(\delta(q, w))$. We say $p, q \in Q$ to be *indistinguishable* and we write $p \sim q$ if for each $w \in \Sigma^*$ that is applicable for both, we have $\lambda(\delta(p, w)) = \lambda(\delta(q, w))$. It is easy to prove that the indistinguishability is an equivalence relation in Q. By using such a relation a reduced automaton can be constructed from a given DMA and it is possible to prove that such an automaton is the minimal equivalent. Note that the minimal DMA equivalent to a given DMA is unique (cf. [15]) up to isomorphism. An example of minimization of a DMA can be found also in [3] where Moore's method is described. An approach by using another equivalence relation is proposed in [18]. Very recently, an implementation of a minimization algorithm based on an operation of gluing two states and on a representation by transition list is considered (cf. [17]).

3   A Variant of Brzozowski's Algorithm on Nondeterministic Moore Automata

The main goal of this paper is to address the problem of *minimizing a coherent nondeterministic Moore automaton* that means to search for the minimal equivalent DMA. Since such a problem is significant for coherent NMA's, in the rest of the paper we will simply denote a coherent automaton by NMA.

Algorithm 2: Reverse operation on $\mathcal{G} = (N_{\mathcal{G}}, E_{\mathcal{G}}, \lambda_{\mathcal{G}})$

1   $N_r = N_{\mathcal{G}}$ $\lambda_r = \lambda_{\mathcal{G}}$ $E_r = \emptyset$
2   **for** $(p, a, q) \in E_{\mathcal{G}}$ **do**
3     $\lfloor$ $E_r = E_r \cup (q, a, p)$
4   **return** $rev(\mathcal{G}) = (N_r, E_r, \lambda_r)$

Fig. 3. Algorithm to compute the reverse of a colored graph

Algorithm 3: Minimization of $\mathcal{A} = (\Sigma, \Gamma, Q, I, \Delta, \lambda)$

1   $\mathcal{G}$ = **state graph of** $\mathcal{A}$
2   $\mathcal{R} = (N_{\mathcal{R}}, E_{\mathcal{R}}, \lambda_{\mathcal{R}}) \leftarrow$ **Reverse operation on** $\mathcal{G}$
3   **for** $j = 1, \ldots, |\Gamma|$ **do**
4     $\lfloor$ $I_j = \{q \in N_{\mathcal{R}} \mid \lambda_{\mathcal{R}}(q) = \gamma_j\}$
5   $P = \{I_j\}_{j=1}^{|\Gamma|}$
6   $(N_{\mathcal{D}}, E_{\mathcal{D}}) \leftarrow$ **Subset Construction on** $(\mathcal{R}, P)$
7   **for** $j = 1, \ldots, |\Gamma|$ **do**
8     $\lfloor$ $\lambda_{\mathcal{D}}(I_j) = \gamma_j$
9   $\mathcal{F} = (N_{\mathcal{F}}, E_{\mathcal{F}}, \lambda_{\mathcal{F}}) \leftarrow$ **Reverse operation on** $\mathcal{D} = (N_{\mathcal{D}}, E_{\mathcal{D}}, \lambda_{\mathcal{D}})$
10   $\mathbf{m}_0 = \{\mathbf{q} \in N_{\mathcal{F}} \mid \mathbf{q} \cap I \neq \emptyset\}$
11   $(N_{\mathcal{M}}, E_{\mathcal{M}}) \leftarrow$ **Subset Construction on** $(\mathcal{F}, \mathbf{m}_0)$
12   **for** $\mathbf{p} \in N_{\mathcal{M}}$ **do**
13 | **if** $\mathbf{p} \cap P = I_j$ **then**
14 | $\lfloor$ $\lambda_{\mathcal{M}}(\mathbf{p}) = \gamma_j$
15   $\mathcal{A}_M = (\Sigma, \Gamma, N_{\mathcal{M}}, \mathbf{m}_0, E_{\mathcal{M}}, \lambda_{\mathcal{M}})$
16   **return** $\mathcal{A}_M$

Fig. 4. Algorithm to minimize an NMA $\mathcal{A}$

Let $\mathcal{A} = (\Sigma, \Gamma, Q, I, \Delta, \lambda)$ be an NMA. We propose an algorithm, inspired by Brzozowski's algorithm (cf. [2,11]), to minimize an NMA. We consider the labeled colored state graph $\mathcal{G} = (N_{\mathcal{G}}, E_{\mathcal{G}}, \lambda_{\mathcal{G}})$ associate to $\mathcal{A}$.

In previous section we defined the subset construction of a labeled graph and a set P of subsets of states. Such an operation, together with another operation defined in this section, will be fundamental steps of the algorithm.

Given a labeled graph $\mathcal{G}$ we call *reverse* of $\mathcal{G}$ (and denoted by $rev(\mathcal{G})$) the graph obtained by inverting the edges of $\mathcal{G}$. If $\mathcal{G}$ is colored, $rev(\mathcal{G})$ inherits the same coloring. Such an operation on a colored graph is described in Fig. 3.

We describe now the algorithm to minimize the NMA $\mathcal{A}$. As well as for Brzozowski's algorithm applied to an NFA, our algorithm is based on four phases that use reverse operation and subset construction that are variants of operations defined on the acceptors. Note that, the intermediate steps of the algorithm produce graphs whose nodes are subsets or set of subsets of states that we denote by $\mathbf{p}, \mathbf{q}, \mathbf{s}, \ldots$ and $\mathbf{p}, \mathbf{q}, \mathbf{s}, \ldots$, respectively. The algorithm is described in Fig. 4.

The first step (line 2) of the algorithm takes as input the labeled colored state graph $\mathcal{G}$ associated to $\mathcal{A}$ and produces the colored labeled graph $\mathcal{R} = (N_\mathcal{R}, E_\mathcal{R}, \lambda_\mathcal{R})$ that is the reverse of $\mathcal{G}$.

The second step (lines 3 - 8) consists of the *subset construction* on the pair $(\mathcal{R}, \{I_j\}_{j=1}^k)$, where $I_j = \{q \in N_\mathcal{R} | \lambda_\mathcal{R}(q) = \gamma_j\}$, followed by a coloring operation. We obtain a colored labeled graph $\mathcal{D} = (N_\mathcal{D}, E_\mathcal{D}, \lambda_\mathcal{D})$. The coloring operation is called *initial coloring* and it is defined by $\lambda_\mathcal{D} : N_\mathcal{D} \rightarrow \Gamma$ that is a partial coloring function with $dom(\lambda_\mathcal{D}) = \{I_1, I_2, ..., I_k\}$, $\lambda_\mathcal{D}(I_j) = \gamma_j$, for each $1 \leq j \leq k$.

The third step (line 9) takes as input the labeled colored graph $\mathcal{D}$ and produces its labeled colored reverse graph of $\mathcal{D}$ named $\mathcal{F} = (N_\mathcal{F}, E_\mathcal{F}, \lambda_\mathcal{F})$.

The last step (lines 10 - 14) consists of the subset construction on the pair $(\mathcal{F}, \{\mathbf{m}_0\})$, where $\mathbf{m}_0 = \{\mathbb{p} \in N_\mathcal{F} | \ \mathbb{p} \cap I \neq \emptyset\}$, and a coloring operation. It produces a colored labeled graph $\mathcal{M} = (N_\mathcal{M}, E_\mathcal{M}, \lambda_\mathcal{M})$ in which the coloring operation, called *final coloring*, is defined by $\lambda_\mathcal{M} : N_\mathcal{M} \rightarrow \Gamma$ that is a total coloring function defined as follows. In Lemma 3 we prove that each $\mathbf{p} \in N_\mathcal{M}$ contains exactly one set I_j, then we pose $\lambda_\mathcal{M}(\mathbf{p}) = \gamma_j$. The line 15 defines the automaton returned by the algorithm.

The following lemmas state some properties regarding the graphs involved in the algorithm.

Lemma 1. *In the graph $\mathcal{D}$, for each j and for each $w \in L(\mathcal{A})$ there exists at most a path from I_j labeled by the reverse of w.*

Proof. The thesis follows from the fact that in the graph the accessible part from each I_j is deterministic by construction.   □

The following lemma can be deduced by the previous one.

Lemma 2. *In the graph $\mathcal{F}$, for each j and for each $w \in L(\mathcal{A})$ there exists at most a unique state $\mathbb{p}$ such that there exists a path from $\mathbb{p}$ to I_j labeled by w.*

Lemma 3. *For each node $\mathbf{p} \in N_\mathcal{M}$ there exists exactly a unique j, ranging from 1 to k, such that $I_j \in \mathbf{p}$.*

Proof. Remind that each node of $\mathcal{M}$ is obtained by a subset construction, so it is a set of nodes of $\mathcal{F}$. Let γ_i the color of the colored initial states. By construction, $\mathbf{m}_0$ contains I_i and no other sets I_h's with $h \neq i$. Let $\mathbf{p}$ an accessible state and let w be the label of the path from $\mathbf{m}_0$ to $\mathbf{q}$. This means that in the NMA $\mathcal{A}$ there is a path from an initial state p to a state q labeled by w. Let γ_l the color of such a path. In the graph $\mathcal{D}$ there is a unique path from I_l labeled by the reverse of w to a set $\mathbb{p}$ that contains p. So, in $\mathcal{F}$ there is a path from $\mathbb{p}$ to I_l labeled by w. Since $\mathbb{p}$ contains a initial state p, then $\mathbb{p}$ belongs to $\mathbf{m}_0$. So, the set $\mathbf{q}$ contains I_l. Moreover $\mathbf{q}$ does not contain any other set $I_h \neq I_l$. In fact, if so, there would exist $\mathbb{p}' \in \mathbf{m}_0$ such that the graph $\mathcal{F}$ contains a path from $\mathbb{p}'$ to I_h labeled by w. This means that there exists in $\mathcal{A}$ two paths having different colors from an initial state labeled by w. This fact contradicts the property of coherency of the NMA.   □

Remark 1. Note that, by previous lemma, the set of colors Γ_M is equal to Γ.

From the graph $\mathcal{M}$ we can, naturally, obtain the deterministic Moore automaton $\mathcal{A}_M = (\Sigma, \Gamma, Q, q_0, \Delta, \lambda)$ where $Q = N_{\mathcal{M}}$, $q_0 = \mathbf{m}_0$, $\Delta = E_{\mathcal{M}}$, $\lambda = \lambda_{\mathcal{M}}$.

Remark 2. It is easy to see that $\mathcal{A}_M$ is a DMA because it is obtained by a subset construction starting from a unique state.

The following theorems state that $\mathcal{A}_M$ is the minimal automaton equivalent to $\mathcal{A}$.

Theorem 1. *The deterministic Moore automaton $\mathcal{A}_M$ is minimal.*

Proof. We have to prove that for each pair of states $\mathbf{p}$ and $\mathbf{q}$, they are distinguishable, i.e. either there exists $w \in \Sigma^*$ that is applicable for $\mathbf{p}$ or for $\mathbf{q}$ but not for both, or there exists w applicable for both such that $\lambda(\delta(\mathbf{p}, w)) \neq \lambda(\delta(\mathbf{q}, w))$. Let $w \in \Sigma^*$, if w is not applicable for one of them then $\mathbf{p}$ and $\mathbf{q}$ are distinguishable. Let us suppose that w is applicable for both. We consider the paths in $\mathcal{M}$ labeled by w from $\mathbf{p}$ to a state $\mathbf{p}'$ and from $\mathbf{q}$ to a state $\mathbf{q}'$. By Lemma 3, $\mathbf{p}'$ contains the set I_h and $\mathbf{q}'$ contains I_j. We prove that $I_h \neq I_j$. This fact follows by using Lemma 2, because in the graph $\mathcal{F}$ there exists a path from $\mathbb{p} \in \mathbf{p}$ to I_h and a path from $\mathbb{q} \in \mathbf{q}$ to I_j both labeled by w and with $\mathbb{p} \neq \mathbb{q}$. □

Theorem 2. *The automata $\mathcal{A}$ and $\mathcal{A}_M$ are equivalent.*

Proof. We prove that for each i, ranging from 1 to k, $w \in L_i(\mathcal{A})$ if and only if $w \in L_i(\mathcal{A}_M)$. Let $w \in L_i(\mathcal{A})$ then there exists a path (p_1, w, p_n) such that $p_1 \in I$ and $\lambda(p_n) = \gamma_i$. There exists in $\mathcal{D}$ a path from I_j to a state $\mathbb{p}$ containing p_1 labeled by the reverse of w. Then there exists a path from $\mathbb{p}$ to I_j in $\mathcal{F}$ labeled by w. This means that there exists in $\mathcal{A}_M$ a path from the initial state $\mathbf{m}_0$ containing $\mathbb{p}$ to $\mathbf{q}$ containing I_j labeled by w. By Lemma 3, $\lambda_M(\mathbf{q}) = \gamma_i$, so $w \in L_i(\mathcal{A}_M)$. The same reasoning in reverse order can be used to prove the vice-versa. □

In the following example the execution of the minimization algorithm is described.

Example 2. In Fig. 5(a) the states graph $\mathcal{G}$ of $\mathcal{A} = (\Sigma, \Gamma, Q, I, \Delta, \lambda)$, in which $Q = \{1, 2, 3, 4, 5, 6, 7\}$, $I = \{2\}$, $\Sigma = \{a, b, c\}$, $\Gamma = \{Red, Green, Yellow, Blue\}$ and coloring function $\lambda(1) = \lambda(2) = Red$, $\lambda(4) = Yellow$, $\lambda(5) = Blue$ and $\lambda(6) = Green$. The language of applicable words is $L(\mathcal{A}) = a\Sigma^* + \epsilon$. The automaton induces the partition of $L(\mathcal{A})$ in $L_{Red} = \{w \in a\Sigma^*c \mid |w| \text{ is even}\} \cup \{\epsilon\}$, $L_{Green} = \{w \in a\Sigma^*a \mid |w| \text{ is even}\}$, $L_{Yellow} = \{w \in a\Sigma^*b \mid |w| \text{ is even}\}$, $L_{Blue} = \{w \in a\Sigma^* \mid |w| \text{ is odd}\}$. We apply the algorithm in order to obtain the minimal equivalent DMA. The first step produces the colored labeled graph $\mathcal{R}$ depicted in Fig. 5(b). In the second step we determine four sets $I_{Green} = \{6\}$, $I_{Red} = \{1, 2\}$, $I_{Yellow} = \{4\}$ and $I_{Blue} = \{5\}$ and we compute the subset construction on the pair $(\mathcal{R}, \{I_{Green}, I_{Red}, I_{Yellow}, I_{Blue}\})$. After the initial coloring,

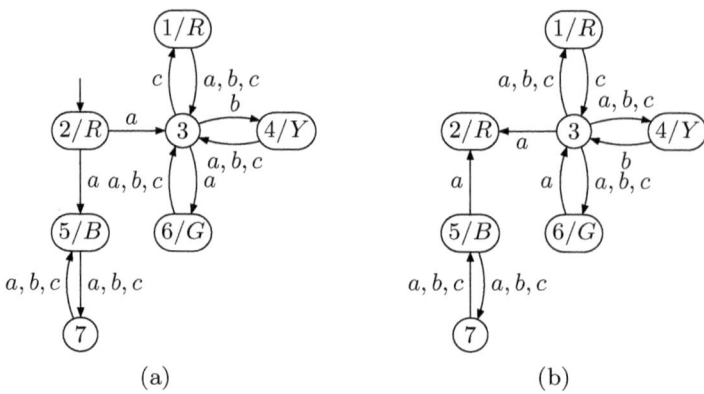

Fig. 5. An NMA $\mathcal{A}$ with three colors (a) and the reverse of the state graph $\mathcal{G}$ of $\mathcal{A}$ (b)

the colored labeled graph $\mathcal{D}$ depicted in Fig. 6(a) is obtained. The third step consists of a reverse operation on $\mathcal{D}$ and produces the colored labeled graph $\mathcal{F}$ depicted in Fig. 6(b). We renamed the states as follows, $A = \{6\}, B = \{1, 2, 4, 6\}, C = \{1, 2\}, D = \{3\}, E = \{1, 4, 6\}, F = \{4\}, G = \{2, 7\}, H = \{5\}, I = \{7\}$. Finally, in the fourth step the graph $\mathcal{M}$ is obtained by a subset construction on $(\mathcal{F}, \{B, C, G\})$ and the final coloring. In this graph, the states are denoted as follows: $\mathbf{1} = \{C, B, E, G, I\}, \mathbf{2} = \{B, C, G\}, \mathbf{3} = \{D, H\}, \mathbf{4} = \{A, B, E, G, I\}, \mathbf{5} = \{E, B, F, G, I\}$. The coloring function is $\lambda_{\mathcal{M}}(\mathbf{1}) = \lambda_{\mathcal{M}}(\mathbf{2}) = Red$ because the only colored set they contain is C that has color Red in $\mathcal{F}$, $\lambda_{\mathcal{M}}(\mathbf{3}) = Blue$, $\lambda_{\mathcal{M}}(\mathbf{4}) = Yellow$, $\lambda_{\mathcal{M}}(\mathbf{5}) = Green$, analogously. The minimal DMA $\mathcal{A}_M = (\Sigma, \Gamma, Q_M, q_0, \Delta_M, \lambda_M)$ obtained by such a graph is depicted in Fig. 7, where $Q_M = N_{\mathcal{M}}, q_0 = \mathbf{2}, \Delta_M = E_{\mathcal{M}}, \lambda_M = \lambda_{\mathcal{M}}$.

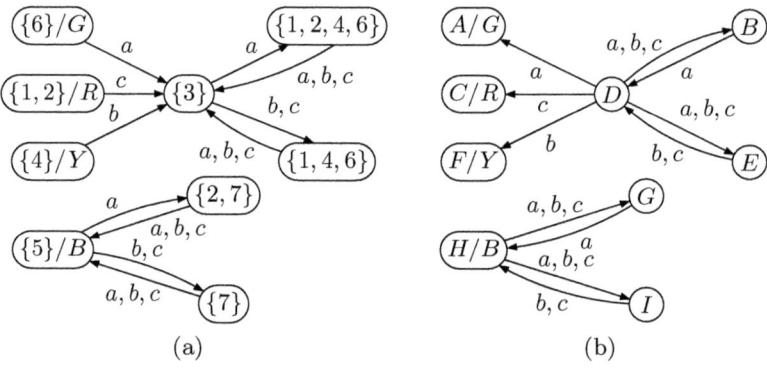

Fig. 6. The subset graph $\mathcal{D}$ of $\mathcal{R}$ (a) and the reverse graph $\mathcal{F}$ of $\mathcal{D}$ (b)

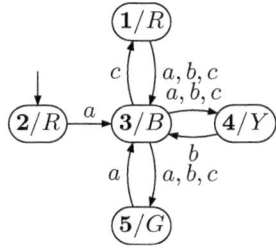

Fig. 7. The minimal DMA $\mathcal{A}_M$ equivalent to $\mathcal{A}$

4   Conclusions and Further work

In this paper we define a nondeterministic notion of Moore automata equipped with the coherence property (here simply denoted by NMA). Such a model could be thought as a particular case of the nondeterministic models defined in the literature when the output alphabet is equipped with a commutative and associative operator. It would be interesting to extend the results shown in this paper to a more general model. Here we propose an algorithm to construct the minimal deterministic Moore automaton equivalent to a nondeterministic one. Such an algorithm sounds like Brzozowski's method that works on acceptors in the sense that it is essentially structured on the operations of reverse and subset construction performed twice but such operations in the context of Moore automata assume a different meaning in particular regarding the coloring.

Recall that Brzozowski's algorithm applied to an NFA has a time complexity that is exponential in the worst case due to the subset constructions. Analogously, for NMA's the time complexity of Brzozowski's method described in this paper is exponential in the worst case. For instance, the Fig. 8 describes a Moore automaton, which falls in such a situation. It would be interesting to study also the time complexity in the average case. Such problems are related to the analysis of the scalability, with respect to the size of a given NMA, of the size of the minimal equivalent DMA. It would be useful to investigate how the transition density and the color density of a given NMA affect the size of the minimal DMA.

In the literature, there exists a model of nondeterministic acceptors called *self verifying* automata (see for instance [10]) that are a particular case of nondeterministic Moore automata, obtained when the set of output symbols is binary. Such automata are a variant of nondeterministic acceptors in which computation paths can give three types of answers: *yes, no* and *I do not know*. Moreover for each input string, at least one path must give answer *yes* or *no* and for the same string two paths cannot give contradictory answers. In [10] a conversion of a self-verifying automaton to a DFA is shown together with the exact cost of such a simulation, in terms of the number of states. Such a deterministic automaton is not necessarily the minimal one. Our method can be also applied to directly simulate a self-verifying automaton by the minimal equivalent DFA. One can observe that, after the process of conversion to a DFA, a classical minimization algorithm could be applied to obtain the minimal DFA. Recall that some experimental results provided in [19] show that in order to construct the minimal DFA equivalent to a given NFA, Brzozowski's algorithm is better in terms of running time for NFA's with high transition densities than the subset construction followed by Hopcroft's algorithm (cf. [9]). Such results could be confirmed also in case of self-verifying automata. It would be useful to find and compare the exact costs (or their upper bounds) of the two transformations with reference to the transition densities and acceptance or rejection densities.

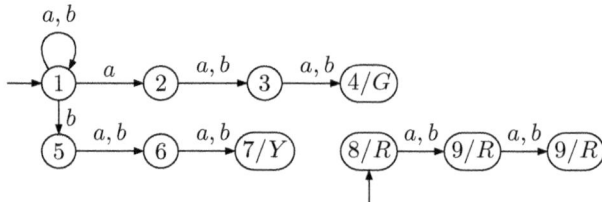

Fig. 8. A nondeterministic Moore automaton for which the size of the minimal equivalent DMA is exponential

Remark that Hopcroft's algorithm could be easily extended to the deterministic Moore automata. Recall that the classical Hopcroft's algorithm starts from a partition of the states of a DFA into accepting and rejecting states and by using splitting operations refines the partitions leading to the coarsest partition compatible with the set of accepting states. In case of Moore automata it would be enough to start from the partition of the states into the the sets of states having the same color. The splitting operation could be defined in a similar way as those used for the DFA's. The running time should be optimized by using the techniques provided in [1]. In this regard, it is worthwhile to recall that, in the case of DFA, by encoding by b each acceptance state and by a the rejection states, an infinite family of automata that are the worst cases of Hopcroft's algorithm has been defined starting from particular families of binary words with special and balanced distributions of the two symbols [6,5]. Such families of automata are challenging also for other classical minimization algorithm [4]. It would be interesting to define new combinatorial properties of families of words over alphabets with cardinality greater than 2 and relate them to the worst cases of Hopcroft's algorithm on Moore automata.

We would like to thank the referees for their helpful suggestions and comments.

References

1. Béal, M.-P., Crochemore, M.: Minimizing local automata. In: IEEE International Symposium on Information Theory (ISIT 2007), pp. 1376–1380 (2007)
2. Brzozowski, J.A.: Canonical regular expressions and minimal state graphs for definite events. Mathematical Theory of Automata 12, 529–561 (1962)

3. Calude, E., Lipponen, M.: Minimal deterministic incomplete automata. Journal of Universal Computer Science 3(11), 1180–1193 (1997)
4. Castiglione, G., Nicaud, C., Sciortino, M.: A challenging family of automata for classical minimization algorithms. In: Domaratzki, M., Salomaa, K. (eds.) CIAA 2010. LNCS, vol. 6482, pp. 251–260. Springer, Heidelberg (2011)
5. Castiglione, G., Restivo, A., Sciortino, M.: Circular sturmian words and Hopcroft's algorithm. Theor. Comput. Sci. 410, 4372–4381 (2009)
6. Castiglione, G., Restivo, A., Sciortino, M.: On extremal cases of Hopcroft's algorithm. Theor. Comput. Sci. 411(38-39), 3414–3422 (2010)
7. Cortes, C., Mohri, M.: Learning with weighted transducers. In: Piskorski, J., Watson, B.W., Anssi, Y.-J. (eds.) Frontiers in Artificial Intelligence and Applications. FSMNLP, vol. 19, pp. 14–22. IOS Press, Amsterdam (2008)
8. García, P., Ruíz, J., Cano, A., Alvarez, G.: Inference improvement by enlarging the training set while learning dFAs. In: Sanfeliu, A., Cortés, M.L. (eds.) CIARP 2005. LNCS, vol. 3773, pp. 59–70. Springer, Heidelberg (2005)
9. Hopcroft, J.E.: An $n \log n$ algorithm for mimimizing the states in a finite automaton. In: Proc. Internat. Sympos. Technion, Haifa,Theory of machines and computations, pp. 189–196. Academic Press, New York (1971)
10. Jirásková, G., Pighizzini, G.: Optimal simulation of self-verifying automata by deterministic automata. Inf. Comput. 209(3), 528–535 (2011)
11. Kameda, T., Weiner, P.: On the state minimization of nondeterministic finite automata. IEEE Trans. Comput. 19, 617–627 (1970)
12. Kupferman, O., Vardi, M.Y.: Robust satisfaction. In: Baeten, J.C.M., Mauw, S. (eds.) CONCUR 1999. LNCS, vol. 1664, pp. 383–398. Springer, Heidelberg (1999)
13. Mohri, M.: Finite-state transducers in language and speech processing. Computational Linguistics 23(2), 269–311 (1997)
14. Mohri, M.: Minimization algorithms for sequential transducers. Theor. Comput. Sci. 234(1-2), 177–201 (2000)
15. Moore, E.F.: Gedaken experiments on sequential machines, pp. 129–153. Princeton University Press, Princeton (1956)
16. Rabin, M., Scott, D.: Finite automata and their decision problems. IBM Journal of Research and Development 3, 114–125 (1969)
17. Solovev, V.: Minimization of Moore finite automata by internal state gluing. Journal of Communications Technology and Electronics 55, 584–592 (2010)
18. Spivak, M.A.: Minimization of a Moore automaton. Cybernetics and Systems Analysis 3, 4–5 (1967)
19. Tabakov, D., Vardi, M.Y.: Experimental evaluation of classical automata constructions. In: Sutcliffe, G., Voronkov, A. (eds.) LPAR 2005. LNCS (LNAI), vol. 3835, pp. 396–411. Springer, Heidelberg (2005)
20. Takahashi, K., Fujiyoshi, A., Kasai, T.: A polynomial time algorithm to infer sequential machines. Systems and Computers in Japan 34(1), 59–67 (2003)
21. Watson, B.W.: Taxonomies and toolkits of regular language algorithms. PhD thesis, Dep. Math. Comput. Sci. Technische Universiteit Eindhoven (1995)

Building Phylogeny with Minimal Absent Words

Supaporn Chairungsee[1] and Maxime Crochemore[1,2]

[1] King's College London, London, WC2R 2LS, United Kingdom
[2] Université Paris-Est, France
{supaporn.chairungsee,maxime.crochemore}@kcl.ac.uk

Abstract. An absent word in a sequence is a segment that does not occur in the given sequence. It is a *minimal absent word* if all its proper factors occur in the given sequence.

In this paper, we review the concept of minimal absent words, which includes the notion of shortest absent words but is much stronger. We present an efficient method for computing the minimal absent words of bounded length for DNA sequence using a Suffix Trie of bounded depth, representing bounded length factors. This method outputs the whole set of minimal absent words and furthermore our technique provides a linear-time algorithm with less memory usage than previous solutions.

We also present an approach to distinguish sequences of different organisms using their minimal absent words. Our solution applies a length-weighted index to discriminate sequences and the results show that we can build phylogenetic tree based on the collected information.

Keywords: minimal absent words, forbidden words, suffix trie of bounded depth, indexing; string similarity, phylogeny construction.

1 Introduction

Processing DNA sequences in an efficient way is a fundamental precondition for the study and analysis of biological molecules, see for example [4]. Sequence alignment is a procedure conducted in any biological study that compares two or more biological sequences [13]. It is the procedure to infer which positions within sequences are homologous, that is, which sites share a common evolutionary history. Alignment is often viewed as a necessary step for further study, for instance, for the identification and quantification of conserved regions or functional motifs, for profiling of genetic disease, for phylogenetic analysis, and for sequence profiling and prediction. In this article we show that missing information can also be used to infer phylogenetic trees in an alignment-free manner.

The availability of complete genome sequences plays an important role for the analysis of similarities and differences between genomes. In biological research, the sequence similarity between different species provides an important source of information to construct phylogenetic tree. Phylogenetics is the tool for studying the evolutionary relationship among different taxa. The sequence similarity between different species provides an important reference for the phylogenetic

B. Bouchou-Markhoff et al. (Eds.): CIAA 2011, LNCS 6807, pp. 100–109, 2011.

analysis although it does not decide the final result of the phylogenetic analysis completely. Selecting suitable invariants/descriptors to characterize DNA sequences to compare sequences effectively instead of using whole genomes is an interesting problem. The first exon of the β-globin gene is often used as a standard example in many DNA-based methods. The gene family of β-globin varies between 86 and 105 bases and has a significant biological role in oxygen transport. In 2005, Liu and Wang [10] presented a relative similarity measure to analyze the similarity of DNA sequences. Their solution applies LZ-complexity to compute similar regions in given sequences to construct phylogenetic trees. They analyze the similarity/dissimilarity of the first exon sequences of β-globin genes of 11 species, which are bovinae, chimpanzee, gallus, gorilla, capra, human, lemur, mouse, opossum, rabbit and rat.

An absent word in a DNA sequence (also called an unword or a forbidden word in other contexts) is a word that does not occur in the given sequence. An absent word is assumed to refer to negative selection. These words can be used as biomarkers for preventive and curative medical applications that derived from personal genomics efforts. When absent words can be identified, this information will be useful for sequence evolution, comparative genomics and genetic engineering.

The idea of using absent words to analyse sequences comes from the field of Symbolic Dynamics and has been initiated by Béal et al. [3]. This powerful concept has been later on at the origin of a new type of successful text compression methods [7], which have been the object of a series of improvements. It has been shown how to compute all the forbidden words of a sequence in linear time and linear memory space [6] and this has been even extended to regular languages [2].

In 2007, Hampikian and Andersen [8] defined the term nullomer to denote the shortest words that do not occur in a given genome and the term prime to refer to the shortest words that are absent from the entire known genetic data. Their motivation was to discover the constraints on natural DNA and protein sequences. The algorithm used by Hampikian and Andersen to obtain the absent words tracks the occurrence of all possible words up to a user-specified length limit n, using a set of 4^n counters for the 4^n possible words of length n. This yields the existing absent words up to the given length limit n. In the same year, Acquisti et al. [1] studied nullomers and the cause, natural selection, of absent words in human.

The fourth approach for solving the absent words problem was presented by Herold et al. [9] and they used the term unword to define the shortest absent words. Their approach has a limitation since it can produce only the shortest absent words. In 2009, an algorithm to find minimal absent words was presented by Armando et al. [11]. They coined the term minimal absent words to define a new and larger class of absent words, including the shortest absent words, independently of previous works. They applied a Suffix Array technique to do the work but the running time of the algorithm is not linear. Recently, Wu et al. [15] presented an algorithm to compute shortest absent words using a probabilistic method. Their algorithm runs in linear time and uses less memory than existing algorithms.

In this paper, we present two efficient approaches. First, we review the concept of minimal absent words, which includes the notion of shortest absent words. In order to compute efficiently minimal absent words we present a new approach to find minimal absent words of given length of the input sequence using the Suffix Trie of its bounded length factors. This approach consumes less memory space than previous approaches while it can compute the whole set of minimal absent words. Moreover, our method runs in linear time according to the sequence length. Second, we introduce a solution to discriminate genomic sequences using their minimal absent words. We define the notion of a length-weighted index to compute the similarity/dissimilarity between sequences based on minimal absent words.

Finally, we apply the technique to the first exon of β-globin. It is used to build a phylogeny of the 11 organisms aforementioned. It confirms the result obtained in [10] and proves that our approach is valid.

2   Basic Definition

A word x is a factor of a word y if there exist two words u and v such that $y = uxv$ [5]. For example consider the word $y = aababaabab$ and $x = baa$; the word x is a factor of the word y.

The Suffix Trie of a word is the deterministic automaton that recognises the set of suffixes of the word and in which two different paths with the same source always have distinct ends [5]. It is a search trie constructed for all suffixes of the word [12]. Thus, the graph structure of the automaton is a tree whose arcs are labelled by letters. The Suffix Trie of the word w is denoted by $\mathcal{T}(w)$. Its nodes are the factors of w, the empty word is the initial state (the root), and the suffixes of w are the terminal states. We define $s\ell[q]$ as the suffix link of (nonempty) state q. If $q = au$ for some letter a, then $s\ell[q] = u$. For instance word $y = $ aabab and the Suffix Trie of this word is presented in Figure 1.

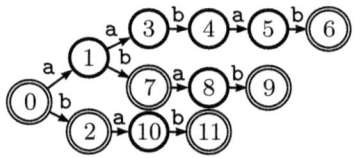

Fig. 1. Suffix Trie of aabab

The Suffix Trie of bounded length factor of word is the deterministic automaton that recognises the set of fixed length suffixes of the word. For example word $y = $ cagaccgttt and the length of the bounded factors equal to 4. The Suffix Trie of bounded length factor of this word is shown in Figure 2.

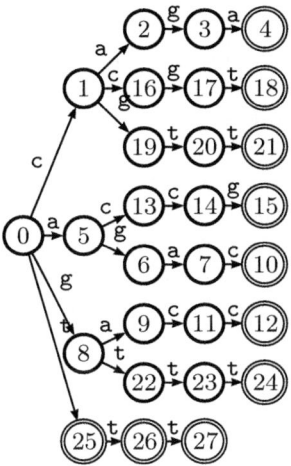

Fig. 2. Suffix Trie of `cagaccgttt` with the length of the bounded factors equal to 4

A word u is said to be absent in the word y if it is not a factor of y. The absent word u is said to be a minimal absent word if all its proper factors are factors of y. For example word $y = aabab$ then the minimal absent words of the word y are $aaa, baa, baba, bb$.

The length weighted index is a technique to find the similarity between sample sets and this solution consider the length of each member in the different$(A\Delta B)$ of sample set. The formal definition is $\sum_{w \in A\Delta B} 1/|w|^2$ where A and B are sample sets and w is the member of $A\Delta B$. For example, given two sets $A = \{aaa, aabb, aba, bbab, bbb\}$ and $B = \{aa, aba, baba, bbb\}$. We can find $A\Delta B = \{aa, aaa, aabb, baba, bbab\}$ and the length weighted index of this example is in the following:

$$\sum_{w \in A\Delta B} = 1/4 + 1/9 + 1/16 + 1/16 + 1/16$$
$$= 0.548611$$

Given a distance matrix M of a set S of n taxa, the Unweighted Pair Group Method with Arithmetic Mean (UPGMA) is a technique for reconstructing the phylogenetic tree T for S [14]. The basic principle of UPGMA is that similar taxa should be closer in the phylogenetic tree. Hence, it builds the tree by clustering similar taxa iteratively. The method works by building the phylogenetic tree bottom up from its leaves.

3   Method

3.1   Minimal Absent Words Trie Computation

In this subsection, we present how to compute minimal absent words of a word in a linear-time with the bounded length factors Suffix Trie.

The code of the algorithm below uses the Suffix Trie of the word y, $\mathcal{T}(y, \ell)$. The algorithm works as follows.

At a given step, *List* is a queue to store pairs of nodes of $\mathcal{T}(y, \ell)$ and nodes of the minimal absent words trie of the word y, δ denotes the transition function of the trie, *initial* is the root node of the trie, $L[q]$ is the maximum length of labels of paths from the root node to the target node q, p is the current node of the trie, p' is the current node of the minimal absent words trie, q' is the target node of transition function of the minimal absent words trie, $s\ell$ is the suffix link of (nonempty) state p and *reach* is a reach status of state p which the value of reach is either equal to 0, if that state has not been reached, or equal to 1, in case the state has been reached. ℓ is a bounded length.

There are two conditions for creating node in a minimal absent words trie. The first condition is $\delta(p, a)$ is not defined and either p is initial node or $s\ell$ is defined. The second condition is $\delta(p, a)$ is defined and $\delta(p, a)$ has not been reached.

AWT($\mathcal{T}(y, \ell)$)
```
 1   M ← NEW-AUTOMATON()
 2   List ← EMPTY-QUEUE()
 3   List ← ENQUEUE(List, (initial[T(y, ℓ)], initial[M]))
 4   while List ≠ 0 do
 5            (p, p') ← DEQUEUE(L)ist
 6            for a ∈ A do
 7                     if (δ(p, a) = NULL) and
                          ((p = initial[T(y, ℓ)]) or
                          (δ(sℓ[p], a) ≠ NULL)) then
 8                            q' ← NEW-STATE()
 9                            terminal[q'] ← TRUE
10                            Succ[p'] ← Succ[p'] ∪ {(a, q')}
11                     elseif (δ(p, a) ≠ NULL) and
                          (List[δ(p, a)] < ℓ) and
                          (reach[δ(p, a)] ≠ 1) then
12                            q' ← NEW-STATE()
13                            Succ[p'] ← Succ[p'] ∪ {(a, q')}
14                            List ← ENQUEUE(List, (δ(p, a), q'))
15   return M
```

Figure 3 displays the minimal absent words trie of the word `cagaccgttt` that is computed by this algorithm. All terminal states represents the value of minimal absent words correspond to the word and outputs are aa, at, ct, gc, gg, ta, tc, tg, aca, acg, agt, cac, cca, ccc, cga, gag and tttt.

Theorem 1. *The algorithm AWT computes the minimal absent words of a word of length n in time $O(n \times card A)$ where A is the size of alphabet set.*

Proof. The operations of the main loop, except the **for** loop in line 6, execute in constant time, this gives a time $O(n)$ for their global execution. In line 14, each operations to enqueue does not corresponds to every node in the Suffix Trie therefore it is not a function of the word size. Each operation in the **for** loop in

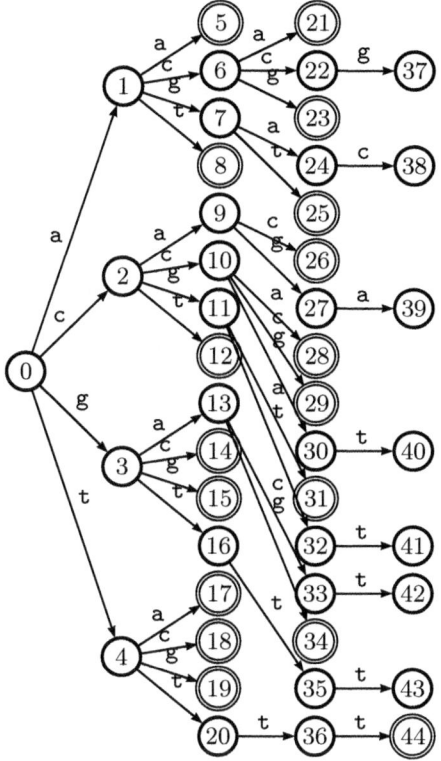

Fig. 3. Suffix Trie of minimal absent words of `cagaccgttt`

line 6 has the total number of targets being bounded by the size of alphabet(card A) and the cumulated time of all the executions of line 6 is O(card A). Therefore the total time of the minimal absent words construction is $O(n \times cardA)$.

3.2   Similarity/Dissimilarity Measures

In this subsection, we present how to compute similarity/dissimilarity of genomic sequences and discriminate sequences using minimal absent words. We propose a method that is called length-weighted index technique.

We present how to find length-weighted index of minimal absent words between two sequences. We will describe by the example, given two sequences: $A = $ `atgagtgatagacc` and $B = $ `gtggctatgttaac`. Then we compute minimal absent words of these sequences and we get two sets of minimal absent words that are $A = $ aa, agag, agat, atgat, ca, ccc, cg, ct, gaga, gatg, gc, gg, gta, gtgag, tac, tagt, tat, tc, tgac, tgt, tt $B = $ aaa, aat, act, ag, ata, atgg, att, ca, cc, cg, ctaa, ctg, ctt, ga, ggg, ggt, gta, gtgt, tac, tc, tgc, tgtg, ttat, ttg, ttt. After that we find the difference between A and B that is $A \triangle B = $ aa, aaa, aat, act, ag, agag, agat, ata, atgat, atgg, att, cc, ccc, ct, ctaa, ctg, ctt, ga, gaga, gatg, gc, gg, ggg, ggt, gtgag, gtgt, tagt, tat, tgac, tgc, tgt, tgtg, tt, ttat, ttg, ttt. Next we compute the length weighted index of this example that is in the following:

$$\sum_{w \in A \triangle B} = 1/4 + 1/9 + 1/9 + 1/9 + 1/4 + 1/16 + 1/16 + 1/9 + 1/25 +$$
$$1/16 + 1/9 + 1/4 + 1/9 + 1/4 + 1/16 + 1/9 + 1/9 + 1/16 +$$
$$1/4 + 1/16 + 1/4 + 1/4 + 1/9 + 1/9 + 1/25 + 1/16 + 1/16 +$$
$$1/9 + 1/16 + 1/9 + 1/9 + 1/16 + 1/4 + 1/16 + 1/9 + 1/9$$
$$= 4.158056$$

4   Results and Discussion

4.1   Minimal Absent Words Trie Computation

In this subsection, we present some experimental results with the first exon sequences of β-globin genes from 11 species that are Human, Capra, Gallus, Opossum, Lemur, Mouse, Rabbit, Rat, Bovinae, Gorilla and Chimpanzee and coding sequences are listed in Table 1.

Figure 4 presents the growth of minimal absent words of the first exon sequences of β-globin genes from 11 genomes. Results show that the range of minimal absent words length is between 2 and 10 and the maximum length of minimal absent words for each genomes is either 4 or 5. Figure 5 displays the trend of memory size for minimal absent words computation that is a linear function with the word length.

4.2   Phylogeny Building from Minimal Absent Words

In order to examine the validity of our new similarity/dissimilarity measure, we apply length weighted index to analyze the similarity/dissimilarity of minimal absent words from the sequence in Table 1 and we present similarity/dissimilarity matrix between each organism based on minimal absent words in Table 2. Take

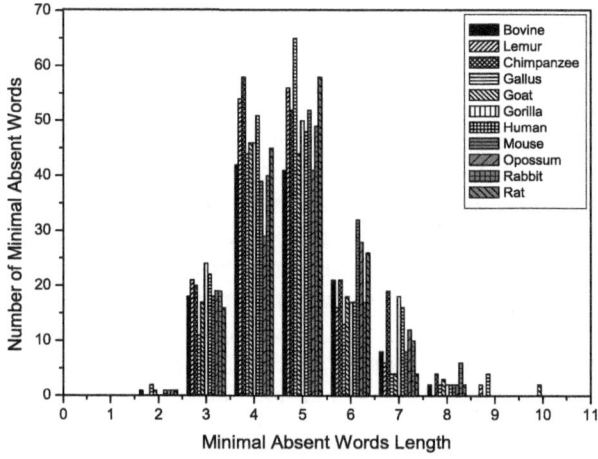

Fig. 4. Growth of minimal absent words

Table 1. Coding sequences of the first exon sequences of β-globin genes from Human, Capra, Gallus, Opossum, Lemur, Mouse, Rabbit, Rat, Bovinae, Gorilla and Chimpanzee [10]

Species	Coding Sequences
Human	ATGGTGCACCTGACTCCTGAGGAGAAGTCTGCCGTTACTGCCCTGTG GGGCAAGGTGAACGTGGATTAAGTTGGTGGTGAGGCCCTGGGCAG
Capra	ATGCTGACTGCTGAGGAGAAGGCTGCCGTCACCGGCTTCTGGGGCAA GGTGAAAGTGGATGAAGTTGGTGCTGAGGCCCTGGGCAG
Opossum	ATGGTGCACTTGACTTCTGAGGAGAAGAACTGCATCACTACCATCTG GTCTAAGGTGCAGGTTGACCAGACTGGTGGTGAGGCCCTTGGCAG
Gallus	ATGGTGCACTGGACTGCTGAGGAGAAGCAGCTCATCACCGGCCTCTG GGGCAAGGTCAATGTGGCCGAATGTGGGGCCGAAGCCCTGGCCAG
Lemur	ATGACTTTGCTGAGTGCTGAGGAGAATGCTCATGTCACCTCTCTGTG GGGCAAGGTGGATGTAGAGAAAGTTGGTGGCGAGGCCTTGGGCAG
Mouse	ATGGTTGCACCTGACTGATGCTGAGAAGTCTGCTGTCTCTTGCCTGT GGGCAAAGGTGAACCCCGATGAAGTTGGTGGTGAGGCCCTGGGCAGG
Rabbit	ATGGTGCATCTGTCCAGTGAGGAGAAGTCTGCGGTCACTGCCCTGTG GGGCAAGGTGAATGTGGAAGAAGTTGGTGGTGAGGCCCTGGGC
Rat	ATGGTGCACCTAACTGATGCTGAGAAGGCTACTGTTAGTGGCCTGTG GGGAAAGGTGAACCCTGATAATGTTGGCGCTGAGGCCCTGGGCAG
Gorilla	ATGGTGCACCTGACTCCTGAGGAGAAGTCTGCCGTTACTGCCCTGTG GGGCAAGGTGAACGTGGATGAAGTTGGTGGTGAGGCCCTGGGCAGG
Bovinae	ATGCTGACTGCTGAGGAGAAGGCTGCCGTCACCGCCTTTTGGGGCAA GGTGAAAGTGGATGAAGTTGGTGGTGAGGCCCTGGGCAG
Chimpanzee	ATGGTGCACCTGACTCCTGAGGAGAAGTCTGCCGTTACTGCCCTGTG GGGCAAGGTGAACGTGGATGAAGTTGGTGGTGAGGCCCTGGGCAGG TTGGTATCAAGG

the first row in Table 2 for example, the element 11.9599 represents the value of dissimilarity between capra and human. Note that 8.82943 and 9.8847 in this row are more close to 12.7895, so we say gorilla and chimpanzee are most similar to human in terms of the coding sequences of the first exon of β-globin genes.

Among these species, human and gorilla, human and chimpanzee (from the first row), capra and bovinae, capra and rabbit, capra and mouse, capra and gorilla (from the 2nd row), mouse and bovinae, mouse and gorilla, mouse and capra (from the 6th row), rabbit and capra, rabbit and bovinae (from the 7th row), rat and capra, rat and bovinae, rat and rabbit, rat and mouse (from the 8th row), gorilla and human, gorilla and chimpanzee (from the 9th row), bovinae and capra, bovinae and mouse (from the 10th row), chimpanzee and gorilla, chimpanzee and human (from the 11th row) are of the most similar.

Gallus and opossum are always the most remote from the other species in most cases, perhaps for gallus is the only nonmammalian representative and opossum is the most remote species from the remaining mammals. These coincide with real biological phenomenon. Besides gallus and opossum, lemur is more remote from the other species relatively. We also apply UPGMA technique to build phylogeny from similarity/dissimilarity matrix. The result is similar to results that present in the work of [10].

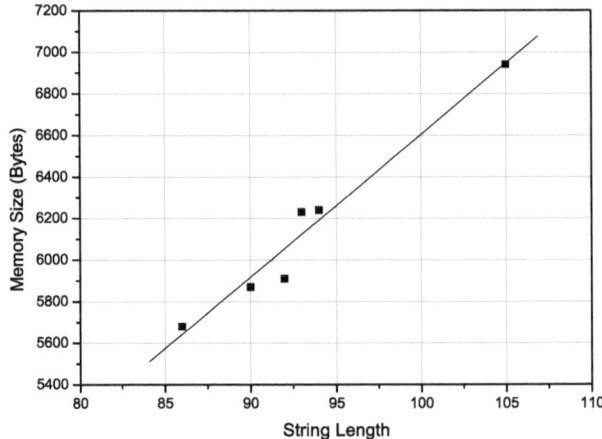

Fig. 5. Memory size of minimal absent words computation

Table 2. Similarity/Dissimilarity measure of analyzed genomes

Species	Human	Capra	Opossum	Gallus	Lemur	Mouse	Rabbit	Rat	Gorilla	Bovinae	Chimpanzee
Human	0	11.9599	12.7895	13.327	13.703	12.1115	12.0944	13.3468	8.82943	11.906	9.8847
Capra		0	12.2199	11.8727	12.4204	11.7775	11.7086	12.7014	11.8341	8.70512	12.4935
Opossum			0	12.4851	13.8097	13.55	12.2393	13.359	12.754	12.4736	13.304
Gallus				0	13.5477	13.0776	12.0693	13.9615	13.6214	12.5143	14.0447
Lemur					0	13.4286	13.4183	13.8036	13.4852	12.2545	14.1932
Mouse						0	12.6632	12.9735	11.7346	11.5871	12.5451
Rabbit							0	12.9485	12.1789	11.9315	12.6647
Rat								0	13.3912	12.8082	14.2242
Gorilla									0	11.7801	9.39664
Bovinae										0	12.3684
Chimpanzee											0

5   Conclusion

Minimal absent words in genomic sequences are interesting and important area for studying and they are useful information for further studies for instance phylogeny building. In this paper, we define the term minimal absent words as a set of all possible minimal absent words and we provide a linear-time algorithm for minimal absent words computation by Suffix Trie of bounded length factors. The memory size of our approach is less than previous solutions. We also apply length-weighted index to compute similarity/dissimilarity between genomic sequences. We present some properties of minimal absent words from first exon of β- globin that are useful information and can be applied to construct phylogeny.

References

1. Acquisti, C., Poste, G., Curtiss, D., Kumar, S.: Nullomers: really a matter of natural selection? PLoS ONE. 10 (2007)
2. Béal, M.P., Crochemore, M., Mignosi, F., Restivo, A., Sciortino, M.: Forbidden words of regular languages. Fundamenta Informaticae 56, 121–135 (2003)
3. Béal, M.P., Mignosi, F., Restivo, A.: Minimal Forbidden Words and Symbolic Dynamics. In: Puech, C., Reischuk, R. (eds.) STACS 1996. LNCS, vol. 1046, pp. 555–566. Springer, Heidelberg (1996)
4. Böckenhauer, H.J., Bongartz, D.: Algorithmic Aspects of Bioinformatics. Springer, Berlin (2007)
5. Crochemore, M., Hancart, C., Lecroq, T.: Algorithms on Strings. Cambridge University Press, Cambridge (2007)
6. Crochemore, M., Mignosi, F., Restivo, A.: Automata and Forbidden Words. Information Processing Letters 67, 111–117 (1998)
7. Crochemore, M., Mignosi, F., Restivo, A., Salemi, S.: Data compression using antidictonaries. Proceedings of the IEEE 88, 1756–1768 (2000)
8. Hampikian, G., Andersen, T.: Absent sequences: Nullomers and primes. In: Pacific Symposium on Biocomputing, vol. 12, pp. 355–366 (2007)
9. Herold, J., Kurtz, S., Giegerich, R.: Efficient computation of absent words in genomic sequences. BMC Bioinformatics 9 (2008)
10. Liu, N., Wang, T.M.: A relative similarity measure for the similarity analysis of DNA sequences. Chemical Physics Letters 408, 307–311 (2005)
11. Pinho, A.J., Ferreira, P.J., Garcia, S.P., Rodrigues, J.M.: On finding minimal absent words. BMC Bioinformatics 10 (2009)
12. Polanski, A., Kimmel, M.: Bioinformatics. Springer, Berlin (2007)
13. Rosenberg, M.S.: Sequence Alignment: Methods, Models,Concepts, and Strategies. University of California Press, California (2009)
14. Sung, W.K.: Algorithms in Bioinformatics: a practical intoduction. CRC Press, New York (2009)
15. Wu, Z.D., Jiang, T., Su, W.J.: Efficient computation of shortest absent words in a genomic sequence. Information Processing Letters 110, 596–601 (2010)

On the Hardness of Priority Synthesis

Chih-Hong Cheng[1], Barbara Jobstmann[2], Christian Buckl[3], and Alois Knoll[1]

[1] Department of Informatics, Technischen Universität München
Boltzmann Str. 3, Garching 85748, Germany
[2] Verimag Laboratory, 2, avenue de Vignate, 38610 Gières, France
[3] fortiss GmbH, Munich, Germany
{chengch,knoll}@in.tum.de, barbara.jobstmann@imag.fr, buckl@fortiss.org

Abstract. We study properties of priority synthesis [2], an automatic method to ensure desired safety properties in component-based systems using priorities. Priorities are a powerful concept to orchestrate components [3], e.g., the BIP[1] framework [1] for designing and modeling embedded and autonomous systems is based on this concept.

We formulate priority synthesis for BIP systems using the automata-theoretic framework proposed by Ramadge and Wonham [5]. In this framework, priority synthesis results in searching for a supervisor from the restricted class of supervisors, in which each is solidly expressible using priorities. While priority-based supervisors are easier to use, e.g., they support the construction of distributed protocols, they are harder to compute. In this paper, we focus on the hardness of synthesizing priorities and show that finding a supervisor based on priorities that ensures deadlock freedom of the supervised system is NP-complete.

1 Introduction

In this paper, we discuss methods to ensure *safety and deadlock avoidance* on component-based systems modeled using the BIP[1] language [1]. In BIP, a system can be modeled using three ingredients: (a) *Behaviors*, an extended automaton using labeled transitions, (b) *Interactions* defining synchronizations between two or more transitions of different components, and (c) *Priorities*, which are used to choose amongst possible interactions [1,2].

In our recent work [2], we present a tool called VISSBIP, which includes a technique called *priority synthesis* for BIP systems. The goal of priority synthesis is to automatically add a set of priorities that enforce a desired safety property of the composed systems. We consider priority synthesis as an instance of *controller synthesis*, which was first presented by Ramadge and Wonham [5]. In their seminal work, they proposed an automata-theoretic framework to constrain the behavior of a system via supervisory control. In priority synthesis, we restrict the supervisor to use only priorities. Constraining a system behavior using priorities has the following benefits.

---

[1] BIP is a shortcut for **B**ehavior-**I**nteraction-**P**riority.

B. Bouchou-Markhoff et al. (Eds.): CIAA 2011, LNCS 6807, pp. 110–117, 2011.
© Springer-Verlag Berlin Heidelberg 2011

- Existing safety properties as well as deadlock freedom is preserved under adding priorities.
- Priorities facilitate distributed control. E.g., by allowing components to coordinate temporarily, priorities can be implemented efficiently [4].

We first formulate priority synthesis under BIP systems using an automata-theoretic framework similar to [5]. Then, we focus on the *hardness of synthesizing priorities*, which constitutes our main contribution. We prove that, given a labeled transition system, finding a set of priorities that ensures safety and deadlock freedom is NP-complete in the size of the system. Our result is in contrast to the work in [5], where a general (monolithic) supervisor, which is usually difficult to distribute, can be found in polynomial-time in the size of the system. Our priority-based supervisors are easier to distribute but harder to compute.

2   Example: Simple BIP Models

Figure 1 shows a BIP model with two components represented in VISSBIP. Using this model, we illustrate in the following the different parts of a BIP system.

- **(Behavior).** The system has two components (Process1 and Process2), and each component has two places (**high** and **low**). A green circle indicates that this place is an initial location of a behavioral component. E.g., place **low** is marked as initial in both Process1 and Process2. Edges between two locations represent *transitions*, and they are labeled with *interaction alphabets*.
- **(Interaction).** For simplicity we use **alphabet bindings** to construct interactions between components, i.e., transitions using the same interaction

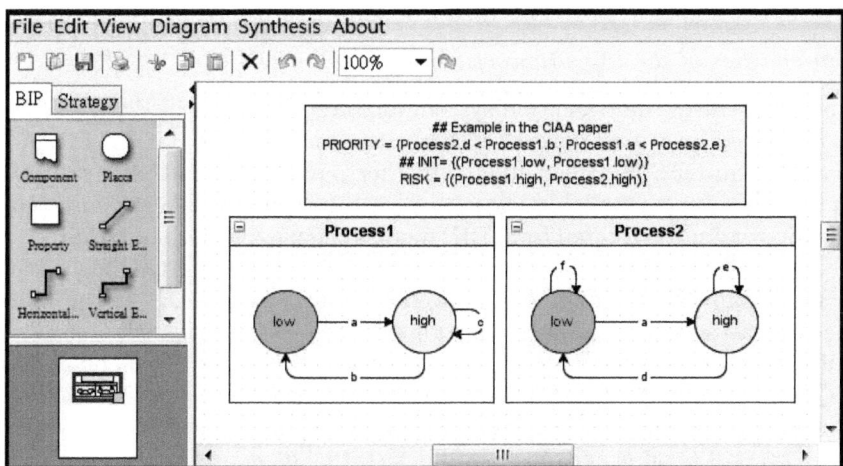

Fig. 1. Constructing BIP models using VISSBIP

alphabet are automatically grouped to a single interaction and are executed jointly. In the following, we refer to an interaction by its interaction alphabet.

- **(Priority).** We use the keyword PRIORITY to state priorities. E.g., the statement Process1.a < Process2.e means that whenever interactions a and e are available, the BIP engine always executes e.
- **(Safety property).** The condition RISK = {(Process1.high, Process2.high)} states that the combined location pair (Process1.high, Process2.high) should never be reached. Also, we implicitly require that the system is deadlock-free, i.e., at anytime, at least one interaction is enabled.

3 Formulating BIP Models and Priority Synthesis Based on Transition Systems

In this section, we first translate simple BIP models (Section 2) into automata, i.e., the logical discrete-event system (DES) model in [5]. Given a simple BIP model, we can always construct the transition system representing the asynchronous product of its components. We follow the definitions in [5] to simplify a comparison between priority synthesis and the controller synthesis technique.

Definition 1 (Transition System). *We define a transition system (called a logical DES model or generator in [5]) as a tuple $G = (Q, \Sigma, q_0, \delta)$, where*

- *Q is a finite set of* states,
- *Σ is a finite set of* event or interaction labels, *called* interaction alphabet,
- *q_0 is the* initial state, *i.e., $q_0 \in Q$,*
- *$\delta : Q \times \Sigma \to Q \cup \{\bot\}$ is a transition function* mapping a state and an interaction label to a successor state or a distinguished symbol $\bot$ that indicates that the given state and interaction pair has no successor. If $\delta(q, \sigma) = \bot$ for some $q \in Q$ and $\sigma \in \Sigma$, then we say $\delta(q, \sigma)$ is undefined. We slightly abuse the notation and extend δ to sequences of interactions in the usual way, i.e., $\delta(q, \epsilon) = q$ and $\delta(q, w\sigma) = \delta(\delta(q, w), \sigma)$ with $w \in \Sigma^*$ and $\sigma \in \Sigma$.

Denote the size of the transition system to be $|Q| + |\Sigma| + |\delta|$.

Figure 2 illustrates the transition system for the BIP model in Figure 1. Transitions in dashed lines are blocked by the priorities. Note that for the formulation in [5], a logical DES model is able to further partition Σ into Σ_c (controllable input) and Σ_u (uncontrollable input), i.e., a transition system can also model a game. For systems translated from BIP models the partition is not required. However, our hardness result of cause applies to the alphabet-partitioned setting as well. We define the **run of G on a word** $w = w_0 \ldots w_n \in \Sigma^*$ as the finite sequence of states $q_0 q_1 \ldots q_{n+1}$ such that for all $i, 0 \le i \le n$, $\delta(q_i, w_i) = q_{i+1}$. Note that if $\delta(q_i, w_i)$ is undefined for some i, then there exists no run of G on w. A state $q \in Q$ with no outgoing transitions, i.e., $\forall \sigma \in \Sigma, \delta(q, \sigma) = \bot$, is called **deadlock state**. A system G has a **deadlock** if there exists a word w such that the run $q_0 \ldots q_{|w|}$ of G on w ends in a deadlock state, i.e., $q_{|w|}$ is a deadlock state.

We now define the concept of **supervisor**, i.e., machinery that controls the execution of the system by suppressing transitions.

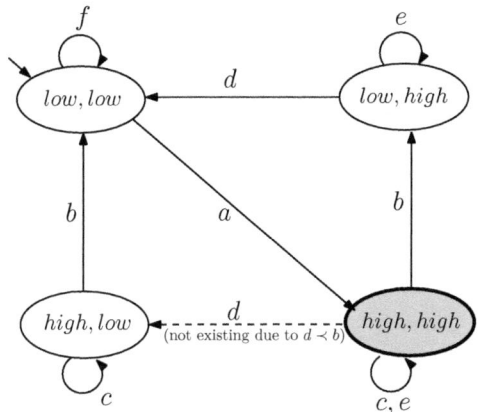

Fig. 2. The transition system for the BIP model (without variables) in Figure 1

Definition 2 (Supervisor). *Given* $G = (Q, \Sigma, q_0, \delta)$, *a supervisor for G is a function* $C : Q \times \Sigma \rightarrow \{\textbf{True}, \textbf{False}\}$. *The transition system* G_C *obtained from G under the supervision of C is defined as follows:* $G_C = (Q, \Sigma, q_0, \delta_C)$ *with* $\delta_C(q, \sigma) = \delta(q, \sigma) \neq \bot$, *if* $C(q, \sigma) = \textbf{True}$, *and* $\delta_C(q, \sigma) = \bot$ *otherwise.*

Definition 3. *Given* $G = (Q, \Sigma, q_0, \delta)$, *a* ***zero-effect supervisor*** $C_\emptyset$ *is a supervisor that disables all undefined interactions, i.e., interactions leading to $\bot$. Formally, for all states $q \in Q$ and interactions $\sigma \in \Sigma$, $C_\emptyset(q, \sigma) = \textbf{False}$ iff $\delta(q, \sigma) = \bot$. Note that $C_\emptyset$ has no effect on G, i.e., $G_{C_\emptyset} = G$.*

Given a transition system, adding priorities to the system can be viewed as masking some transitions. The masking can be formulated using supervisors.

Definition 4 (Priorities). *Given an interaction alphabet Σ, a set of priorities $\mathcal{P}$ is a finite set of interaction pairs defining a relation $\prec \subseteq \Sigma \times \Sigma$ between the interactions. We called a priority set legal, if the relation $\prec$ is (1) transitive and (2) non-reflexive (i.e., there are no circular dependencies) [3].*

We are only interested in legal sets, as a supervisor from a non-legal set of priorities may induce more deadlocks over the existing system. Note that given an arbitrary set, we can easily check if there exists a corresponding legal set.

Definition 5 (Priority Supervisor). *Given a transition system* $G = (Q, \Sigma, q_0, \delta)$ *and a legal priority set* $\mathcal{P} = \bigcup_{i=0}^{n} \sigma_i \prec \sigma_i'^2$ *with* $\sigma_i, \sigma_i' \in \Sigma$, *we define the corresponding supervisor $C_\mathcal{P}$ inductively over the number of priority pairs as follows:*

- *Base case:* $C_\mathcal{P} = C_\emptyset$, *if* $\mathcal{P} = \{\}$
- *Inductive step: Let* $\mathcal{P}' = \mathcal{P} \cup \{\sigma_k \prec \sigma_k'\}$, *then for all state $q \in Q$, if $C_\mathcal{P}(q, \sigma_k) = C_\mathcal{P}(q, \sigma_k') = \textbf{True}$, then $C_{\mathcal{P}'}(q, \sigma_k) = \textbf{False}$ and for all interactions $\sigma \neq \sigma_k' : C_{\mathcal{P}'}(q, \sigma) = C_\mathcal{P}(q, \sigma)$, otherwise for all $\sigma \in \Sigma : C_{\mathcal{P}'}(q, \sigma) = C_\mathcal{P}(q, \sigma)$.*

---

[2] We write $\sigma_i \prec \sigma_i'$ instead of (σ_i, σ_i') to emphasize that priorities are not symmetric.

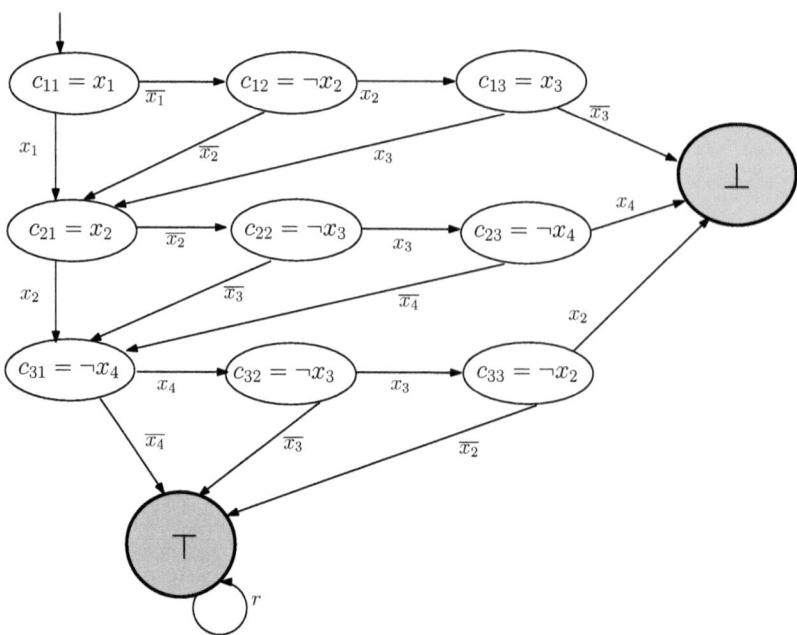

Fig. 3. The reduced system from the 3SAT instance $\phi = c_1 \wedge c_2 \wedge c_3$, where $c_1 :=$ $(x_1 \vee \neg x_2 \vee x_3), c_2 := (x_2 \vee \neg x_3 \vee \neg x_4), c_3 := (\neg x_4 \vee \neg x_3 \vee \neg x_2)$

Definition 6 (Safety). *Given a transition system* $G = (Q, \Sigma, q_0, \delta)$ *and the set of risk states* $Q_{risk} \subseteq Q$, *the system is **safe** if the following conditions holds.*

- **(Risk-free)** $\forall w \in \Sigma^*$, *if* $\delta(q_0, w) \neq \bot$, *then* $\delta(q_0, w) \notin Q_{risk}$
- **(Deadlock-free)** $\forall w \in \Sigma^*, \exists \sigma \in \Sigma$ *s.t. if* $\delta(q_0, w) \neq \bot$, *then* $\delta(q_0, w\sigma) \neq \bot$.

*A system that is not safe is called **unsafe**.*

Note that by removing all outgoing transitions for risk states every risk state is also a deadlock state. Therefore, risk-freeness reduces to deadlock-freeness and there is no need to handled it separately.

Definition 7 (Priority Synthesis). *Given a transition system* $G = (Q, \Sigma, q_0, \delta)$, *and the set of risk states* $Q_{risk} \subseteq Q$, *priority synthesis searches for a set of priorities* $\mathcal{P}$ *such that* G *supervised by* $\mathcal{C}_\mathcal{P}$ *is safe.*

4   Priority Synthesis Is NP-Complete

We now state the main result, i.e., the problem of priority synthesis is NP-complete.

Theorem 1. *Given a transition system* $G = (Q, \Sigma, q_0, \delta)$, *finding a set* $\mathcal{P}$ *of priorities such that* G *under* $\mathcal{C}_\mathcal{P}$ *is safe is NP-complete in the size of* G.

Proof. Given a set of a priorities $\mathcal{P}$, checking if $G_{\mathcal{C}_{\mathcal{P}}}$ is safe can be done in polynomial time by a simple graph search in $G_{\mathcal{C}_{\mathcal{P}}}$ for reachable states that have no outgoing edges. Therefore, the problem is in NP.

For the NP-hardness, we give a polynomial-time reduction from Boolean 3-Satisfiability (3-SAT) to Priority Synthesis. Consider a 3-SAT formula ϕ with the set of variables $X = \{x_1, \ldots, x_n\}$ and the set of clauses $C = \{c_1, \ldots, c_m\}$, where each clause c_i consists of the literals c_{i1}, c_{i2}, and c_{i3}. We construct a transition system $G_{\phi} = (Q, \Sigma, q_0, \delta)$ using Algorithm 1. The transition system has one state for each literal c_{ji} and two designated states $\top$ and $\bot$, indicating if an assignment satisfies or does not satisfy the formula. For each variable x, the alphabet Σ of G includes two interactions x_i and $\overline{x_i}$ indicating if x is set to true or false, respectively. The transition system consists of m layers. Each layer corresponds to one clause. The transitions allows one to move from layer i to the layer $i+1$ iff the corresponding clause is satisfied. E.g., consider the 3SAT formula $\phi = c_1 \wedge c_2 \wedge c_3$ with $c_1 := (x_1 \vee \neg x_2 \vee x_3)$, $c_2 := (x_2 \vee \neg x_3 \vee \neg x_4)$, $c_3 := (\neg x_4 \vee \neg x_3 \vee \neg x_2)$, Figure 3 shows the corresponding transition system.

We prove that ϕ is satisfiable iff there exists a set of priorities $\mathcal{P}$ such that G_{ϕ} supervised by $\mathcal{C}_{\mathcal{P}}$ is safe, i.e., in G_{ϕ} supervised by $\mathcal{C}_{\mathcal{P}}$ the state $\bot$ is unreachable.

($\rightarrow$) Assume that ϕ is satisfiable, and let $v : X \rightarrow \{0, 1\}$ be a satisfying assignment. Then, we create the priority set $\mathcal{P}$ as follows:

$$\mathcal{P} := \{\overline{x} \prec x \mid v(x) = 1\} \cup \{x \prec \overline{x} \mid v(x) = 0\}$$

E.g., consider the example in Figure 3, a satisfying assignment for ϕ is $v(x_1) = 1$ and $v(x_2) = v(x_3) = v(x_4) = 0$, then we obtain $\mathcal{P} = \{\overline{x_1} \prec x_1, x_2 \prec \overline{x_2}, x_3 \prec \overline{x_3}, x_4 \prec \overline{x_4}\}$.

Recall that G_{ϕ} under $\mathcal{C}_{\mathcal{P}}$ is safe iff it never reaches the state $\bot$. In G_{ϕ}, we can only reach the state $\bot$, if the priorities allows us, in some layer i, to move from c_{i1} to c_{i2} to c_{i3} and from there to $\bot$. This path corresponds to an unsatisfied clause. Since the priorities are generated from a satisfying assignment, in which all clauses are satisfied, there is no layer in which we can move from c_{i1} to $\bot$.

($\leftarrow$) For the other direction, consider a set of priorities $\mathcal{P}$. Let $\mathcal{P}'$ be the set of all priorities in $\mathcal{P}$ that refer to the same variable, i.e., $\mathcal{P}' = \{p \prec q \in \mathcal{P} \mid \exists x \in X : (p = x \wedge q = \overline{x}) \vee (p = \overline{x} \wedge q = x)\}$. Since $\mathcal{P}$ is a valid set of priorities (no circular dependencies), the transition system G_{ϕ} has the same set of reachable states under $\mathcal{C}_{\mathcal{P}}$ and under $\mathcal{C}_{\mathcal{P}'}$. There, the state $\bot$ is also avoided with using the set $\mathcal{P}'$. Given $\mathcal{P}'$, we construct a corresponding satisfying assignment as follows:

$$v(x) = \begin{cases} 0 & x \prec \overline{x} \in \mathcal{P}' \\ 1 & \overline{x} \prec x \in \mathcal{P}' \\ 0 & \text{otherwise.} \end{cases}$$

The size the transition system G_{ϕ} is polynomial in n and m. In particular, the transition system G_{ϕ} has $3 \cdot m + 2$ states, $2 \cdot n + 1$ interaction letters, and $2 \cdot 3 \cdot m + 1$ transitions.

Algorithm 1. Transition System Construction Algorithm

Data: 3SAT Boolean formula ϕ with n variables and m clauses
Result: Transition System $G_\phi = (Q, \Sigma, q_0, \delta)$
begin

$\quad Q := \{\top, \bot\}$
$\quad$**for** *clause* $c_i = (c_{i1} \vee c_{i2} \vee c_{i3})$, $i = 1, \ldots, m$ **do**
$\quad\quad Q := Q \cup \{c_{i1}, c_{i2}, c_{i3}\}$
$\quad \Sigma = \bigcup_{i=1\ldots n}\{x_i, \overline{x_i}\} \cup \{r\}$
$\quad$**for** *clause* $c_i = (c_{i1} \vee c_{i2} \vee c_{i3})$ *with variables* x_{i1}, x_{i2}, x_{i3}, $i = 1, \ldots, m$ **do**
$\quad\quad$**if** $i \neq m$ **then**
$\quad\quad\quad$/* Connect the truth assignment to state $c_{(i+1)1}$ */
$\quad\quad\quad$**if** x_{i1} appears positive in c_{i1} **then**
$\quad\quad\quad\quad \delta(c_{i1}, x_{i1}) := c_{(i+1)1}; \delta(c_{i1}, \overline{x_{i1}}) := c_{i2}$
$\quad\quad\quad$**else**
$\quad\quad\quad\quad \delta(c_{i1}, \overline{x_{i1}}) := c_{(i+1)1}; \delta(c_{i1}, x_{i1}) := c_{i2}$
$\quad\quad\quad$**if** x_{i2} appears positive in c_{i2} **then**
$\quad\quad\quad\quad \delta(c_{i2}, x_{i2}) := c_{(i+1)1}; \delta(c_{i2}, \overline{x_{i2}}) := c_{i3}$
$\quad\quad\quad$**else**
$\quad\quad\quad\quad \delta(c_{i2}, \overline{x_{i2}}) := c_{(i+1)1}; \delta(c_{i2}, x_{i2}) := c_{i3}$
$\quad\quad\quad$**if** x_{i3} appears positive in c_{i3} **then**
$\quad\quad\quad\quad \delta(c_{i3}, x_{i3}) := c_{(i+1)1}; \delta(c_{i3}, \overline{x_{i3}}) := \bot$
$\quad\quad\quad$**else**
$\quad\quad\quad\quad \delta(c_{i2}, \overline{x_{i3}}) := c_{(i+1)1}; \delta(c_{i3}, x_{i3}) := \bot$
$\quad\quad$**else**
$\quad\quad\quad$/* Connect the truth assignment to $\top$ */
$\quad\quad\quad$**if** x_{i1} appears positive in c_{i1} **then**
$\quad\quad\quad\quad \delta(c_{i1}, x_{i1}) := \top; \delta(c_{i1}, \overline{x_{i1}}) := c_{i2}$
$\quad\quad\quad$**else**
$\quad\quad\quad\quad \delta(c_{i1}, \overline{x_{i1}}) := \top; \delta(c_{i1}, x_{i1}) := c_{i2}$
$\quad\quad\quad$**if** x_{i2} appears positive in c_{i2} **then**
$\quad\quad\quad\quad \delta(c_{i2}, x_{i2}) := \top; \delta(c_{i2}, \overline{x_{i2}}) := c_{i3}$
$\quad\quad\quad$**else**
$\quad\quad\quad\quad \delta(c_{i2}, \overline{x_{i2}}) := \top; \delta(c_{i2}, x_{i2}) := c_{i3}$
$\quad\quad\quad$**if** x_{i3} appears positive in c_{i3} **then**
$\quad\quad\quad\quad \delta(c_{i3}, x_{i3}) := \top; \delta(c_{i3}, \overline{x_{i3}}) := \bot$
$\quad\quad\quad$**else**
$\quad\quad\quad\quad \delta(c_{i2}, \overline{x_{i3}}) := \top; \delta(c_{i3}, x_{i3}) := \bot$
$\quad \delta(\top, r) := \top$
$\quad q_0 := c_{11}$
$\quad$**return** (Q, Σ, q_0, δ)

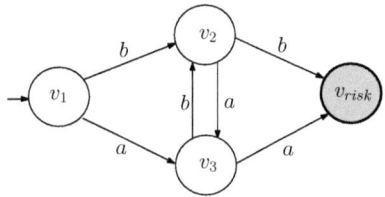

Fig. 4. An example where priority synthesis is unable to find a set of priorities

5   Discussion

The framework of priority systems in [3,1,2] offers a methodology to incrementally construct a system satisfying safety properties while maintaining deadlock freedom. In this paper, we use an automata-theoretic approach to formulate the problem of priority synthesis, followed by giving an NP-completeness proof. We conclude that, although using priorities to control the system has several benefits, the price to take is the hardness of an automatic method which finds appropriate priorities. Also, based on the formulation, it is not difficult to show that it is possible to find a supervisor in the framework of Ramadge and Wonham [5] while priority synthesis is unable to find one. This is because priorities are stateless properties, and sometimes to achieve safety, executing interactions conditionally based on states is required. E.g., for the transition system in Figure 4, applying priority $a \prec b$ or $b \prec a$ is unable to ensure system safety, but there exists a supervisor (for safety) which disables b at state v_2 and a at v_3.

References

1. Basu, A., Bozga, M., Sifakis, J.: Modeling heterogeneous real-time components in BIP. In: Proceedings of the 4th IEEE International Conference on Software Engineering and Formal Methods (SEFM 2006), pp. 3–12. IEEE Computer Society Press, New York (2006)
2. Cheng, C.-H., Bensalem, S., Jobstmann, B., Yan, R., Knoll, A., Ruess, H.: Model construction and priority synthesis for simple interaction systems. In: Bobaru, M., Havelund, K., Holzmann, G.J., Joshi, R. (eds.) NFM 2011. LNCS, vol. 6617, pp. 466–471. Springer, Heidelberg (2011)
3. Gößler, G., Sifakis, J.: Priority systems. In: de Boer, F.S., Bonsangue, M.M., Graf, S., de Roever, W.-P. (eds.) FMCO 2003. LNCS, vol. 3188, pp. 314–329. Springer, Heidelberg (2004)
4. Graf, S., Peled, D., Quinton, S.: Achieving distributed control through model checking. In: Touili, T., Cook, B., Jackson, P. (eds.) CAV 2010. LNCS, vol. 6174, pp. 396–409. Springer, Heidelberg (2010)
5. Ramadge, P., Wonham, W.: The control of discrete event systems. Proceedings of the IEEE 77(1), 81–98 (1989)

Smaller Representation of Finite State Automata

Jan Daciuk[1] and Dawid Weiss[2]

[1] Knowledge Engineering Department, Gdańsk University of Technology, Poland
[2] Institute of Computing Science, Poznan University of Technology, Poland

Abstract. This paper is a follow-up to Jan Daciuk's experiments on space-efficient finite state automata representation that can be used directly for traversals in main memory [4]. We investigate several techniques of reducing the memory footprint of minimal automata, mainly exploiting the fact that transition labels and transition pointer offset values are not evenly distributed and so are suitable for compression. We achieve a size gain of around 20–30% compared to the original representation given in [4]. This result is comparable to the state-of-the-art dictionary compression techniques like the LZ-trie [10] method, but remains memory and CPU efficient during construction.

1   Introduction

Minimal, deterministic, finite-state automata are a good data structure for representing natural language dictionaries [6]. They are not only fast in construction and traversals, but also take little space. Small memory footprint stems from minimality, but it is possible to reduce it even further using various compression and bit-packing schemes. It is also possible to change the definition of an automaton so that transitions, and not the states, can become final [3]. Ciura and Deorowicz [2] call such an automaton a *Mealy's acceptor* to underline the parallel with Moore's transducers and Mealy's transducers. Moore's transducers store their output in their states, Mealy's transducers – in their transitions.

One compression technique is universal in all implementations. Fields like a transition's label, target state's address (a pointer), or various flags can be packed into a minimal bit field required for their representation. Packing the fields so that they occupy as few bytes or bits as possible greatly reduces memory requirements. Decoding bit-aligned representation on modern hardware does not impose a large overhead on processing time and compact memory representation contributes nicely to reuse of CPU cache lines.

Packing fields bit- and byte-wise may be done in two manners: the fields can maintain fixed length, or their length may become variable, that is different instances of the field may have different lengths. The latter can be implemented using Huffman coding or other variable-bit representation schemes. While it can lead to greater savings, such compression requires additional memory lookups for decoding and can even lead to increased overall size if addresses need to be bit- or byte-aligned instead of being multiplications of a node's fixed size.

In most efficient implementations, an automaton is a vector of transitions. States are represented implicitly. There are two major methods of representing

B. Bouchou-Markhoff et al. (Eds.): CIAA 2011, LNCS 6807, pp. 118–129, 2011.
© Springer-Verlag Berlin Heidelberg 2011

states. In the first one, a state is a list of outgoing transitions. In the other one, a state is a vector of possible outgoing transitions with an allocated place for a transition for every transition label from the alphabet. The vector for each state is put into a larger vector of transitions so that states overlap whenever possible without conflicts between transitions. The latter method is called *superimposed coding* [8]. It is faster for recognition, as each time we traverse a transition, we go to it directly without looking at any other transition going out from the same state (a transition is indexed directly by its label), and it is slower for exploration, as we need to check which transitions exist. That method allows for fewer compression techniques, so we will focus on the state-as-a-list representation.

In the state-as-a-list representation, it is possible to link subsequent transitions with pointers, but using a vector is more economical. The next transition in the vector is the next transition on the list. The problem of knowing what the last transition is can be solved by either storing an outgoing transition counter in an incoming transition, or by using a flag [7] (we call it L for **LAST**) to mark the last transition on the list. The latter approach saves more space.

A transition connects two states. Since we group transitions going out from a state, we need to specify the target of a transition, that is the address of the target state, which in state-as-a-list representation is the address of its first outgoing transition. There are several methods of reducing the size of that field. When the target state is placed directly after the current transition, it is possible to omit the field altogether at the cost of adding a new flag that we call N for **NEXT**. When this flag is set, the transition has no target address field, and the target state begins right after the current transition. It is also possible to vary the size of the address field so that there are local (short) and global (long) pointers as in [9] at the cost of an additional flag. In a US patent 5 551 027 granted on August 7th, 1996 to Xerox, frequently used addresses are put into a vector of full length pointers, and the addresses are replaced with shorter indexes to the pointers in the vector.

Since a state is stored as a list of outgoing transitions, it is possible to share transitions between states. When all transitions of one state are also present as transitions of another state (that has more transitions), then the "smaller" state can be stored inside the "bigger" one. When we use the L flag, the transitions of the smaller state have to be the last transitions of the bigger state. If it is not the case, the transitions need to be rearranged to conform to this condition. There may be many combinations of smaller states fitting into some larger ones, so heuristics have to be used. Note that once a state is stored inside another one, there is no speed or memory penalty for using this type of compression, it just reuses the same memory regions.

Another technique of reusing states' transitions is based on the fact that two states may share a subset of their transitions, but are not subsets of each other (each of the states has transitions that the other one does not have). In such case, one state is stored intact, the unique transitions of the second state are stored as usual, but the last transition has a flag we call T for **TAIL**, followed by the address of the common set of transition stored in the first state. Reordering of transitions inside individual states may lead to greater savings.

A generalization of transition sharing is presented in the LZ-trie method by Strahil Ristov [10]. The LZ-trie method treats an automaton as a sequence of transitions and applies compression to this sequence. A suffix tree (or array) is used for finding all subsequences of transitions, storing them once, and replacing redundant instances with pointers to their previous occurrences. This gives state-of-the-art compression ratios [1]. Note that combining LZ-trie with other methods described above gives much poorer results [5].

Some research has been devoted to finding substructures in an automaton – subautomata [12,11]. Although conceptually different from the LZ-trie method, these methods can be seen as a variant of the LZ-trie method with some restrictions that limit compression efficiency. On the other hand, subautomata can have applications other than mere reduction of representation size.

The remaining part of this work is structured as follows. Our motivation and goals are given in Section 2. Section 3 introduces the data sets used in evaluating various methods described later in the paper. Section 4 describes our attempts to reduce the size of automata representation in memory. Section 5 provides an overview of computational experiments and their results, comparing them to the known state of the art. Section 6 concludes the paper.

2   Motivation and Goals

Many of the compression techniques described in the introduction are implemented in Jan Daciuk's fsa package [4]: transition-based representation, accepting transitions (Mealy's recognizers), optimizations of pointers in the form of the N bit or bit-packing of the target address with the rest of the flags. These tricks allow for direct, incremental construction in the compressed format, suitable for immediate serialization to disk or storage in memory, and implementation of traversals over the packed format with very little overhead. The goals of this work were to investigate the following open problems:

1. Is it possible to construct a more space-efficient automaton representation that would retain the features present in the fsa package?
2. There is a trade-off between compressing representation and traversal efficiency. Is there a representation that would balance small size with an efficient (read: simple) automaton traversals?

3   Test Data

The research presented in this paper was mostly trial-and-error driven, where the baseline was acquired by comparing the output to the equivalent automata compiled using the fsa_build command from the fsa package. The choice of test data was thus important. The test files, their size and number of terms, are given in Table 1. The first five files on that list were collected by the authors of this

Table 1. File size (bytes), number of terms (lines) and an average number of bits per term for all the files used in experiments

Name	Size (bytes)	Terms	BPT		*continued*		
pl	165 767 147	3 672 200	361		esp	8 001 052	642 014	100
streets	706 187	59 174	95		files	212 761 171	2 744 641	620
streets2	203 590	17 144	95		fr	2 697 825	221 376	97
wikipedia	105 316 228	9 803 311	86		ifiles	212 761 171	2 744 641	620
wikipedia2	504 322 111	38 092 045	106		polish	18 412 441	1 365 467	108
	—				random	1 151 303	100 000	92
deutsch	2 945 114	219 862	107		russian	9 933 320	808 310	98
dimacs	7 303 884	309 360	189		scrable	1 916 186	172 823	89
enable	1 749 989	173 528	81		unix	235 236	25 481	74
english	778 340	74 317	84		unix_m	191 786	20 497	75
eo	12 432 197	957 965	104		webster	985 786	92 342	85

work and the remaining files come from [2]. The `pl` data set is a morphological dictionary of inflected forms and their encoded lexemes and morphological annotations. It has highly repeatable suffixes (a limited set of inflection frames and morphological tags). The two data sets named `wikipedia` and `wikipedia2` contain terms from an inverted index of English Wikipedia (`wikipedia` is a sample, `wikipedia2` is an index of full content). Data sets called `streets` and `streets2` carry street and city names covering the area of Poland and have been acquired from a proprietary industrial application. The first five files in Table 1 contained UTF-8 encoded text. We did not alter the original character encoding used to represent the remaining data sets – they all used single-byte encodings of their respective languages (ISO8859-2 for Polish, for example). Our automata implementation was byte-based, so input character encoding was simply preserved in the automaton structure.

4   Size Reduction Techniques

Figure 1 shows a binary data layout of fields in a single transition in Jan Daciuk's fsa package. Recall this was the baseline representation we started from. A single transition is composed of the initial label, then a byte with three flags – (F for FINAL, acceptor transition), N (no address, the target state follows this state's last transition) and L (this is the last transition of the current state). If the N bit is not set, partial address is bit-packed into the remaining five bits of the flags

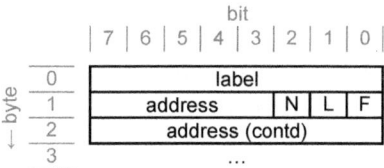

Fig. 1. Binary layout of data fields in a single transition. fsa package compiled with N and L options. N, F and L are bit flags, address field's length is as large, as the largest state offset in the automaton (but constant for every transition).

byte and as many bytes, as are needed to encode the largest integer offset in this automaton.

Starting with the baseline above, we tried numerous variations to decrease the representation size of each transition (and the transition graph as a whole). We describe these that yielded maximum gains in the paragraphs below.

V-coding of target addresses. The fsa package uses fixed-length address encoding integrated with the flags byte. This has an effect of abrupt increases of automaton size once 1, 2, 3 or more bytes are needed to encode the largest state's offset. We used a simple form of variable length encoding for non-negative integers (*v-coding*), where the most significant bit of each byte is an indicator whether this is the last byte of the encoded integer and the remaining bits carry the integer's data. For example, 0 is encoded as (binary representation) 0000 0000, 127 as 0111 1111, 128 using two bytes: 1000 0001 and 0000 0000, and so on. Encoding and decoding of v-coded integers can be implemented efficiently without bit rotations if we have them in consecutive bytes, so we moved the transition's target address to separate bytes, which left us with 5 unused bits in the flags byte.

Transitions with index-coded label. We assumed each transition's label is a single byte. Each transition's label can be therefore an integer between 0 and 255. For multi-byte or variable-byte character encoding schemes (such as UTF-8) the automaton stores their raw binary representation. When performing traversals or lookups, the automaton's encoding must be respected – the input text must be converted to the automaton's code page, for example. Another side effect is that certain transitions can lead to incomplete character codes, but we never had a problem with this in real applications (even with multibyte planes from Unicode).

In reality, for automata created on non-degenerate input, and in particular on text, the distribution of label values is often skewed. Figure 2 illustrates the distribution of labels in the pl data set, for example – there are many transitions with a small subset of the label range and a few transitions outside this range.

The observation that labels have uneven distribution leads to an optimization that has a profound effect on automaton size: we can integrate the 31 most frequent labels ($2^5 - 1$) into the flags byte as an index to a static lookup table. Zeros on all these bits would indicate the label is not indexed and is stored separately. Note that we tried to avoid any complex form of encoding (like Huffman trees); a fixed-length table with 31 most frequent labels is a balanced tradeoff between auxiliary lookup structures and label decoding overhead at runtime.

Combining v-coding of the target address and table lookup for the most frequent labels yields two alternative transition formats, as shown in Figure 3. With such encoding most transitions take $1 + length(address)$ bytes. In an extreme case when the N bit is also set (target follows the current state's last transition), the entire transition is encoded in a single byte.

Rearranging states to minimize the total length of address fields. By default states (actually a list of transitions of each state) in an automaton are

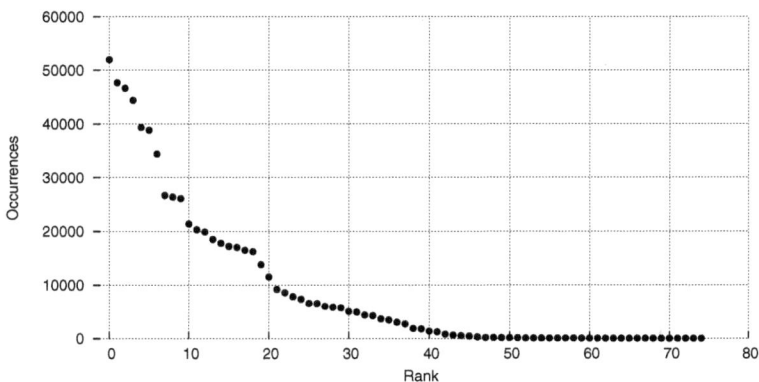

Fig. 2. Number of occurrences of 75 most frequent labels in the `pl` data set

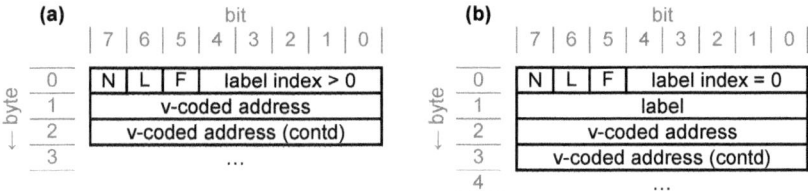

Fig. 3. Binary layout of data fields in a single transition with v-coding of target address and indexed labels. Two variants of each transition are possible: (a) with the index to the label, (b) with the label directly embedded in the transition structure.

serialized in a depth-first order to maximize the number of occurrences of the N flag and hence the gain from not having to emit the target address for such transitions. For these transitions where N is not set, the target address must be emitted and the amount of space taken by such an address depends on its absolute value (recall addresses are v-coded and thus take a variable number of bytes). If we move certain states (these to which there are a lot of incoming transitions) to the beginning of the automaton, the global amount of space for address encoding should be smaller than if we leave these states somewhere farther in the serialized automaton structure. The question is which states we should move and in what order they should appear in the automaton structure.

The problem of rearranging states to minimize the global sum of bytes required for encoding target addresses is complicated. There are several things to consider:

– States located at offsets 0–127 require only one byte for target address code, states located at offsets 128–16 383 two bytes, and so on. But then, a single state may have many transitions, so it occupies a variable number of bytes. We can move to the front a single large state with many incoming transitions

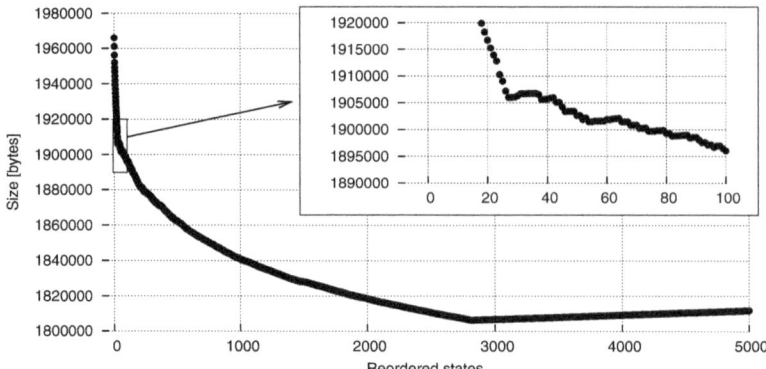

Fig. 4. Automaton size in relation to the number of moved states, `pl` data set (first 5 000 state reorderings shown). The zoomed-in section of the chart shows the relation is not monotonic, even at the very beginning.

or, alternatively, many smaller states with fewer incoming transitions but all fitting in the "one-byte" offset range.

- By moving a state from its original location we also shift the offset of other states, possibly rearranging the fields across the entire automaton.
- We may lose the gain from applying the N flag optimization if we move a state (or its predecessor) to which the N flag applies.

The question if there is an "optimal" arrangement of states to minimize the global serialized automaton length remains open. The problem itself seems to be equivalent to bin-packing (in terms of computational complexity) and thus not have a solution working in a reasonable (polynomial) time.

Our first attempt to solve this issue was a simple heuristic: in the first step, we determine the serialization order for all states as to maximize the number of N bits (depth-first traversal). Then, we create a priority queue of states in the decreasing number of their incoming transitions and keep moving states from the top of the queue to the start of the automaton as long as the serialized automaton is smaller than before.

This heuristic has a serious flaw because serialized automaton size does not decrease monotonically with the number of moved states. For example, Figure 4 depicts automaton size in relation to the number of moved states for the `pl` data set. The minimum size is reached at around 2 900 reordered states with the largest number of incoming transitions, but a closer look at the beginning of this chart shows that the function is not monotonic – see the zoomed rectangle inside Figure 4. At around 29 reordered states the size goes up from 1 911 758 to 1 912 027, only to drop further down after more states are reordered.

The second take at the state reordering heuristic was a simulated-annealing like process that worked similar to the first approach (initial states order to maximize the use of the N bit, then a queue of states with most inlinks), and then probed at various subsets of the states' queue, decaying over time and focusing on ranges that promised the smallest output.

To compute the resulting automaton size after each state reordering, both heuristics performed its full serialization. This was the key factor slowing down automaton compression and took vastly more time than automaton construction itself. Nonetheless, if size is of major importance, even these simple heuristics provide significant gain: as shown in Figure 4, for the pl data set the serialized size decreases from 1 966 038 bytes achieved by depth-first order traversal of states to 1 806 621 bytes (8% gain) with around 2 900 states moved to the front.

5   Experiments and Results

Table 2 shows the results of compressing the test data sets into finite state automata using fsa_build (version 0.50, with patches), LZ-trie and a binary format utilizing optimizations presented in this paper, called cfsa2, implemented as part of the Morfologik project. The data sets and software used to compress them are available at: https://github.com/dweiss/paper-fsa-compression.

Automata packed using cfsa2 were on average 29% smaller compared to the result of fsa_build, regardless of the nature of the input file ($\sigma = 3.31\%$). LZ-trie produced files smaller by 13.7% on average (compared to cfsa2), but here the standard deviation is $\sigma = 11.86$ and there is a notable exception of the pl data set, smaller by 27% when packed using cfsa2. We do not have an explanation for

Table 2. The size of on-disk automaton representation and bits per byte and term ratios for the input files compressed with fsa_build (fsa), cfsa2 and LZ-trie (LZ). The $\%^1$ column shows size drop from fsa to cfsa2, $\%^2$ from cfsa2 to LZ-trie. The smallest compressed size of each data set is marked with a ▪ symbol.

Name	Output size (KB)			%		Bits per byte			Bits per term		
	fsa	cfsa2	LZ	$\%^1$	$\%^2$	fsa	cfsa2	LZ	fsa	cfsa2	LZ
pl	2 655	1 764 ▪	2 245	34	−27	0.13	0.09	0.11	5.9	3.9	5.0
streets	334	244	217 ▪	27	11	3.87	2.83	2.52	46.2	33.8	30.1
streets2	128	93	86 ▪	28	7	5.16	3.73	3.46	61.3	44.3	41.1
wikipedia	—	40 362	36 413 ▪	—	10	—	3.14	2.83	—	33.7	30.4
wikipedia2	—	168 683	157 126 ▪	—	7	—	2.74	2.55	—	36.3	33.8
deutsch	285	215	188 ▪	24	13	0.79	0.60	0.52	10.6	8.0	7.0
dimacs	2 436	1 487	1 299 ▪	39	13	2.73	1.67	1.46	64.5	39.4	34.4
enable	401	290	264 ▪	28	9	1.88	1.36	1.23	18.9	13.7	12.4
english	243	173	145 ▪	29	16	2.56	1.82	1.53	26.8	19.1	16.0
eo	211	147	109 ▪	31	26	0.14	0.10	0.07	1.8	1.3	0.9
esp	385	268	187 ▪	30	30	0.39	0.27	0.19	4.9	3.4	2.4
files	12 425	9 205	7 120 ▪	26	23	0.48	0.35	0.27	37.1	27.5	21.3
fr	220	153	120 ▪	30	22	0.67	0.47	0.36	8.2	5.7	4.4
ifiles	12 770	9 748	8 147 ▪	24	16	0.49	0.38	0.31	38.1	29.1	24.3
polish	676	477	352 ▪	29	26	0.30	0.21	0.16	4.1	2.9	2.1
random	1 162	832	798 ▪	28	4	8.27	5.92	5.68	95.2	68.2	65.4
russian	505	354	262 ▪	30	26	0.42	0.29	0.22	5.1	3.6	2.7
scrable	435	310	263 ▪	29	15	1.86	1.33	1.12	20.6	14.7	12.4
unix	132	95	83 ▪	28	13	4.61	3.30	2.88	42.6	30.5	26.6
unix_m	104	72	63 ▪	30	12	4.42	3.09	2.70	41.4	28.9	25.3
webster	417	298	248 ▪	28	17	3.46	2.48	2.06	37.0	26.4	22.0

Table 3. Automata compression times (in seconds). Experiments were performed on the following hardware: **cfsa2** and **fsa5** – Intel Core i7 CPU 860 @ 2.80GHz, 8GB RAM, Ubuntu Linux; LZ-trie – Intel Xeon W3550 @ 3.07 Ghz, 12GB RAM, CentOS. Ratios are shown only for compression times greater than a few seconds (**cfsa2** is written in Java and the timings include HotSpot warm-up time, so times for really short input data are not directly comparable).

Name	Compression time (s)			Ratio (%)	
	fsa	cfsa2	LZ	cfsa2/fsa	cfsa2/LZ
pl	40.01	20.15	6 000.00	50	0.34
streets	0.16	1.15	0.84		
streets2	0.13	3.57	0.21		
wikipedia		226.90	1 860.00		12
wikipedia2		1 556.84	57 600.00		3
deutsch	0.22	1.10	1.00		
dimacs	1.66	8.38	100.00		
enable	0.21	1.20	0.89		
english	0.12	0.90	0.50		
eo	0.84	1.10	5.00		
esp	0.57	1.14	3.00		
files	645.17	99.77	7 200.00	15	1
fr	0.21	0.85	1.00		
ifiles	453.53	102.10	25 200.00	23	0.41
polish	1.33	2.46	9.00		
random	0.61	4.12	3.00		
russian	0.66	1.50	4.00		
scrable	0.25	1.16	0.84		
unix	0.06	0.50	0.11		
unix_m	0.04	0.50	0.72		
webster	0.20	1.21	0.68		

this at the time of writing, but we suspect that this difference is caused by the fact that the `pl` data set has a huge number of repetitive suffixes (morphological tags); it is likely that the transitions to these repetitive suffixes ended up moved to the front of the automaton and thus resulted in small sizes of target address pointers of many arcs, whereas in LZ-trie each such pointer is represented as a constant-size data structure.

Yet, smaller files produced by LZ-trie come at a much longer compression time – for example, `wikipedia2` took 16 hours, while the (Java-based) `cfsa2` compressed it in 25 minutes (of which 42 seconds were spent in constructing the FSA and the rest seeking for the optimum number of states to reorder, which yet again proves the point of improving this heuristic somehow). Table 3 shows a complete list of compression times for the three methods used. Note that these times are only roughly comparable because LZ-trie compression was performed on a different hardware (CPUs computational performance is nearly identical though, according to `cpubenchmark.net`) and `cfsa2` timings included the time to launch Java VM, HotSpot JITting, etc.

The largest size reduction is achieved by integrating transition labels with the flags byte (see Table 4) – most data sets did not even use transitions with separate label byte. Note that even a truly random byte sequence would still benefit from

Table 4. Ratios of integrated and separate labels and lengths of v-coded target state
addresses. V-code zero is equivalent to the presence of the N flag.

Name	Labels (%)		V-code length (%)				
	int.	sep.	0	1	2	3	4
pl	94	6	30	9	32	29	
streets	98	2	39	17	23	20	
streets2	98	2	46	13	20	21	
wikipedia	79	21	38	18	14	13	16
wikipedia2	85	15	45	13	11	14	17
deutsch	100	0	31	15	35	19	
dimacs	98	2	48	12	19	21	
enable	100	0	25	27	30	19	
english	100	0	26	29	27	17	
eo	100	0	19	34	33	14	
esp	100	0	17	34	35	14	
files	86	14	64	3	6	12	15
fr	99	1	25	33	26	16	
ifiles	88	12	65	1	4	11	18
polish	99	1	20	33	29	18	
random	100	0	60	5	13	23	
russian	100	0	20	33	30	16	
scrable	100	0	23	30	29	17	
unix	99	1	26	31	21	22	
unix_m	100	0	26	33	23	18	
webster	100	0	24	30	28	18	
$\mu =$	96	4	34	22	23	18	3

integrated labels at around 12% (even if label distribution is uniform, 31 labels
would still be integrated in the flags field). Table 4 also shows the benefit of using
the N bit (34% of transitions on average) and v-coding of transition pointers (an
average of 45% of transitions used one or two bytes for the address).

6   Conclusions

We have shown that three basic techniques:

- table-lookup encoded labels, exploiting their uneven distribution,
- variable-length coding of transition target addresses, and
- state ordering to minimize the global size of encoded target addresses

make it possible to compress (already compact) dictionaries considerably, in
some cases even better than the LZ-trie method, whose results were so far considered the best in the field. Not of less importance is the fact that the representation presented in this paper retains simple automaton structure and allows very
efficient, non-recursive traversals. There is a considerable space for further research in how to efficiently determine an optimal or nearly-optimal arrangement
of states to minimize their global representation length, but even the presented
naïve heuristic implemented in Java turns out to be much faster than fsa_build
or LZ-trie, especially on large data sets.

Comparing our method to the LZ-trie method, the main difference is that we
do not search for repeatable substructures. By finding subautomata, we might

possibly boost the compression ratio at the cost of slightly increased traversal time and moderately increased construction time. It is worth mentioning that the LZ-trie method could also benefit from the improvements we introduced here, mainly of variable-length coding of transition target addresses and table-lookup encoded labels. How large this gain can be and what exactly could be borrowed from the ideas presented here is a matter of further study and we plan to address it in a follow-up paper.

Another interesting aspect that requires attention is automaton traversal speeds. All methods exercized in this paper represent a state's transitions in a form that requires a linear lookup scan to find a matching label. This is highly in-effective when traversing highly fanning-out states, which unfortunately usually happen to be close to (and including) the automaton root. We created a simple benchmark where the same traversal routine was executing a simple hit/miss test using a mix of random and matching sequences. The traversal speed (same hardware as in Table 3) on an automaton in fsa5 format averaged around 1.9 million checks per second, on cfsa2 – around 800 thousand checks per second (variable transition length requires partial decoding hence the slowdown). These figures compare favorably to the speed achieved by LZ-trie, which, as reported by the author, achieves around 1 million checks per second.

A few simple improvements can be made to make the traversal much, much faster at a slight size penalty. The most obvious improvement is to expand states with a larger fan-out into a form allowing direct table-lookup (or binary search) of a given label. This has been implemented in Apache Lucene recently and yields nearly 4 million terms/ second check speed. Another optimization hint is related to utilizing CPU caches better – we can clump together the representation of states reachable from the root state so that they fit in as few cache lines as possible. This can be easily done by breadth-first traversal to a given depth and even combined with state reordering mentioned earlier. We plan to tackle these ideas in our future work on the subject.

As a concluding remark, let us note that morphological dictionaries com-pressed very well in our experiments, achieving incredible compression ratios (1.3 bits per entry for eo or 3.9 bits per entry for the pl data set). Knowing that finite state automata can be used for calculating perfect hashes (or with minor modifications as transducers) it is somewhat surprising to learn that quite a few tools for natural language processing still opt for using traditional databases to store and search for linguistic data.

Acknowledgments. Strahil Ristov and Damir Korencic kindly responded to our request to run LZ-trie on the provided data sets and passed back valuable comments and suggestions. Michael McCandless provided a list of terms ex-tracted from Wikipedia and was always ready for long (and fruitful) discussions on the subject of automata compression and traversal speeds. Finally, we really appreciate all the corrections, comments and remarks given by three anonymous reviewers of this work. Thank you all.

References

1. Budiscak, I., Piskorski, J., Ristov, S.: Compressing Gazetteers Revisited. In: Yli-Jyrä, A., Kornai, A., Sakarovitch, J., Watson, B. (eds.) FSMNLP 2009. LNCS, vol. 6062, Springer, Heidelberg (2010)
2. Ciura, M., Deorowicz, S.: How to squeeze a lexicon. Software – Practice and Experience 31(11), 1077–1090 (2001)
3. Daciuk, J.: Incremental Construction of Finite-State Automata and Transducers, and their Use in the Natural Language Processing. Ph.D. thesis, Technical University of Gdańsk (1998)
4. Daciuk, J.: Experiments with automata compression. In: Yu, S., Păun, A. (eds.) CIAA 2000. LNCS, vol. 2088, pp. 105–119. Springer, Heidelberg (2001)
5. Daciuk, J., Piskorski, J.: Gazetteer Compression Technique Based on Substructure Recognition. In: Proceedings of the International Conference on Intelligent Information Systems, Ustroń, Poland, pp. 87–95 (2006)
6. Daciuk, J., Piskorski, J., Ristov, S.: Scientific Applications of Language Methods. In: Mathematics, Computing, Language, and Life: Frontiers in Mathematical Linguistics and Language Theory, pp. 133–204. World Scientific Publishing, Singapore (2010)
7. Kowaltowski, T., Lucchesi, C.L., Stolfi, J.: Minimization of binary automata. In: 1st South American String Processing Workshop, Belo Horizonte, Brasil (1993)
8. Liang, F.M.: Word Hyphenation by Computer. Ph.D. thesis, Stanford University (1983)
9. Lucchiesi, C., Kowaltowski, T.: Applications of finite automata representing large vocabularies. Software Practice and Experience 23(1), 15–30 (1993)
10. Ristov, S., Laporte, É.: Ziv Lempel Compression of Huge Natural Language Data Tries Using Suffix Arrays. In: Crochemore, M., Paterson, M. (eds.) CPM 1999. LNCS, vol. 1645, pp. 196–211. Springer, Heidelberg (1999)
11. Tounsi, L.: Sous-automates à nombre fini d'états. Application à la compression de dictionnaires électroniques. Ph.D. thesis, Université François Rabelais Tours (2008)
12. Tounsi, L., Bouchou, B., Maurel, D.: A Compression Method for Natural Language Automata. In: Proceedings of the 7th International Workshop on Finite-State Methods and Natural Language Processing, Ispra, Italy, pp. 146–157 (2008)

Compositional Failure Detection in Structured Transition Systems

Ingo Felscher and Wolfgang Thomas

RWTH Aachen University

Abstract. In model-checking, systems are often given as products. We propose an approach that is built on a preprocessing of specifications in terms of appropriate automata. This allows to incorporate information about the local behaviour and synchronization of the system components into the specification. We develop a framework of (partially) synchronized automaton products and a format of corresponding specification automata that allows for a compositional failure detection of linear regular properties (either for finite or for infinite behaviour). As a result we obtain an algorithm which separates the local and the non-local segments of system runs, resulting in improved complexity bounds in typical specifications.

Keywords: model-checking, finitely synchronized products, compositional failure detection.

1 Introduction

In model-checking we examine whether a given system, normally modelled as a transition system, satisfies a specification, modelled as a logic formula. The systems under investigation often arise as products composed of several components – again transition systems – that may interact with some or all other components and may also perform actions independently of the other components. The main problem in this scenario is the question of state space explosion, studied in a large body of literature, see e. g. [2].

The basic problem is to separate aspects of the specification that are local (to the components) from each other and from synchronizing features. This is a natural idea which is also familiar from the "composition method" of algorithmic model theory. The method allows to deduce the truth of a formula in a product from information about the truth value of formulas in the components. In model theory this approach was initiated in the pioneering paper [9] of Feferman and Vaught and further developed by numerous authors [5,12,13,14,15,17,18]. For more recent results, now in the field of model-checking, see [10,16,20]. The complexity of this compositional approach is excessive (in fact, non-elementary in the size of the given formula – even for modal logic and first-order logic), due to the large number of auxiliary formulas that have to be constructed. At least for first-order logic this effect is known to be unavoidable [7]. (Apart from this, the classical approach is restricted to (variants of) first-order logic. Already for

B. Bouchou-Markhoff et al. (Eds.): CIAA 2011, LNCS 6807, pp. 130–141, 2011.

modal logic extended by the logical operator EG and for first-order logic with regular reachability predicates the composition method fails [16,20].)

In the present paper we offer a compositional analysis of reachability (motivated by failure detection) that may lead to a considerable reduction of the high complexity as known from the logical framework. Our approach relies on automata theoretic specifications. (For a full paper with corrections see [11].)

Another advantage of the insistence on automata as specification formalism is the avoidance of the initial conversion of a given logic formula to an automaton. In most cases (e.g., for temporal, first-order or monadic second-order logic), the costs of this conversion of formulas are exponential (or much more) in time complexity. (It is well-known that an $MSO(<)$ formula can be translated into an equivalent automaton [4,8] and that the complexity of this translation is non-elementary [1,19].)

In the present work we start with the description of undesired behaviour using a "complement specification", denoted $\overline{Spec}$. The given system is a partially synchronized product Sys with (binary, labeled) relations and (unary) predicates. We split $\overline{Spec}$ into parts which can be checked in the individual components. For purpose of exposition, we first consider the case of $\overline{Spec}$ where unary predicates are missing, and then treat the general case.

The general idea is to split the complement specification automaton into parts (called "local blocks") each of which has only labels and predicates from a fixed set of components. As result we then get the local blocks (as mentioned above, as specification automata which can be checked in the individual components) and a "global specification automaton" $Glob$ which describes the possible concatenations of these local block automata. For this, information about the synchronization behaviour of a transition is used: the "synchronization profile". This profile specifies which of the components are synchronized via the transition's label. In the runs of $Glob$, sequences of transitions with the same synchronization profile are grouped together.

In the main result, first stated for the case without unary predicates, the question whether a product of transition systems and a given complement specification automaton have paths with a common labeling is reduced to the question whether a path in the global specification automaton exists such that the components and parts of the complement specification which are described by the local blocks of the global specification automaton have paths with common labeling.

To generalize the result to specifications with predicates we first linearize the transition system Sys and the complement specification $\overline{Spec}$: We code the predicates of states in labeled self-loops of these states and thus dissolve, for example, the fulfillment of predicates $p_1, \neg p_2, p_3$ at a state s into the subsequent execution of self loops at s, labeled $p_1, \neg p_2, p_3$. If an action move c is executable at s, the corresponding c-transition may be taken after the mentioned self-loops.

The terminological complexity of a compositional framework as developed here is considerable – an unavoidable feature also known from the literature above. As a gain of this effort, we will show that the algorithm derived from this automaton composition method will only be single exponential in the number of states and predicates that the components and the complement specification have.

Of course, a drawback is the necessity of preprocessing when a logical specification is given. However, in many practical situations, when specifications are short, this preprocessing can often be done efficiently in spite of the exponential standard algorithms [21]. In other cases, one might be able to use automata theoretic specifications directly.

The applicability of our method depends on an appropriate set-up of the specification automaton: It should offer as much as possible the potential to separate the various local and synchronized computations. Of course, in the worst case as represented by always fully synchronized transitions, the decomposition does not pay (since only blocks of length one are formed).

The paper is structured as follows: After this introduction we present in Sect. 2 technical preliminaries. For this, we show our notion of a synchronized product and the complement specification automaton. These definitions are then used in Sect. 3 for the main result. We further add a sketch how this result can be generalized to infinite behaviours captured by Büchi automata. In Sect. 4 we treat the case of specifications with unary predicates. We conclude the paper in Sect. 5 with a summary and some remarks on open problems.

2   Technical Preliminaries

In this section we introduce the basic definitions: In Sect. 2.1 we treat products of transition systems and in Sect. 2.2 the automaton models used for complement specification and its transformation into the global automaton.

2.1   Products of Transition Systems

A *transition system* is a labeled graph $K = (S, \{R_a \mid a \in \Sigma\}, \{P_v \mid v \in V\})$ with state set S, transition relations $R_a \subseteq S \times S$ and predicates $P_v \subset S$.

We introduce our notion of a synchronized product with asynchronous and synchronous behaviour: *Synchronized transitions* are transitions which are taken at the same time in a subset of the components – captured by the "synchronization profile" – and independently of the transitions of the other components. *Asynchronous transitions* are taken independently of all transitions in the other components, i. e. they can be seen as synchronized with a synchronization profile which contains only one component. Therefore, we restrict ourselves to synchronized transitions to simplify the constructions.

From now on, we use $[m]$ for $m \in \mathbb{N}$ as an abbreviation for the set $\{1, \ldots, m\}$.

Definition 1 (Synchronized product). *Let $I = [n]$ be a finite set of indices and Σ an alphabet of labels (of the transitions) and $V := \{v_1, \ldots, v_l\}$ a set of names of unary predicates. For $i \in I$ let a component transition system K_i be of the form $K_i = (S_i, \{R_c^i \mid c \in \Sigma\}, \{P_v^i \mid v \in V\})$ as mentioned above.*

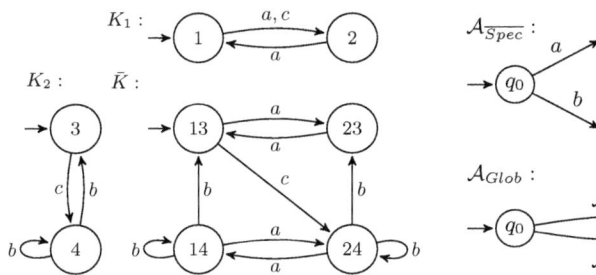
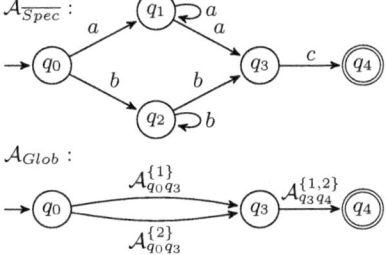

Fig. 1. Components K_1, K_2 and their product $\overline{K}$

Fig. 2. Complement specification and Global automaton

A synchronization profile $sp(c)$ for $c \in \Sigma$ defines which components are synchronized via c-transitions and is formally defined as $sp : \Sigma \to Pot(I), c \mapsto \{i \mid |R_c^i| \neq \emptyset\}$.

The synchronized product $\mathcal{A}_{Sys}$ of the components K_i is defined as the transition system $\overline{K} := (\bar{S}, \{\bar{R}_c \mid c \in \Sigma\}, \{\bar{P}_{v^i} \mid v \in V\})$ where

- the state set $\bar{S}$ is the product of the component state sets: $\bar{S} := \prod_{i \in I} S_i$. (We write $\bar{s}[i]$ for the state of the i-th component of $\bar{s} \in \bar{S}$.)
- the synchronized transition relation $\bar{R}_c$ is defined by $(\bar{x}, \bar{y}) \in \bar{R}_c$ iff $\forall i \in sp(c)$: $(\bar{x}[i], \bar{y}[i]) \in R_c^i$ and $\forall j \in I$ with $j \notin sp(c)$: $\bar{x}[j] = \bar{y}[j]$.
- the predicate $\bar{P}_{v^i}$ is the set $\{\bar{s} \mid \bar{s}[i] \in P_v^i\}$.

Example 1. In Fig. 1 we show a synchronized product $\overline{K}$ of two components K_1, K_2 with asynchronous a- and b-transitions in K_1, respectively K_2, and synchronized c-transitions with synchronization profile $\{1, 2\}$. For better readability, the state names of K_1, K_2 are chosen differently.

2.2 Automata

In this section we introduce the format of complement specification automata for a given synchronized product. They are used to express properties that lead to a failure in the product. Afterwards, we translate the complement specification automaton into a "global specification automaton". For this, the complement specification is split into parts that can be checked in the synchronization profiles.

Note that the definitions in this section do not treat unary predicates of a product yet. How to cope with the predicates will be shown in Sect. 4.

Let us recall usual finite automata to fix notation. A *(non-deterministic) finite automaton* is defined by $\mathcal{A} := (Q, \Sigma, \Delta, q_0, F)$ with finite state set Q, input alphabet Σ, transition relation $\Delta \subseteq Q \times \Sigma \times Q$, initial state $q_0 \in Q$ and final state set $F \subseteq Q$. A *complement specification automaton* (without predicates) for a synchronized product is a finite automaton $\mathcal{A}_{\overline{Spec}} = (Q, \Sigma, \Delta_{\overline{Spec}}, q_0, F)$ which is compatible with the action alphabet of the synchronized product.

The *global specification automaton* of a complement specification combines subsequent transitions of the same synchronization profile. Such a combination will result in "super"-transitions labeled from a *local block alphabet*: the alphabet of all sub-automata $\mathcal{A}^I_{q,q'}$ of $\mathcal{A}_{\overline{Spec}}$ where $\mathcal{A}^I_{q,q'}$ contains only transitions from the components of I and q' is reachable from q.

Definition 2 (Global specification automaton). *Given a complement specification automaton* $\mathcal{A}_{\overline{Spec}} = (Q, \Sigma, \Delta_{\overline{Spec}}, q_0, F)$, *let the global specification automaton of* $\mathcal{A}_{\overline{Spec}}$ *be* $\mathcal{A}_{Glob} := (G, \Sigma_B, \Delta_{Glob}, q_0, F)$ *where:*

- *the state set G contains all states of Q such that in $\mathcal{A}_{\overline{Spec}}$ there are out-going and in-coming transitions that belong to different synchronization profiles:*
  $G := \{q_0\} \cup F \cup \{q \in Q \mid \exists q_1, q_2 \in Q \exists (q_1, c, q), (q, d, q_2) \in \Delta_{\overline{Spec}} \text{ with } sp(c) \neq sp(d)\}$,
- *the set Σ_B is the local block alphabet of letters $\mathcal{A}^I_{qq'}$ with $q, q' \in G$ and $I = sp(c)$ for $c \in \Sigma$ and*
- *the transition relation Δ_{Glob} is defined as the set $\{(q, \mathcal{A}^I_{qq'}, q') \mid q, q' \in G$ such that there exists a path from q to q' in $\mathcal{A}_{\overline{Spec}}$ containing only labels of the components of I.\}*.

For a given $z = t_1 \ldots t_u \in L(\mathcal{A}_I)$ let the *projection of z to component i*, denoted by $z\!\restriction_i$, be the restriction of z to all $t_j = \mathcal{A}^{I_j}_{q_j q'_j}$ with $i \in I_j$.

Example 2. Figure 2 shows a complement specification and its transformation into a global automaton for the product from Fig. 1. Each letter from the local block alphabet can be interpreted as an automaton $\mathcal{A}^I_{q,q'}$, e. g. the letter $\mathcal{A}^{\{1\}}_{q_0 q_3}$ corresponds to the automaton $\mathcal{A}_{\overline{Spec}}$ with initial state q_0, final state set $\{q_3\}$, and transitions $(q_0, a, q_1), (q_1, a, q_1)$ and (q_1, a, q_3).

3   Composition: Simple Case

In this section we present the result that reduces the question whether a given synchronized product and given complement specification have common labeling sequences to checking whether the components of this product and certain parts of the complement specification have common labeling sequences.

Theorem 1. *For a given complement specification automaton $\mathcal{A}_{\overline{Spec}}$ without predicates and any synchronized product $\mathcal{A}_{Sys}$ of components K_i for $i \in I$, compatible with $\mathcal{A}_{\overline{Spec}}$, we have:*

$$L(\mathcal{A}_{Sys}) \cap L(\mathcal{A}_{\overline{Spec}}) \neq \emptyset \Leftrightarrow \exists z \in L(\mathcal{A}_{Glob}) \text{ such that } \forall i \in I: \ L(z\!\restriction_i) \cap L(K_i) \neq \emptyset.$$

Let us mention that the length of the word z can be restricted. A complexity analysis is deferred to the treatment of the general case in Sect. 4.

Example 3. The complement specification $\mathcal{A}_{\overline{Spec}}$ from Fig. 2 expresses that a synchronized transition should never be taken after any component has taken more than two asynchronous transitions. Obviously, in the product from Fig. 1 there is a path which conflicts with this property, namely $(12, a, 13, a, 12, c, 24)$.

In $\mathcal{A}_{Glob}$ there exists a path with label $z = \mathcal{A}_{q_0 q_3}^{\{1\}} \mathcal{A}_{q_3 q_4}^{\{1,2\}}$ and for $z \upharpoonright_1 = \mathcal{A}_{q_0 q_3}^{\{1\}} \mathcal{A}_{q_3 q_4}^{\{1,2\}}$ there exists the label sequence aac in K_1 and for $z \upharpoonright_2 = \mathcal{A}_{q_3 q_4}^{1,2}$ there exists the label sequence c in K_2. These sequences lead together to the failure via aac to state 24 in the product.

We can generalize Theorem 1 to complement specifications given as Büchi automata. Thus, we can capture any linear time property if it is converted into a Büchi automaton via the standard techniques.

Corollary 1. *For a given complement specification Büchi automaton $\mathcal{B}_{\overline{Spec}}$ and any synchronized product $\mathcal{A}_{Sys}$ compatible with $\mathcal{A}_{\overline{Spec}}$: $L(\mathcal{A}_{Sys}) \cap L(\mathcal{B}_{\overline{Spec}}) \neq \emptyset$ holds iff there exists a word $z \in L(\mathcal{A}_{Glob})$ such that $\forall i \in [n]$: $L(z \upharpoonright_i) \cap L(K_i) \neq \emptyset$, where z is*

- *either a finite word and at least one word $z \upharpoonright_i$ ends with Büchi automaton as local block*
- *or an ω-word and all local blocks of $z \upharpoonright_i$ are finite automata.*

4   Extension to Specifications with Predicates

In this section we discuss specifications with unary predicates. For this, we encode the predicates (respectively their negation) in the components (as well as in the product) as self-loop transitions. For the complement specification we introduce a sequential projection which allows us to check which predicates hold at a state in the product by checking that all (possibly negated) "predicate" transitions exist before taking a "normal" transition. Further, we analyse the transition structure of the complement specification to reduce the number of checks of "predicate" transitions, if successive "normal" transitions belong to the same transitions profiles.

This section is structured as follows: after the modification of the product, we fix the format we use for a complement specification automaton that is compatible with the actions and the predicates of a product. Then, we introduce its sequential projection in which the predicates are checked via transitions. We conclude by splitting this sequential projection into the local blocks of a global automaton as in Sect. 3.

To store the predicates as self loop transitions, we modify the components by adding transition relations $R_v^i/R_{\neg v}^i$ with $(x, x) \in R_v^i/R_{\neg v}^i$ iff $x \in P_v/x \notin P_v$, and we modify the synchronized product by adding transition relations $\bar{R}_{(\neg)v^i}$ by $(\bar{x}, \bar{y}) \in \bar{R}_{(\neg)v^i}$ iff $\bar{x} = \bar{y}$ and $(\bar{x}[i], \bar{x}[i]) \in R_{(\neg)v}^i$.

Example 4. In Fig. 3 three component transition systems K_1, K_2, K_3 and their synchronized product $\bar{K}$ are shown. Again, the state names are chosen differently

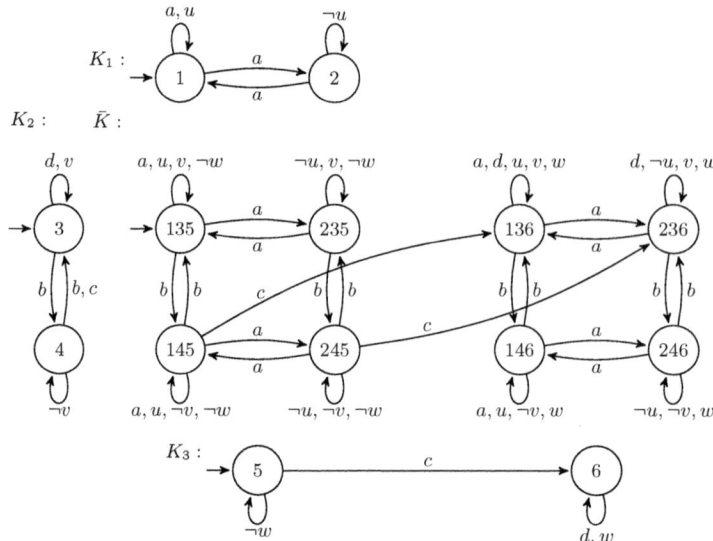

Fig. 3. Components K_1, K_2, K_3 and their synchronized product $\bar{K}$

for the components: $S_1 := \{1,2\}$, $S_2 := \{3,4\}$ and $S_3 := \{5,6\}$. We use different letters for the predicates and only show the corresponding self-loop transitions $(\neg)u, (\neg)v, (\neg)w$ in K_1, K_2 respectively K_3 (and not the predicates itselves). The labels a/b are used for asynchronous transitions of K_1/K_2, i.e. $sp(a) = \{1\}$ and $sp(b) = \{2\}$ and the labels c, d are used for synchronized transitions with synchronization profile $sp(c) = sp(d) = \{2,3\}$.

For a given complement specification automaton we use a small modification to improve the results later: we double each state s which has self loop transitions if all of these transitions are non-switching w.r.t. each other. Fig. 4 shows a complement specification automaton and Fig. 5 this modification.

To compare the paths of a synchronized product with the complement specification, we translate $\mathcal{A}_{Sys}$ in an expanded form $\mathcal{A}_{ESys}$, where the values of the predicates are added to the transition labels, e.g. we have $s \xrightarrow{(a,1,1,0)} s'$ in $\mathcal{A}_{ESys}$ iff $\mathcal{A}_{Sys}$ contains the transition $s \xrightarrow{a} s'$ and P_u, P_v hold at state s, whereas P_w does not. Formally, $\mathcal{A}_{ESys}$ of $\mathcal{A}_{Sys}$ is defined as $(\bar{S}, \bar{R})$ with $\bar{S}$ as in Definition 1 and for $c \in \Sigma$: $(s, (c, b_1^1, \ldots, b_l^n)^T, s') \in \bar{R}$ holds iff $(s, s') \in \bar{R}_c$ and $(b_j^i = 1$ iff $\bar{P}_{v_j^i}$ holds at state s).

Now, we introduce a complement specification automaton with predicates.

Definition 3. *A complement specification automaton with predicates for a synchronized product $\bar{K}$ is an automaton $\mathcal{A}_{\overline{Spec}}$, compatible with the action and predicate alphabet of $\bar{K}$. Formally, $\mathcal{A}_{\overline{Spec}} := (Q, \Sigma \times \mathbb{B}^{l \cdot n}, \Delta, q_0, F)$ with $l := |V|$. A transition has the form $(q, (c, B^1, \ldots, B^n)^T, q')$ with $c \in \Sigma$ and $B^i := (b_1^i, \ldots, b_l^i)$ specifies the truth values of the predicates $\bar{P}_{v^i}$ for $v \in V = \{v_1, \ldots, v_l\}$ at the state q.*

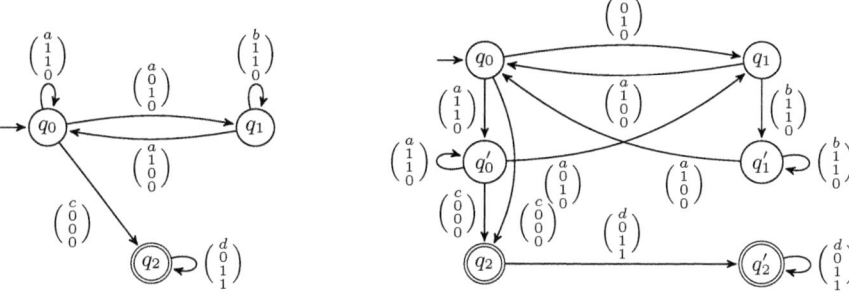

Fig. 4. Complement specification   **Fig. 5.** Modified complement specification

As preparation for the sequential projection of the complement specification and to reduce the number of checks of "predicate" transitions in it, we introduce the notion of *switching* transitions: for subsequent transitions we distinguish between transitions that use labels of the same synchronization profile as the transition before and those which switch to another component.

We call a transition with label $B = (c, b_1^1, b_2^1, \ldots, b_l^n)$ *switching with respect to a transition with label* $B' = (c', b_1'^1, \ldots, b_l'^n)$ if c has a synchronization profile different from c' $(sp(c) \neq sp(c'))$ or if there exists at least one predicate valuation of the other components which does not coincide $(\exists j \in [l]$ with $b_j^k \neq b_j'^k$ for $k \notin sp(c))$. A transition t is called *switching* if there exists a predecessor t' such that t is switching with respect to t'. A transition is called *non-switching* if it is not switching with respect to all predecessors.

The *sequential projection* of a complement specification automaton $\mathcal{A}_{\overline{Spec}} = (Q, \Sigma \times \mathbb{B}^{l \cdot n}, \Delta_{\overline{Spec}}, q_0, F)$ transfers the truth value of the predicates into "predicate transitions" which are checked before the "normal" transitions. It is defined by the automaton $\mathcal{A}_{Proj} := (Q \cup R, \Sigma \cup (V \times [n]), \Delta_{Proj}, q_0, F)$ with $R := (\Sigma \times \mathbb{B}^{l \cdot n}) \times [l \cdot n] \times Q$. We explain the definition of the transition relation Δ_{Proj}: for a transition $t = (q, (c, B^1, \ldots, B^n), q') \in \Delta_{\overline{Spec}}$ we check the predicates – corresponding to $B^1, \ldots, B^n$ – one after the other and afterwards the label c of the transition t. Note that the order in which the predicates have to be checked can be chosen freely. For each synchronized transition with $c \in \Sigma$ we first verify that for all components different from $sp(c)$ there exist transitions for the predicates corresponding to the sets B^j of t before verifying this for the components of $sp(c)$ and before taking the c-labeled transition. If the transition is non-switching with respect to all predecessor transitions, we only check the predicates of the components of $sp(c)$ before taking the c-labeled transition.

Example 5. In Fig. 6 we see the sequential projection of the modified complement specification from Fig. 5. For readability the states that were added to the complement specification automaton are abbreviated with r_i $(1 \leq i \leq 16)$, where e.g. $r_{10} := ((a, 1, 1, 0)^T, 1, q_1)$ and $r_6 := ((c, 0, 0, 0)^T, 3, q_2)$. The transitions for the predicates P_u, P_v, P_w are labeled with u, v, w, respectively their negation. To indicate the assigned component, the transitions with labels of K_1, K_2, respec-

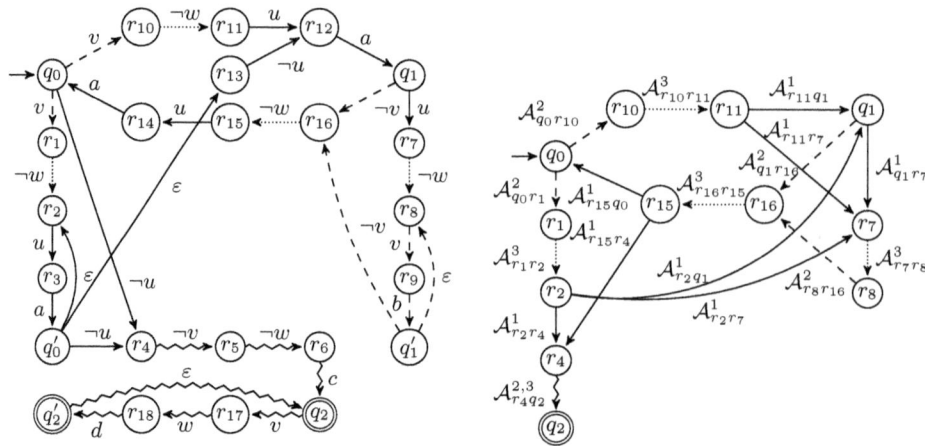

Fig. 6. Complement specification projection   **Fig. 7.** Global specification

tively K_3, are drawn as normal, dashed respectively dotted lines. The transitions of the synchronization profile $sp(c) = sp(d) = \{2,3\}$ are drawn as zigzag lines.

The *global specification automaton* $\mathcal{A}_{Glob} := (G, \Sigma_B, \Delta_{Glob}, q_0, F)$ for a sequential projection $\mathcal{A}_{Proj} = (Q \cup R, \Sigma, \Delta_{Proj}, q_0, F)$ of a complement specification automaton is defined as in Definition 2, but the state set G is the union of the sets

- $\{q_0\} \cup F$
- $\{q \in Q \mid \exists r_1, r_2 \in R\ \exists (r_1, c, q), (q, v^i, r_2) \in \Delta_{Proj} : c \in \Sigma \wedge i \notin sp(c)\}$
- $\{r = (c, B^1, \ldots, B^n, k, q) \in R \mid \exists r_1, r_2 \in Q \cup R\ \exists (r_1, v^i, r), (r, w^j, r_2) \in \Delta_{Proj} : i \neq j \wedge (i \notin sp(c) \vee j \notin sp(c))\}$

For a local block $\mathcal{A}^I_{r,r'}$ and $i \in I$ let $\mathcal{A}^I_{r,r'}\!\restriction_i := \mathcal{B}^i_{r,r'}$ where $\mathcal{B}^i_{r,r'}$ is the automaton which results from $\mathcal{A}^I_{r,r'}$ if we replace all transitions from components different from i by ε-transitions. For a given $z = t_1 \ldots t_u \in L(\mathcal{A}_I)$ we define the *projection of z to component i*, denoted by $z\!\restriction_i$, as $t^i_1 \ldots t^i_{len(i)}$ with $len(i)$ maximal such that for $j \in [len(i)] : t^i_j = \mathcal{A}^{I_j}_{r_j, r'_j}\!\restriction_i$ (for states $r_j, r'_j \in G$) and $i \in I_j$ in the order in which the $\mathcal{A}^{I_j}_{r_j, r'_j}$ appear in z.

Example 6. In Fig. 7 we show the global specification automaton of the sequential projection from Fig. 6. The local blocks are defined as in Sect. 3, e.g. $\mathcal{A}^1_{r_{11}, r_7} = (\{r_{11}, r_{12}, q_1\}, \{u, a\}, \Delta, r_{11}, \{q_1\})$ where Δ contains only the transitions (r_{11}, u, r_{12}) and (r_{12}, a, q_1).

With these preliminaries we generalize Theorem 1 to specifications which can also check the predicates of a product. Further, we give an upper bound for the induced algorithm.

Theorem 2. *For a given complement specification automaton $\mathcal{A}_{\overline{Spec}}$ and any synchronized product $\mathcal{A}_{ESys}$ of components K_i for $i \in I$, compatible with $\mathcal{A}_{\overline{Spec}}$ we have:*

$$L(\mathcal{A}_{ESys}) \cap L(\mathcal{A}_{\overline{Spec}}) \neq \emptyset \Leftrightarrow \exists z \in L(\mathcal{A}_{Glob}) \text{ such that } \forall i \in I : L(z\!\restriction_i) \cap L(K_i) \neq \emptyset.$$

The size of $\mathcal{A}_{Glob}$ is quadratic in the size of $\mathcal{A}_{\overline{Spec}}$ and linear in the number of predicates and components. The length of z is exponential only in the maximal number of states a component has. The tests whether $L(z\!\restriction_i) \cap L(K_i) \neq \emptyset$ need a precalculation which is exponential in the number of components, predicates and states of the complement specification, and in the number of states the synchronization profiles have.

Example 7. In $\mathcal{A}_{ESys}$ (which is $\mathcal{A}_{Sys}$ from Fig. 3 with the predicate valuations of the current state on the outgoing transitions) and in $\mathcal{A}_{\overline{Spec}}$ from Fig. 5 there exist the paths $\pi_{\overline{Spec}} = (q_0, q_0', q_1, q_1', q_0, q_2, q_2')$ and $\pi_{ESys} = (s_{135}, s_{235},$ $s_{135}, s_{145}, s_{245}, s_{236}, s_{236})$ labeled with $(a,1,1,0)^T (a,0,1,0)^T (b,1,1,0)^T (a,1,0,0)^T$ $(c,0,0,0)^T (d,0,1,1)^T$. In $\mathcal{A}_{Sys}$ there exists a path π_{Sys} with the same state sequence like π_{ESys}, but with each state repeated four times and the label sequence $v \neg wua\varepsilon \neg uau \neg wvb \neg v \neg wua \neg u \neg v \neg wcvwd$. From the path $\pi_{\overline{Spec}}$ we get a path $\pi_{Proj} = (q_0, r_1, r_2, r_3, q_0', r_{13}, r_{12}, q_1, r_7, r_8, r_9, q_1', r_{16}, r_{15}, r_{14}, q_0, r_4, r_5, r_6,$ $q_2, r_{17}, r_{18}, q_2')$ in $\mathcal{A}_{Proj}$ of Fig. 6 with the same label sequence.

From π_{Proj} we get $\pi_{Glob} = (q_0, r_1, r_2, r_7, r_8, r_{16}, r_{15}, r_4, q_2)$ in $\mathcal{A}_{Glob}$ from Fig. 7 for the word $z = \mathcal{A}_{q_0, r_1}^2 \mathcal{A}_{r_1, r_2}^3 \mathcal{A}_{r_2, r_7}^1 \mathcal{A}_{r_7, r_8}^3 \mathcal{A}_{r_8, r_{16}}^2 \mathcal{A}_{r_{16}, r_{15}}^3 \mathcal{A}_{r_{15}, r_4}^1 \mathcal{A}_{r_4, q_2}^{2,3}$. Thus, for $i \in \{1, 2, 3\}$ there exist a word in $L(z\!\restriction_i) \cap L(K_i)$, e.g. for $i = 2$: $z\!\restriction_2 = \mathcal{A}_{q_0, r_1}^2 \cdot \mathcal{A}_{r_8, r_{16}}^2 \cdot (\mathcal{A}_{r_4, q_2}^{2,3}\!\restriction_2)$ and $v \cdot vb \neg v \cdot \neg v\varepsilon cv\varepsilon d \in L(z\!\restriction_2) \cap L(K_2)$ with the path $\pi_2 = (3, 3, 3, 4, 4, 4, 3, 3, 3, 3)$ in K_2.

We now justify the complexity claims, by giving more precise complexity bounds. For a synchronized product let l be the number of predicates, n the number of components, n_i the number of states of component K_i and N the maximum over all n_i for $i \in [n]$. Further, call q the number of states for a complement specification automaton.

Given a synchronization profile $sp = \{i_1, \ldots, i_f\} \subseteq [n]$ for synchronized transition labels $c_1, \ldots, c_e$ we consider the synchronous product which contains only these transitions. Let n_{sp} denote the number of states with adjacent transitions of this synchronous product.

Then the size of $\mathcal{A}_{Glob}$ is $\leq (2q)^2 \cdot l \cdot n$ and the length of the word z can be restricted to $(2q)^2 \cdot l \cdot n \cdot N \cdot 2^N$. The complexity of the precalculation is at most $q \cdot (l \cdot n + 1) \cdot 2^{(q \cdot (l \cdot n + 1) \cdot p) + 1}$ where p is the maximum over all n_{sp} for synchronization profiles sp.

The generalization to Büchi automata as in Sect. 3 also works in the case with predicates. The complexity differs only by a constant factor.

5 Further Results and Conclusion

We have presented a compositional approach for reducing failure detection in a product of transition systems to the components, working in an automata theoretic rather than a logical framework. The method allows us to reduce the question whether a product of transition systems and a given complement specification automaton have paths with a common labeling to the question whether a path in the global specification automaton exists such that the components and parts of the complement specification which are described by the local blocks of the global specification automaton have paths with common labeling. The composition method uses information about the transitions in the product – their synchronization profiles – to split the complement specification automaton into parts. Further, we have shown that the complexity of the induced algorithm is at most exponential in the number of components, in the number of states and predicates the complement specification has, and in the number of states and predicates the largest synchronization profile has.

These results complement research on synchronized state/event systems [3] in which the descriptional framework is modal logic, and where model-checking is done by a reduction of the product index set while transforming the given specification (formula). As another related paper we mention [6] where a different set-up for specifying synchronization is used (via "interface processes").

We mention that the present technique can be improved further in appropriate scenarios: E. g., one could use the fact that in the complement specification successive transitions of the same component must have the same valuation of the predicates of the other components, to reduce the number of transitions by deleting transitions where this is not the case. A second improvement would be to duplicate states with incoming transitions of different components and thereby to split the different paths. However, one would have to ensure that this procedure does terminate by considering the decomposition of the complement specification automaton into strongly connected components and aborting the procedure if we reach the same state of a loop again.

Let us mention a possible generalization: One should get a deeper understanding of the technique by looking at how the decomposition of the complement specification automaton can be translated to a decomposition of a logical formula. Therefore, one could consider e. g. a variation of linear time temporal logic (LTL) with additional information about the components, respectively the synchronization profile on parts of the formula. Linear time temporal logic is here a better candidate than classical first-order logic.

Acknowledgment

Thanks are due to an anonymous critical referee whose remarks have led to a clearer presentation.

References

1. Aho, A.V., Hopcroft, J.E., Ullman, J.D.: The Design and Analysis of Computer Algorithms. Addison Wesley, Reading (1974)
2. Baier, C., Katoen, J.P.: Principles of Model Checking. MIT Press, Cambridge (2008)

3. Bodentien, N.O., Vestergaard, J., Friis, J., Kristoffersen, K.J., Larsen, K.G.: Verification of state/event systems by quotienting. BRICS RS-99-41 (December 1999) Nordic Workshop in Programming Theory, Uppsala, Sweden, October 6–8 (1999)
4. Büchi, J.R.: Weak second-order arithmetic and finite automata. Zeitschrift für Mathematische Logik und Grundladen Der Mathematik 6, 66–92 (1960)
5. Chang, C.C., Keisler, H.J.: Model Theory. North Holland, Amsterdam (1990)
6. Cheung, S.C., Kramer, J.: Context constraints for compositional reachability analysis. ACM Trans. Softw. Eng. Methodol. 5, 334–377 (1996)
7. Dawar, A., Grohe, M., Kreutzer, S., Schweikardt, N.: Model theory makes formulas large. In: Arge, L., Cachin, C., Jurdziński, T., Tarlecki, A. (eds.) ICALP 2007. LNCS, vol. 4596, pp. 913–924. Springer, Heidelberg (2007)
8. Elgot, C.: Decision problems of finite automata design and related arithmetics. Transactions of the American Mathematical Society 98, 2152 (1961)
9. Feferman, S., Vaught, R.: The first-order properties of products of algebraic systems. Fundamenta Mathematicae 47, 57–103 (1959)
10. Felscher, I.: The compositional method and regular reachability. Electronic Notes in Theoretical Computer Science 223, 103–117 (2008)
11. Felscher, I., Thomas, W.: On compositional failure detection in structured transition systems, RWTH Aachen Unviersity, Department of Computer Science, Tech. Rep.AIB-2011-12
12. Gabbay, D., Shehtman, V.: Products of modal logics. Logic Journal of IGPL 6(1), 73–146 (1998)
13. Hodges, W.: Model theory, Encyclopedia of Mathematics and its Applications, vol. 42. Cambridge University Press, Cambridge (1993)
14. Makowsky, J.A.: Algorithmic uses of the feferman-vaught theorem. Annals of Pure and Applied Logic 126(1-3), 159–213 (2004)
15. Mostowski, A.: On direct products of theories. The Journal of Symbolic Logic 17(1), 1–31 (1952)
16. Rabinovich, A.: On compositionality and its limitations. ACM Transactions on Computational Logic 8(1) (January 2007)
17. Shelah, S.: The monadic theory of order. The Annals of Mathematics 102(3), 379–419 (1975)
18. Thomas, W.: Ehrenfeucht games, the composition method, and the monadic theory of ordinal words. In: Mycielski, J., Rozenberg, G., Salomaa, A. (eds.) Structures in Logic and Computer Science. LNCS, vol. 1261, pp. 118–143. Springer, Heidelberg (1997)
19. Thomas, W.: Languages, automata and logic. In: Rozenberg, G., Salomaa, A. (eds.) Handbook of Formal Languages, Beyond Words, vol. 3, pp. 389–455. Springer, New York (1997)
20. Wöhrle, S., Thomas, W.: Model checking synchronized products of infinite transition systems. In: Proceedings of the 19th Annual IEEE Symposium on Logic in Computer Science. LNCS, pp. 2–11. IEEE Computer Society Press, Los Alamitos (2004)
21. Wolper, P.: Constructing automata from temporal logic formulas: A tutorial. In: Brinksma, E., Hermanns, H., Katoen, J.-P. (eds.) EEF School 2000 and FMPA 2000. LNCS, vol. 2090, pp. 261–277. Springer, Heidelberg (2001)

Chrobak Normal Form Revisited, with Applications*

Paweł Gawrychowski

Institute of Computer Science,
University of Wrocław,
ul. Joliot-Curie 15, 50–383 Wroclaw, Poland
gawry@cs.uni.wroc.pl

Abstract. It is well known that any nondeterministic finite automata over a unary alphabet can be represented in a certain *normal form* called the Chrobak normal form [1]. We present a very simple conversion procedure working in $\mathcal{O}(n^3)$ time. Then we extend the algorithm to improve two trade-offs concerning conversions between different representations of unary regular languages. Given an n-state NFA, we are able to find a regular expression of size $\mathcal{O}(\frac{n^2}{\log^2 n})$ describing the same language (which improves the previously known $\mathcal{O}(n^2)$ size bound [8]) and a context-free grammar in Chomsky normal form with $\mathcal{O}(\sqrt{n \log n})$ nonterminals (which improves the previously known $\mathcal{O}(n^{2/3})$ bound [3]).

Keywords: unary automata, descriptional complexity.

1 Introduction

Finite automata are a simple yet particularly ubiquitous and useful model of computation. There exists a vast amount of research devoted to studying trade-offs between different methods of describing a language recognized by such devices, starting with the classic conversions between deterministic and nondeterministic finite automata [10]. In this paper we focus on the cost of converting an automaton to a regular expression. If there are no restrictions on the size of the input alphabet, the conversion might require an exponential blow-up [4], even if the alphabet is binary [7]. On the other hand, if the alphabet consists of just one letter, it turns out that such exponential blow-up is not necessary. Additionally, it turns out that nondeterministic automata over such an alphabet can be converted into the so-called Chrobak normal form, meaning that there exists a nondeterministic automaton M' such that $L(M) = L(M')$ and M' consists of a path of $\mathcal{O}(n^2)$ states followed by a single nondeterministic choice to a set of disjoint cycles, where the cycles contain at most n states altogether [1] (also see the errata to the original article [2]). The original proof did not address the computational complexity of finding such M' given M. Martinez showed [8] that this

* Supported by MNiSW grant number N N206 492638, 2010–2012.

B. Bouchou-Markhoff et al. (Eds.): CIAA 2011, LNCS 6807, pp. 142–153, 2011.

conversion requires polynomial time, or more precisely, $\mathcal{O}(n^5)$. This has been improved by Sawa to $\mathcal{O}(n^2(n+m))$ [12]. Both the original proof and the Martinez's improvement contained a minor flaw observed and corrected in [13]. In the next section we give a more efficient version of the construction and then show how to extend the method to construct a regular expression of size $\mathcal{O}(\frac{n^2}{\log^2 n})$ describing the same language. While the improvement might seem minor, it requires combining a few ideas and refutes a conjecture of Martinez who asked for $\Omega(n^2)$ lower bound. Furthermore, we give an evidence that a more substantial improvement would require dramatically different ideas: we show that for some automata converting to Chrobak normal form involves a quadratic blow-up. Then we show that using a similar technique we can construct a context-free grammar in Chomsky normal form with $\mathcal{O}(\sqrt{n \log n})$ nonterminals thus improving the previously known bound $\mathcal{O}(n^{2/3})$ [3].

Because of the space constraints, a few simpler proofs are just sketched or completely removed from the conference version.

2 Preliminaries

We are given a nondeterministic finite automaton $M = \langle \Sigma, Q, q_0, \delta, F \rangle$ over a unary alphabet $\Sigma = \{a\}$. Because the automaton is nondeterministic, without loss of generality there is exactly one final state q_f. Similarly, we can assume that there are no edges incoming into q_0. As the alphabet is unary, we can (and will) view the automaton as a directed graph on $n = |Q|$ vertices and $m = |\delta|$ edges, where $m \leq n^2$. The Chrobak normal form of such automaton consists of a path of length $\mathcal{O}(n^2)$ followed by a single nondeterministic choice to a set of disjoint cycles of lengths $c_1, c_2, \ldots, c_\ell$, with $\sum_i c_i \leq n$, see Figure 1.

A *strongly connected component* of a directed graph is a maximal subset of vertices $\{v_1, v_2, \ldots, v_s\}$ such that for any i, j there is a path from v_i to v_j. We call such component nontrivial if there exists at least one edge $v_i \to v_j$ inside, i.e., either $s > 1$ or there is a loop from v_1 to v_1. The *girth* of a directed graph

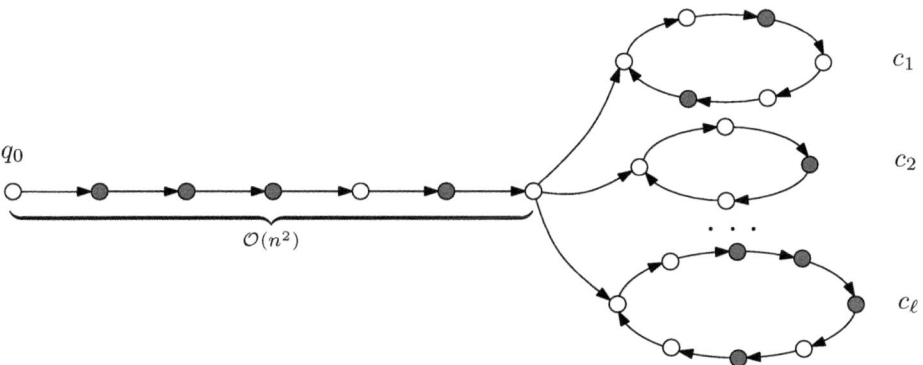

Fig. 1. Automaton in Chrobak normal form

is the length of its shortest cycle. We will use g to denote the girth of the graph corresponding to the given automaton.

Regular expressions considered in this paper are defined in the standard recursive way: a and ϵ are regular expressions, and if R and S are both regular expressions, so are R^*, RS and $R + S$. The size of such expression is simply the number of characters necessary to write it down.

We will also consider languages described by context-free grammars in Chomsky normal form, meaning that all productions are of the form $A \rightarrow a$ or $A \rightarrow BC$ where A, B, C are nonterminals. While it is known that context-free grammars over an unary alphabet describe exactly regular languages [11], they might allow a more succinct description of the language in question that nondeterministic automata.

3   The Algorithm

We are interested in lengths of paths from q_0 to q_f in the corresponding directed graph G. We are going to compute a succinct description of all such paths. First consider the case when G is acyclic. Then any path consists of at most $n - 1$ edges so in time $\mathcal{O}(nm)$ we can easily compute for all vertices v and possible lengths of path $0 \leq \ell < n$ whether there exists a path from q_0 to v of length ℓ. This gives us a description of all paths from q_0 to q_f.

Now consider the case when G is not acyclic, i.e., contains vertices belonging to nontrivial strongly connected components. We can compute all lengths of paths from q_0 to q_f avoiding vertices belonging to nontrivial strongly connected components in the same way as in the acyclic case. Thus now we are only concerned with paths that go through at least one vertex v belonging to a nontrivial strongly connected component. We are going to consider all possible choices of v one by one.

Lemma 1. *Given a vertex v belonging to a nontrivial strongly connected component we can represent all accepting paths through v by a path of length $2n^2$ followed by a cycle of length at most n. The representation can be found in $\mathcal{O}(n^3)$ time.*

Proof. First we need (any) simple cycle $v = v_0 \rightarrow v_1 \rightarrow \cdots \rightarrow v_d = v$ containing v. In fact we can find the shortest such cycle in linear time by replacing v with two vertices v' and v'', all edges of the form $u \rightarrow v$ with $u \rightarrow v''$, and all edges of the form $v \rightarrow u$ with $v' \rightarrow u$, and computing the shortest path from v' to v''. It corresponds to the shortest cycle containing v. As the cycle is simple, $d \leq n$.

Observe that whenever we have a path from q_0 to q_f through v of length ℓ, there is such a path of length $\ell + d$ as well. Thus among all paths with $\ell \equiv r \pmod{d}$ we need to find just the shortest one. Such shortest paths can be computed efficiently in the following way: for each vertex u in the original graph create its $2d$ copies $u(0), u(1), \ldots, u(d-1)$ and $u'(0), u'(1), \ldots, u'(d-1)$. Then add appropriate edges so that there is a path from $q_0(0)$ to $u(r)$ of length ℓ if and only if $\ell \equiv r \pmod{d}$ and there is a path from q_0 to u of the same length in

G. Similarly, there is a path from $q_0(0)$ to $u'(r)$ of length ℓ when $\ell \equiv r \pmod{d}$ and there is a path from q_0 to u going through v and of the same length in G. It is easy to check that the following construction ensures the above conditions: for each edge $x \to y$ and for all possible values of $r = 0, 1, \ldots, d-1$ create edges:

1. $x(r) \to y((r+1) \bmod d)$
2. $x'(r) \to y'((r+1) \bmod d)$
3. if $y = v$, $x(r) \to y'((r+1) \bmod d)$

Then use breadth first search to find shortest paths from $q_0(0)$ in the resulting graph. This requires time $\mathcal{O}(d(n+m)) = \mathcal{O}(nm)$ and gives us a succinct description of all paths from q_0 to q_f going through v. Indeed, there is such path of length ℓ if and only if the distance to $q_f'(\ell \bmod d)$ is finite and does not exceed ℓ. Observe that the new graph contains $2dn$ vertices so all finite distances do not exceed $2dn$. Hence we can represent all those paths by creating a path of length $2dn \leq 2n^2$ followed by a cycle of length $d \leq n$. □

The above lemma gives a description of all paths going through a fixed vertex v. Hence we must consider all possible n choices for v and take the union of the representations found for all of them. As for each of them we create a path of the same length $2n^2$, we can share it among all representations, and then follow with a single nondeterministic choice to a set of n disjoint cycles. This construction works in time $\mathcal{O}(n^2 m) = \mathcal{O}(n^4)$ but it is not enough to match the bounds of the original proof: we must show that the combined size of all the cycles is at most n. Although this can be ensured in the above version, it is more convenient to give an improved algorithm which is faster by an order of magnitude and explicitly guarantees this property.

Theorem 1. *We can represent all accepting paths by a path of length $2n^2$ followed by a nondeterministic choice to a collection of disjoint cycles, with the combined size of all the cycles at most n. Such a representation can be found in $\mathcal{O}(n^3)$ time.*

Proof. To improve the running time of Lemma 1 we try to process vertices in groups instead of one-by-one. Take any simple cycle $v_0 \to v_1 \to \cdots \to v_d = v_0$. We will consider all paths going through at least one of the vertices on this cycle at once. Among all such paths of length ℓ with $\ell \equiv r \pmod{d}$ we need to find just the shortest one. This can be done by a similar construction as in Lemma 1, the only difference being that we create edge $x(r) \to y'((r+1) \bmod d)$ when $y = v_i$ for any $i = 0, 1, \ldots, d-1$. Then there is a path from q_0 to q_f of length ℓ going through the cycle if and only if the distance in the new graph to $q_f'(\ell \bmod d)$ is finite and does not exceed ℓ, so we can represent all such paths by a single path of length $2n^2$ followed by a cycle of length d. As this describes all possible paths going through the cycle, we can then delete all vertices $v_0, v_1, \ldots, v_{d-1}$ and repeat, as long as the graph is not acyclic. Let the lengths of the cycles found in successive iterations be $c_1, c_2, \ldots, c_t$. As they are all disjoint, $\sum c_i \leq n$, so the whole complexity is $\sum_i \mathcal{O}(c_i m) = \mathcal{O}(nm) = \mathcal{O}(n^3)$. Also, the combined size of the cycles in the representation is at most n. □

Using the method from the above theorem we can also prove the following lemma. It will be an important tool in the subsequent sections.

Lemma 2. *We can represent all accepting paths going through at least one vertex contained in some cycle of length at most c by a path of length $2cn$ followed by a nondeterministic choice to a collection of disjoint cycles, with the combined size of all the cycles at most n. The representation can be found in $\mathcal{O}(n^3)$ time.*

As recently shown by Geffert [6], it is always possible to convert an automaton with into the Chrobak normal form so that the path consists of at most $n^2 - 2$ vertices, if $n > 1$. While Theorem 1 gives us a path of length $2n^2$, we can improve this bound to $n^2 - n$. Note that we assumed that there is just one accepting state and there are no edges incoming into q_0, and it might increase n by 2. This was just for the sake of simplicity: all above proofs can be modified to work even without such assumption.

Theorem 2. *We can represent all accepting paths by a path consisting of $n^2 - n$ vertices followed by a nondeterministic choice to a collection of disjoint cycles, with the combined size of all the cycles at most n, assuming $n > 1$. Such a representation can be found in $\mathcal{O}(n^3)$ time.*

Proof. We use the same method as in Theorem 1 but bound the shortest paths lengths more carefully. Assume that for some $0 \leq r < d$ the shortest path from $q_0(0)$ to $q'_f(r)$ contains more than $n^2 - n$ vertices. Then there must be a vertex v in the original graph which appears on this path at least n times. By cutting out parts between two occurrences of v, we get different shorter paths. There are two problems here: by cutting out parts we might remove all vertices from the chosen cycle of length d (and hence get a path to $q_f(r)$ instead of $q'_f(r)$), and we might get a different remainder modulo d of the resulting path length. The former can be removed by reserving one occurrence of v. Hence if $d \leq n - 2$, by the pigeonhole principle we can always find two occurrences such that the distance between them on the path is divisible by d. It remains to deal with the case of $d \geq n - 1$. d can be assumed to be the smallest cycle length possible. Thus if $d = n$ the whole graph consists of just one cycle and the claim is obvious. The case of $d = n - 1$ is slightly more complicated. If the distance between two occurrences of some vertex is $n - 1$, we can shorten the path. Hence the path must of the form $v_1 \rightarrow v_2 \rightarrow \ldots \rightarrow v_n \rightarrow v_1 \rightarrow v_2 \rightarrow \ldots$ If its length exceeds $n^2 - n$, we can remove the first $n(n - 1)$ vertices and get a shorter path with the same length modulo d. $\qquad\square$

4 Application to Regular Expressions Conversion

Given a NFA over a unary alphabet we would like to construct a small regular expression describing the same language. In the regular expression we are allowed to use concatenation, union and Kleene star. A straightforward construction gives an expression of size $\mathcal{O}(n^2)$. We will show that with some number theoretic insight this can be improved to $\mathcal{O}(\frac{n^2}{\log n})$. While the improvement is of only logarithmic magnitude, it requires combining a few ideas, and refutes the conjecture of Martinez who asked for a quadratic lower bound [9].

First we show that for some automata converting into the Chrobak normal form implies a quadratic blow-up. More precisely, we construct an infinite family of automata N_n on n states requiring such an increase in size after the conversion.

Lemma 3. *For any n there exists an automaton N_n on n states such that for any automaton M with $L(N_n) = L(M)$ consisting of a path followed by a nondeterministic choice into a collection of disjoint cycles, the path is of length $\Omega(n^2)$.*

To overcome the quadratic increase we must use a stronger notion than the Chrobak normal form alone. For that to happen we split the set of all accepting paths into acyclic, *strongly cyclic*, and *weakly cyclic*. All paths of a given type will be represented separately as regular expressions of bounded size.

Definition 1. *A path is:*

1. strongly cyclic *if it contains a vertex v such that there is a cycle of length at most $\frac{n}{\alpha \log^2 n}$ through v,*
2. weakly cyclic *if it is not strongly cyclic but contains a vertex belonging to some cycle,*
3. acyclic *otherwise.*

The constant α in the above definition is to be chosen later.

Lemma 4. *A regular expression of size $\mathcal{O}(n)$ describing all acyclic accepting paths can be constructed in $\mathcal{O}(nm)$ time.*

Proof. Computing the lengths of all acyclic accepting paths in the claimed complexity is trivial. To encode them in a regular expression, use the following simple trick: if $x_1 < x_2 < \ldots < x_k$ then $a^{x_1} + (\epsilon + a^{x_2 - x_1} (\ldots (\epsilon + a^{x_k - x_{k-1}}) \ldots))$ generates exactly $\{x_1, x_2, \ldots, x_k\}$ and is of size $\mathcal{O}(x_k)$. □

Lemma 5. *A regular expression of size $\mathcal{O}(\frac{n^2}{\log^2 n})$ describing all strongly cyclic accepting paths can be constructed in $\mathcal{O}(nm)$ time.*

Proof. Apply Lemma 2 with $c = \frac{n}{\alpha \log^2 n}$ and use the trick from Lemma 4 to encode the path and all cycles. □

We still have to construct an expression representing the weakly cyclic paths. Note that we have already described all strongly cyclic paths and so we can safely assume that the girth is at least $\frac{n}{\alpha \log^2 n}$. Nonexistence of smaller cycles implies that there are at most $\alpha \log^2 n$ nontrivial strongly connected components.

We split weakly cyclic paths into two groups. For that we define $C(v) = \{1 \leq \ell \leq n :$ there is a cycle through v of length $\ell\}$ and $C(S) = \bigcup_{v \in S} C(v)$.

Definition 2. *A weakly cyclic path going through strongly connected components $S_1, S_2, \ldots, S_k$ is:*

1. thin *if* $\left| \bigcup_{i=1}^{k} C(S_i) \right| \leq \beta \log n$,
2. fat *otherwise.*

The constant β will be chosen later. Using the above notion and defining the set of *nonnegative combinations* of positive integers $a_1, a_2, \ldots, a_n$ as $N(a_1, \ldots, a_n) = \{\sum_{i=1}^{n} x_i a_i : x_i \geq 0 \text{ for all } i\}$ we can establish a certain normal form of all accepting paths.

Definition 3. *Given an accepting path* $v_0 \rightarrow v_1 \rightarrow \ldots \rightarrow v_\ell$ *we define its skeleton of length* ℓ' *to be an accepting path* $v_0' \rightarrow v_1' \rightarrow \ldots \rightarrow v_{\ell'}'$ *such that* $\ell - \ell' \in N\left(\bigcup_{i=0}^{\ell'} C(v_i') \right)$.

Note that a skeleton of a given accepting path can possibly go through completely different vertices than the original path. The only required condition is on its length and the set of cycles it intersects (which, again, does not have to be anyhow similar to the set of cycles the original path intersects).

Lemma 6. *Any accepting path* $v_0 \rightarrow v_1 \rightarrow \ldots \rightarrow v_\ell$ *has a skeleton of length at most* $n + n \left| \bigcup_{i=0}^{\ell} C(v_i) \right|$.

Proof. Let $C = \bigcup_{i=0}^{\ell} C(v_i)$ be the set of the lengths of all cycles having nonempty intersection with the path. For each element of $c \in C$ we mark the first vertex v_i such that $c \in C(v_i)$. As long as there exist $i < j$ such that $v_i = v_j$ and no vertex v_k with $i < k < j$ is marked we can cut out $v_{i+1}, v_{i+2}, \ldots, v_j$ obtaining a shorter accepting path with the same set C. If such pair of indices does not exists, the distance between any pair of marked vertices must be strictly smaller than n. Thus the total length ℓ' of the final path cannot exceed $n + n |C|$. Because it has been constructed by cutting out cycles, $\ell - \ell'$ can be represented as a nonnegative combination of elements of C. Thus this final path is a skeleton of claimed length. □

Lemma 7. *A regular expression of size* $\mathcal{O}(n^{1+\alpha+\beta} \log n)$ *describing all thin accepting paths can be constructed in polynomial time.*

Proof. We construct a separate expression for each possible choice of the set of strongly connected components $S_1, S_2, \ldots, S_k$ such that $\left| \bigcup_{i=1}^{k} C(S_i) \right| \leq \beta \log n$. Assume such fixed choice and remove all other strongly connected components.

We would like to generate all pairs (ℓ', C) such that there exists a skeleton of length ℓ' and a specified set of cycles $C' \subseteq C = \bigcup_{i=1}^{k} C(S_i)$. There are just n^β subsets of C and by Lemma 6 we can restrict our attention to $\ell' \leq n(1 + \beta \log n)$ so the maximum number of possible pairs is fairly small. To generate the pairs efficiently we define a new graph G' with the following vertices and edges:

$$V' = \{(v, X) : v \in V, X \subseteq C\}$$
$$E' = \{((u, X), (v, X \cup C(v))) : (u, v) \in E\}$$

It is easy to see that there exists a path from $(q_0, C(q_0))$ to (q_f, C') of length ℓ' in G' if and only if there exists a skeleton of length ℓ' and the set of cycles C'. Thus we can generate all valid pairs (ℓ', C') in polynomial time by computing paths of lengths not exceeding $n(1 + \beta \log n)$ from $(q_0, C(q_0))$ in G'. Then we consider all pairs $(\ell'_1, C'), (\ell'_2, C'), \ldots, (\ell'_k, C')$ with the same set of cycles C'. We can construct a regular expression of size $\mathcal{O}(n \log n)$ describing all paths with the corresponding skeletons using the trick from Lemma 4 to encode all $\ell'_1, \ell'_2, \ldots, \ell'_k$ and appending $(a^{c_1} + a^{c_2} + \ldots + a^{c_s})^*$ where $C' = \{c_1, c_2, \ldots, c_s\}$.

For fixed choices of the set of strongly connected components and C' we get a description of all thin paths of size $\mathcal{O}(n^{1+\beta} \log n)$. There are n^{α} choices possible so the total size is $\mathcal{O}(n^{1+\alpha+\beta} \log n)$. □

To deal with fat accepting paths we need to dig deeper into the structure of nonnegative combinations.

Lemma 8. *Let $a_1, a_2, \ldots, a_n$ be a set of different positive integers with $M = \max_i a_i$. Elements of $N(a_1, \ldots, a_n)$ greater than $2\frac{M^2}{n}$ are exactly the multiples of $\gcd(a_1, \ldots, a_n)$.*

Proof. Follows from a result of Erdős and Graham [5]. □

Lemma 9. *If X is a set of positive integers, we can choose its subset $X' \subseteq X$ such that $\gcd(X) = \gcd(X')$ and $|X'| \leq \log \max_{x \in X} x$.*

Proof. Let $X = \{x_1, x_2, \ldots, x_s\}$. Start with $X' = \{x_1\}$, then for $i = 2, 3, \ldots, s$ check if $\gcd(X' \cup \{x_i\}) = \gcd(X')$ holds. If it does, continue. Otherwise add x_i to the current X'. Each time we add something to X', the value of $\gcd(X')$ decreases at least by a factor of 2, and the claim follows. □

Lemma 10. *A regular expression of size $\mathcal{O}(\frac{n^2}{\log n} + n^{1+\alpha})$ describing all fat accepting paths can be constructed in polynomial time.*

Proof. Choose a subset of strongly connected components $S_1, S_2, \ldots, S_k$ such that $\left| \bigcup_{i=1}^{k} C(S_i) \right| > \beta \log n$. If for some $i < j$ there is no path from S_i to S_j nor from S_j to S_i there exists no path hitting all those components at once and we take another subset. Otherwise we can sort all S_i topologically and compute $d = \gcd\left(\bigcup_{i=1}^{k} C(S_i)\right)$. Observe that $d \leq \frac{n}{\beta \log n}$ as a bigger value of d implies that there would be less than $\beta \log n$ different multiplies of d not exceeding n, and each element of $\bigcup_{i=1}^{k} C(S_i)$ must be a multiple of d. We compute for any $0 \leq r < d$ the smallest integer t_r such that there exists an accepting path of length $t_r d + r$ going through all $S_1, S_2, \ldots, S_k$. Note that we require that the path goes through each of those components. This can be done by constructing a new graph G' consisting of d copies of the original G:

$$V' = \{(v, r) : v \in V, 0 \leq r < d\}$$
$$E' = \{((u, r), (v, (r + 1) \bmod d)) : (u, v) \in E\}$$

and repeating a multiple sources shortest path computation k times. First we compute the shortest paths from $(q_0, 0)$ ending in S_1 and avoiding all $S_2, S_3, \ldots, S_k$. Then, assuming we already have shortest paths visiting at least one vertex from each $S_1, S_2, \ldots, S_i$ and ending in S_i, we find the same information for $i + 1$ by a single multiple sources shortest paths computation in G'.

If $t_{\ell \bmod d}$ is not defined, there are no accepting paths of length ℓ, but the converse is not necessarily true, for at least two different reasons. First of all, by computing lengths modulo d we assumed that any multiple of d can be realized as a nonnegative combination of cycles. While it is true for sufficiently large multiples, we might need to combine cycles which are contained in some of the S_i but are completely disjoint with the path.

Assume $\ell > 2(1 + \beta)n \log n + 3\frac{n^2}{\beta \log n}$ and $t_{\ell \bmod d}$ is defined. Then we can find an accepting path of length $\ell' \leq \ell$ visiting all components S_i such that $\ell' \leq \frac{n^2}{\beta \log n}$ and d divides $\ell - \ell'$ (because ℓ' is created by subtracting multiples of d from ℓ). By Lemma 9 we can choose a set of at most $\log n$ vertices $v_1, v_2, \ldots, v_s$ from the strongly connected components S_i such that $d = \gcd\left(\bigcup_{i=1}^{s} C(v_i)\right)$. Because the total number of different cycle lengths in all components $S_1, S_2, \ldots, S_k$ exceeds $\beta \log n$, we can also choose a set of at most $\beta \log n$ vertices $v'_1, v'_2, \ldots, v'_{s'}$ such that $\left|\bigcup_{i=1}^{k} C(v'_i)\right| \geq \beta \log n$ and $s' \leq \beta \log n$. By extending the path to hit all v_i and v'_i we can create another accepting path ℓ'' such that $\ell'' \leq \frac{n^2}{\beta \log n} + 2(1 + \beta)n \log n$, d divides $\ell'' - \ell$ and the path visits all vertices v_i and v'_i. Hence there exists a collection of cycles $c_1, c_2, \ldots, c_t$ having nonempty intersections with this new path of length ℓ'' such that $\gcd(c_1, c_2, \ldots, c_t)$ divides $\ell - \ell'$ and $t \geq \beta \log n$. Then by Lemma 8 the new path is a skeleton of the original path so there exists an accepting path of length $\ell > 2(1 + \beta)n \log n + 3\frac{n^2}{\beta \log n}$ if and only if $t_{\ell \bmod d}$ is defined.

By repeating the above reasoning for all choices of strongly connected components we get a succinct description of all fat accepting paths: we can compute a collection of sets $R_1, R_2, \ldots, R_{n^\alpha}$ such that there is such path of length ℓ exceeding $2(1 + \beta)n \log n + 3\frac{n^2}{\beta \log n}$ if and only if $\ell \bmod d_i \in R_i$ for some $1 \leq i \leq n^\alpha$. Thus we can construct a regular expression of size $\mathcal{O}(\frac{n^2}{\beta \log n} + n^{1+\alpha})$ describing all such paths by considering lengths smaller or equal and greater than $2(1 + \beta)n \log n + 3\frac{n^2}{\beta \log n}$ separately. We write down the former explicitly and to deal with the latter we take a union of the expressions describing all R_i concatenated with $a^{\left\lceil 2(1+\beta)n \log n + 3\frac{n^2}{\beta \log n} \right\rceil}$ which is shared among all i.   □

By choosing $\alpha + \beta < 1$ and combining Lemma 4, 5, 7 and 10 we get:

Theorem 3. *A regular expression of size $\mathcal{O}(\frac{n^2}{\log n})$ describing all accepting paths can be constructed in polynomial time.*

5   Application to Context-Free Grammar Conversion

Given a NFA M over a unary alphabet we would like to construct a small context-free grammar describing the same language. The grammar should be in Chomsky normal form (and thus we relax the problem a little bit by assuming that the empty word is not accepted by M), and we would like to minimize the number of nonterminals. An application of Chrobak normal form results in $\mathcal{O}(n^{2/3})$ bound [3]. In this section we develop a substantially more efficient conversion procedure requiring just $\mathcal{O}(\sqrt{n \log n})$ nonterminals. We start with a simple combinatorial lemma.

Lemma 11. *Given a collection of t sets $A_1, A_2, \ldots, A_t \subseteq U$ we can efficiently find $B \subseteq U$ of cardinality at most $\frac{|U|}{s} \lg t$ such that $A_i \cap B \neq \emptyset$ for all i, where $s = \min_i |A_i|$.*

Proof. We use a simple greedy method: start with $B = \emptyset$ and as long as there exists A_i disjoint with B, select $x \notin B$ maximizing $|\{i : A_i \cap B = \emptyset \text{ and } x \in A_i\}|$. Let $t = t_0, t_1, t_2, \ldots, t_k \geq 1$ be the cardinalities of $\{i : A_i \cap B = \emptyset\}$ in successive steps. We claim that $t_{i+1} \leq t_i - t_i \frac{s}{|U|}$: there are t_i sets left, each of them contains at least s elements, thus there exists x belonging to at least $t_i \frac{s}{|U|}$ sets. Now observe that the claim implies $t_k \leq t \left(1 - \frac{s}{|U|} \right)^k$. Setting $k = \frac{|U|}{s} \lg t$ yields:

$$1 \leq t_k \leq t \left(1 - \frac{1}{\frac{|U|}{s}} \right)^{\frac{|U|}{s} \lg t} < t \left(\frac{1}{e} \right)^{\lg t} = 1$$

so the method terminates after the k-th step, which gives the lemma.   □

We give two different conversion methods, one appropriate for large girth graphs and one which efficiently describes all paths going through at least one vertex contained in a short cycle. Consider the representation found using Lemma 2 (with some c to be chosen later). We call the part of $L(M)$ accepted on the path of length $2cn$ *finite* while and the part accepted on the collection of cycles *infinite*. Dealing with the infinite part is relatively simple, no matter what c is.

Lemma 12. *A context-free grammar with $\mathcal{O}(\sqrt{n})$ nonterminals describing the infinite part of $L(M)$ can be constructed in polynomial time.*

Proof. Let $c_1 < c_2 < \ldots < c_\ell$ be lengths of the cycles in the Chrobak normal form. As $\sum_i c_i \leq n$, ℓ is at most $\sqrt{n}$. First we introduce $\mathcal{O}(\sqrt{n})$ nonterminals $X_0, X_1, \ldots, X_{\lfloor \sqrt{n} \rfloor}$ and $Y_0, Y_1, Y_2, \ldots, Y_{\lfloor \sqrt{n} \rfloor}$ such that X_i derives a^i and Y_i derives $a^{i \lfloor \sqrt{n} \rfloor}$. Using those nonterminals we can express any a^k as $X_i Y_j$ as long as k is at most n. Then we introduce ℓ nonterminals $C_1, C_2, \ldots, C_\ell$ such that C_i describes all words accepted on the i-th cycle. For that we first define D_i which derives exactly a^{c_i} and add production $C_i \to C_i D_i$. Then for any a_k which is accepted on this cycle we add production $C_i \to X_{k \bmod \lfloor \sqrt{n} \rfloor} Y_{\lfloor k / \lfloor \sqrt{n} \rfloor \rfloor}$. We combine all C_i by introducing a special nonterminal P which derives a^{2cn} (this can be done by introducing a logarithmic number of new nonterminals) and productions $S \to P C_i$ for all $i = 1, 2, \ldots, \ell$.   □

First we show how to represent all accepting paths going through at least one vertex contained in a short cycle.

Lemma 13. *A context-free grammar with $\mathcal{O}\left(\sqrt{n} + (cn)^{1/3}\right)$ nonterminals describing accepting paths going through at least one vertex contained in a cycle of length at most c can be constructed in polynomial time.*

Proof. Apply Lemma 2, Lemma 12 and Lemma 2.1 of [3].   □

Note that after applying the above lemma we can assume that there are no short cycles in the graph.

Lemma 14. *If $b \leq g$, a context-free grammar with $\mathcal{O}\left(\sqrt{n} + \frac{n}{b}\log n + b\right)$ nonterminals describing all accepting paths can be constructed in polynomial time, where g is the girth of the underlying graph.*

Proof. Apply Theorem 1 and use Lemma 12 to describe the infinite part of the language found while introducing $\mathcal{O}(\sqrt{n})$ nonterminals. Let its finite part be $\{a^{x_1}, a^{x_2}, \ldots, a^{x_s}\}$ ($s \leq 2n^2$). Choose any accepting computation for each a^{x_i}. Each computation corresponds to a path from q_0 to q_f of length $\ell \leq 2n^2$. Take its prefix of length $\ell - \ell \bmod b$ and split it into blocks of consecutive b states. For each such block we create one set containing the corresponding states. Observe that because $b \leq g$ the states in a single block do not repeat. Then by Lemma 11 we can choose $Q' \subseteq Q$ such that $|Q'| \leq \frac{n}{b}\lg\frac{(2n^2)^2}{b} = \mathcal{O}(\frac{n}{b}\log n)$ and any block contains at least one element of Q'. We add q_0, q_f to Q' and create one nonterminal A_q for any $q \in Q'$. Then we create $2b$ nonterminals $B_1, B_2, \ldots, B_{2b}$ such that B_k derives a^k and for any $q, q' \in Q'$ and $k \leq 2b$ we add production $A_q \rightarrow B_k A_{q'}$ whenever there is a path from q to q' of length k. The total number of introduced nonterminals is $\mathcal{O}(\frac{n}{b}\log n + b)$. Now observe that if we make A_{q_0} the starting state and add production $A_{q_f} \rightarrow \epsilon$, any a^{x_i} can be derived in the resulting grammar. Indeed, consider Figure 2: at least one state from each block belongs to Q', and we can jump between two adjacent blocks using nonterminals B_k. Furthermore, any word derived in the grammar corresponds to an accepting computation of M. The epsilon production can be removed without creating any new nonterminals.   □

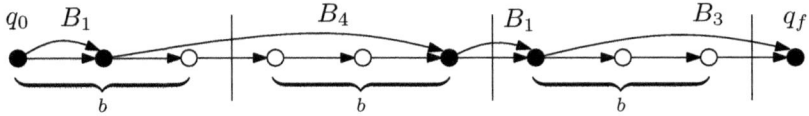

Fig. 2. Jumping between blocks using states from Q'

Combining the two above lemmas gives the claimed bound.

Theorem 4. *A context-free grammar with $\mathcal{O}\left(\sqrt{n\log n}\right)$ nonterminals describing $L(M)$ can be constructed in polynomial time.*

Proof. We apply Lemma 13 to describe all accepting paths going through at least one vertex contained in a cycle of length at most $g = \sqrt{n \log n}$. This requires $\mathcal{O}((gn)^{1/3}) = \mathcal{O}(\sqrt{n} \log^{1/3} n) = \mathcal{O}(\sqrt{n \log n})$ nonterminals. Then we remove all such vertices, which leaves us with a graph of girth at least g, so by Lemma 14 with $b = \sqrt{n \log n}$ we can construct a context free grammar with $\mathcal{O}(\frac{n}{b} \log n + b) = \mathcal{O}(\sqrt{n \log n})$ nonterminals describing all remaining accepting paths.   □

Acknowledgments

I am undoubtedly grateful to Tomasz Jurdziński and Artur Jeż for a thorough reading of an initial version of this paper and many helpful remarks.

References

1. Chrobak, M.: Finite automata and unary languages. Theor. Comput. Sci. 47, 149–158 (1986)
2. Chrobak, M.: Errata to: finite automata and unary languages. Theor. Comput. Sci. 302, 497–498 (2003)
3. Domaratzki, M., Pighizzini, G., Shallit, J.: Simulating finite automata with context-free grammars. Information Processing Letters 84(6), 339–344 (2002)
4. Ehrenfeucht, A., Zeiger, P.: Complexity measures for regular expressions. In: Proceedings of the Sixth Annual ACM Symposium on Theory of Computing (STOC 1974), pp. 75–79. ACM Press, New York (1974)
5. Erdős, P., Graham, R.: On a linear diophantine problem of Frobenius. Acta Arith. 21, 399–408 (1972)
6. Geffert, V.: Magic numbers in the state hierarchy of finite automata. Inf. Comput. 205, 1652–1670 (2007)
7. Gruber, H., Holzer, M.: Finite automata, digraph connectivity, and regular expression size. In: Aceto, L., Damgård, I., Goldberg, L.A., Halldórsson, M.M., Ingólfsdóttir, A., Walukiewicz, I. (eds.) ICALP 2008, Part II. LNCS, vol. 5126, pp. 39–50. Springer, Heidelberg (2008)
8. Martinez, A.: Efficient computation of regular expressions from unary NFAs. In: DFCS 2002, pp. 174–187 (2002)
9. Martinez, A.: Topics in formal languages: String enumeration, unary NFAs and state complexity. M. Math Thesis, University of Waterloo (2002)
10. Meyer, A.R., Fischer, M.J.: Economy of description by automata, grammars, and formal systems. In: Proceedings of the 12th Annual Symposium on Switching and Automata Theory (SWAT 1971), pp. 188–191. IEEE Computer Society Press, Los Alamitos (1971)
11. Parikh, R.J.: On context-free languages. J. ACM 13, 570–581 (1966)
12. Sawa, Z.: Efficient construction of semilinear representations of languages accepted by unary NFA. In: Kučera, A., Potapov, I. (eds.) RP 2010. LNCS, vol. 6227, pp. 176–182. Springer, Heidelberg (2010)
13. To, A.W.: Unary finite automata vs. arithmetic progressions. Information Processing Letters 109(17), 1010–1014 (2009)

A Cellular Automaton Model for Car Traffic with a Form-One-Lane Rule

Yo-Sub Han and Sang-Ki Ko

Department of Computer Science, Yonsei University
Seoul 120-749, Republic of Korea
{emmous,narame7}@cs.yonsei.ac.kr

Abstract. We propose a cellular automaton model that simulates a traffic flow with a junction. We include a 'form-one-lane' rule that decides which car moves ahead when two cars on two different lanes are in front of a junction. We present a fundamental diagram of the proposed model and car distribution examples. We also demonstrate that the proposed model is useful for predicting the real-world traffic flow with a junction.

Keywords: Cellular automata, Car traffic, Traffic junction, Form-one-lane rule.

1 Introduction

Traffic jam is a major problem in most of the major cities in the world. There are several researches that attempt to predict the traffic flow accurately and realistically. One of such approaches is a cellular automaton (CA) model for traffic flow. CA models are intuitive and can simulate a complex behavior with a set of simple CA rules. Wolfram [15] presented a basic one-dimensional CA model for highway traffic flow (R184). Nagel and Schreckenberg [9] proposed another traffic simulation model using CAs, which is a variant of R184 [15]. This model shows a transition from laminar traffic flow to start-stop-waves as the car density increases using Monte-Carlo simulations. Benjamin et al. [1] developed another model (in short, BJH model) that is similar to the Nagel-Schreckenberg (NaSch) model with a 'slow-to-start' rule that reflects the flawed behavior of real drivers. A 'slow-to-start' rule assumes that drivers sometimes lose attentions because of having been stuck in the queue of stopped cars and then start with some delay. Clarridge and Salomaa [2] proposed a 'slow-to-stop' rule and added the new rule into the BJH model. The 'slow-to-stop' rule is based on the following behaviors of drivers: Drivers decelerate before the traffic jam to avoid collision. Note that these models simulate the single-lane highway traffic with one-dimensional CAs. However, in reality, most highways have several junctions where two or more lanes join and they may cause a heavy traffic congestion. See Fig. 1 for example.

Benjamin et al. [1] examined the presence of a junction. They studied the effects of acceleration, disorder and slow-to-start behavior on the queue length at the entrance to the highway. Xiao et al. [17] analyzed a bridge traffic bottleneck

B. Bouchou-Markhoff et al. (Eds.): CIAA 2011, LNCS 6807, pp. 154–165, 2011.
© Springer-Verlag Berlin Heidelberg 2011

Fig. 1. An example of traffic jam around a junction where several lanes join

based on the R184 model [15,16]. Researchers investigated two-lane or multi-lane traffic simulations using CAs and lane changing rules [6,7,13].

We focus on how to simulate a traffic with a junction and how to compute a maximal traffic flow that does not increase the traffic jam while the traffic density varies. We use two one-dimensional CA arrays with various parameters. This helps us to predict the flux of a junction and the length of traffic congestion in front of a junction and to simulate other possible cases with a junction.

2   CA-Based Traffic Simulation Models

A CA is a collection of cells on a grid that evolves through a number of discrete time steps according to a set of rules based on the states of neighboring cells [16]. The NaSch model [9] is the first nontrivial traffic simulation model based on CAs. There are several papers analyzing this model in detail [8,10,11,12] and modifying the model for better simulations [3,4,5]. The NaSch model is defined on a one-dimensional array with periodic boundary conditions. Each cell may either be occupied by a car or be empty. Each car has an integer velocity with values between zero and v_{max}. For an arbitrary configuration, one update of the system consists of four consecutive steps performed in parallel for all cars. These four steps are acceleration, slowing down, randomization and car motion, and respectively reflect the features of cars on highways.

The BJH model [1] is an extension of the NaSch model [9]. Benjamin et al. [1] noticed that drivers have a possibility of starting slowly when they pull away from being in a static queue of cars. This can arise from a driver's loss of attention as a result of having been stuck in the queue. The BJH model introduces p_{slow} to simulate the driver's behavior stochastically. The p_{slow} is the probability of starting slowly from the static queue of cars. When the velocity of a car is 0 and the distance between the next car is long enough, this car stays at velocity 0 on this time step with probability p_{slow} and accelerates to 1 on the next time step. On the other way, this car may accelerate normally with probability $1 - p_{slow}$. This rule is called a 'slow-to-start' rule.

Lastly, there is one more rule for more realistic traffic simulation, a 'slow-to-stop' rule by Clarridge and Salomaa [2]. They observed that the cars following the previous models behave in an unrealistic fashion when approaching a traffic jam. If a car B ahead has velocity 0, then a car A may drive up to B at velocity v_{max} only to brake down to velocity zero in one time step in the cell right behind B. To make it more realistic, they suggested the addition of a 'slow-to-stop' rule. This rule causes drivers to go slower when approaching jams since drivers would slow down beforehand where a small jam is visible from a distance. Clarridge and Salomaa [2] used this rule to the BJH model and demonstrated that there are fewer long jams with many cars at a complete stop, and instead there appear to be many slowdowns to avoid these situations, which is more realistic than the BJH model.

3   Form-One-Lane Rule Model

We propose new CA transition rules for traffic simulation with a highway junction where two lanes become a single lane.

Fig. 2. The left diagram shows a case of merging traffic and the right diagram shows a case of forming one lane. These two cases are the same in the respect to joining two lanes at a junction and becoming a single-lane.

When two lanes join at a junction, there are two types of rules: the first is merging and the second is forming one lane [14]. Merging traffic is where a lane is ending and a driver is required to cross a broken or dotted line to merge with other traffic. In this case, the driver who is about to cross the broken line must give way to traffic in close proximity in another lane regardless of which car is in front. This case is shown in the left side of Fig. 2; car A must give way to car B. The 'form-one-lane' rule requires a driver to give way to a car in another lane if that car is in front of the driver's car when the lanes merge. Hence, "The driver in front has right of way". This case is shown in the right side of Fig. 2; car B must give way to car A. Between these two rules, we consider the second rule, 'form-one-lane', since it gives the same priority to all lanes. This implies that all cars on the road have the same priority regardless of which lane they are in.

In the 'form-one-lane' rule, when there are two cars near a junction, the car that is in front of the car on the other lane goes first. We can adopt this rule to the velocity rule of the BJH model and the 'slow-to-stop' model. In the BJH model, the next velocity of the car is determined based on the current velocity, the maximal velocity and the distance between the next car. The 'slow-to-stop' model uses one more information, the velocity of the next car, for determining the next velocity.

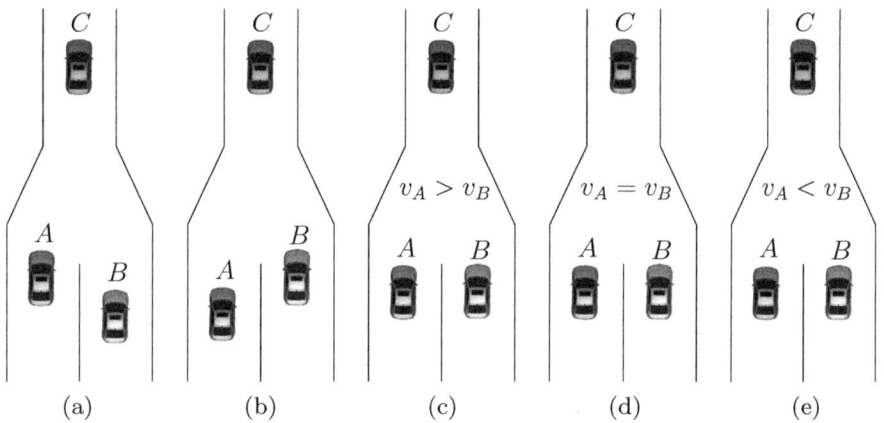

Fig. 3. Five possible cases with a junction where two lanes join. We say that A and B are on the same position in (c), (d) and (e).

There are five possible cases with a junction as illustrated in Fig. 3. In trivial cases, there are two cars approaching to a junction on each lane and they are at different positions. Let L_1 and L_2 be the two lanes that join and make a junction and L_3 be the joined lane after the junction. Let A, B and C be the closest cars to the junction on L_1, L_2 and L_3, respectively. In Fig. 3(a), A goes first because A is in front of B. Let $next(A)$ denote the next car of A. Then $C = next(A)$ and $A = next(B)$ in Fig. 3(a). Similarly B goes first, and $C = next(B)$ and $B = next(A)$ in Fig. 3(b).

Assume that $A = next(B)$. This implies that we can put the distance of B as $pos(A) - pos(B)$, where $pos(A)$ is the index of the cell occupied by A. However, sometimes two cars in front of a junction can be in the same position; namely, $pos(A) - pos(B) = 0$. In the real world, under this condition, the faster car goes first. This is quite reasonable since the faster car is more likely to reach a

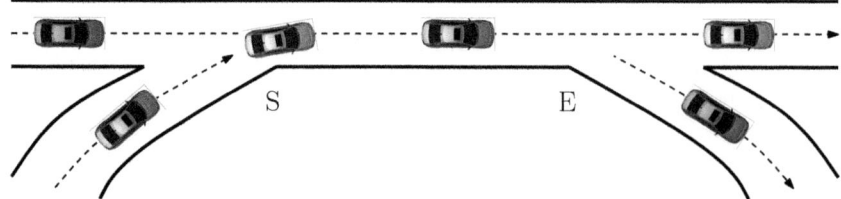

Fig. 4. Our model simulates the traffic where two lanes join at a junction and split into two lanes later. Let S and E be the beginning and the ending of the joint single-lane, respectively.

junction earlier than the other car. These cases are depicted in Fig. 3(c) and (d). In (c), A goes first and $A = next(B)$ since A is faster than B.

Here we consider one more case: Two cars with the same distance from a junction and the same velocity as depicted in Fig. 3(e). We can assume that one of the cars would decrease the velocity because otherwise the two cars would crash into each other. We introduce a simple rule to avoid a crash: We randomly select a car and reduce the velocity of the selected car with probability p_{follow} to mimic the real-world behavior. Since the two lanes have the same priority, we set $p_{follow} \leftarrow 0.5$. Now the two cars have different velocities and thus we can follow one of the two cases in Fig. 3(c) and (d). We say that two cars A, B are on the same position if $pos(A) = pos(B)$.

Fig. 4 illustrates the traffic flow that we consider: Two lanes join as a single-lane and later split again. We use *start* and *end* to denote the beginning and the ending of the joint single-lane. Let $\mathcal{N}(start, i)$ be the car nearest to the junction on lane i. We include a 'form-one-lane' rule as follows:

1. On the joint single-lane: if $pos(start) \leq pos(A) < pos(end)$, then d is the distance from A to the nearest car ahead of A.
2. Closest to a junction: if $pos(\mathcal{N}(start, 1)) = pos(A) > pos(\mathcal{N}(start, 2))$, then d is the distance from A to the nearest car C ahead of A and we set v_{next} as the velocity of C. (Fig. 3(a) case)
3. Behind another car on the other lane: if $pos(\mathcal{N}(start, 1)) = pos(A) < pos(\mathcal{N}(start, 2))$, then $d \leftarrow pos(\mathcal{N}(start, 2)) - pos(A)$ and $v_{next} \leftarrow v(\mathcal{N}(start, 2))$. (Fig. 3(b) case)
4. On the same position: when $pos(\mathcal{N}(start, 1)) = pos(A) = pos(\mathcal{N}(start, 2))$.
   (a) If $v > v(\mathcal{N}(start, 2))$, then d is the distance between the car that is in front and on the single-lane. (Fig. 3(c) case)
   (b) If $v < v(\mathcal{N}(start, 2))$, then we set $d \leftarrow 0$ in order to stop the car and follow the car on the other lane. In this case, we do not need to set v_{next} since this car stops here regardless of the velocity of the next car. (Fig. 3(d) case)
   (c) If $v = v(\mathcal{N}(start, 2))$, then with probability p_{follow} d is the distance between the next car on the single-lane, v_{next} is the velocity of that car and the velocity of the car on the other lane decreases ($v(\mathcal{N}(start, 2)) \leftarrow v(\mathcal{N}(start, 2)) - 1$). (Fig. 3(e) case)

5. Otherwise: the car is not affected by the junction. Let C be the car that is ahead of A. Then d is the distance from A to C and v_{next} is the velocity of C.

These rules determine the values of d and v_{next} of all cars. However, it is impossible to determine the values of all cars in parallel with these rules. When two cars are on the same position and have the same velocity, we reduce the velocity of one car with probability p_{follow}. Then the two cars become to have different velocities and the slower car follows the faster car. If these rules are applied to all cars in parallel, then the velocity of two cars can be reduced at the same time and this causes two cars to stop. Thus, for two cars with the same velocity, we avoid this problem by applying these rules to each car one by one.

1. Slow-to-start: if $v = 0$ and $d > 1$, then the car accelerates on this step or stays there and accelerates on the next step.
2. Deceleration (when the next car is near): if $d \leq v$ and either $v < v_{next}$ or $v \leq 2$, then the next car is either very close or going at a faster speed, and we prevent a collision by setting $v \leftarrow d - 1$ but do not slow down more than is necessary. Otherwise, if $d \leq v$, $v \geq v_{next}$, and $v > 2$ we set $v \leftarrow min(d-1, v-2)$ in order to possibly decelerate slightly more, since the car ahead is slower or the same speed and the velocity of the current car is substantial.
3. Deceleration (when the next car is far): if $v < d \leq 2v$, then if $v \geq v_{next} + 4$, decelerate by 2 ($v \leftarrow v - 2$). Otherwise, if $v_{next} + 2 \leq v \leq v_{next} + 3$ then decelerate by one ($v \leftarrow v - 1$).
4. Acceleration, Randomization, Car motion: these rules are same as in the NaSch model.

These velocity rules calculate the velocities of all cars. Now we design a 'form-one-lane' model by using these two rule sets. The main concern of our model is how to determine the next car when two cars are near a junction, especially when they are on the same position.

Note that when two cars are on the same position, by our rules, the faster car has d as the distance to the car that is in front of the two cars and v_{next} as the velocity of that car while the other car has d as 0. By the deceleration rule, if d is 0, then the velocity of the car becomes -1. However, since the domain of velocity is from 0 to v_{max}, we set velocity as 0. This follows that when two cars are on the same position near a junction, one car stops there and thus two cars cannot advance at the same time.

4   Experiments and Analysis

4.1   Single-Lane Model and the Proposed Model

We simulate the proposed model for traffic with a junction and compare the simulation results with an example of the 'slow-to-stop' model by Clarridge and Salomaa [2]. Fig. 5 is an example of two simulations.

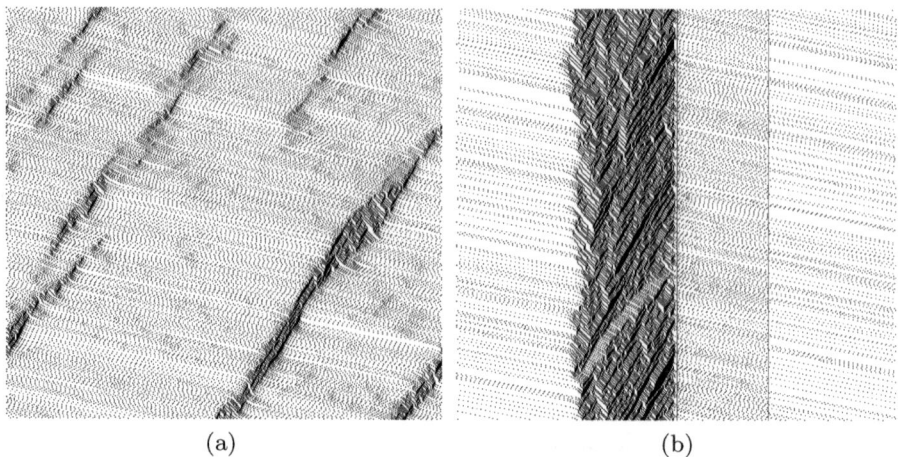

(a)   (b)

Fig. 5. The two pictures depict the distribution of cars over 500 consecutive time steps. (a) is an example of the 'slow-to-stop' model by Clarridge and Salomaa [2] and (b) is an example of the proposed simulation model. Two vertical lines denote the beginning and the ending of a single-lane. Cars are moving from left to right.

In the simulation in Fig. 5, we use the following parameter values: $v_{max} = 5, p_{fault} = 0.1, p_{slow} = 0.5$ and $\rho = 0.15$. v_{max} is the maximal velocity of cars. The cars on cells can move to the right at most v_{max} cells for each time step. p_{fault} and p_{slow} are the probabilities for the disorder rule and the slow-to-start rule. We calculate the traffic density ρ by the ratio of the number of cars to the number of cells. In our model, there are two lanes at first and become a single-lane at a certain point. We have simulated with the same density (0.15) for two lanes.

Notice that the start-stop-waves (traffic jams) in Fig. 5(a) often appear in the real-world traffic. These jams move backwards slowly as the time passes and occur randomly. Note that the locations of traffic jams are different and unpredictable. On the other hand, the location of traffic jams in our model are consistent.

As shown in Fig. 5(b), most start-stop-waves occur in front of the traffic junction. This is quite similar to the real-world traffic flow, where most of traffic jams occur in front of traffic junctions when the number of lanes decreases. The lengths of traffic jams are almost the same when we simulate it with the fixed number of cars. This is because our simulation is carried out using a circular road. In our simulation, there is a fixed number of cars on a fixed number of cells. The influx, which is the number of cars coming into the start of the lane in each time step, is the same as the outflux of the lane after a junction. This means that the flux of a junction goes into the start of the lane along the circular road repeatedly. Thus, the amounts of the influx and the outflux are the same and this is the reason why the lengths of traffic jams remain still as the time passes.

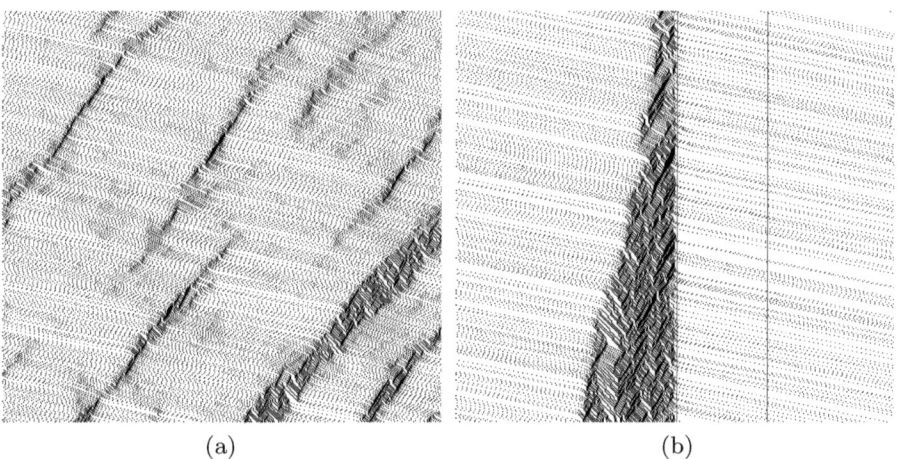

(a)   (b)

Fig. 6. The two pictures depict the distribution of cars over 500 consecutive time steps. We use randomly generated roads to add the cars to a stretch of road. (a) is an example of single-lane model and (b) is an example of form-one-lane model.

However, roads are typically not circular in reality. Furthermore, if the influx is larger than the maximal flux that the junction can process, then the number of cars before the junction would increase. On circular roads, it is impossible to simulate this. Clarridge and Salomaa [2] addressed this problem with an alternative method using Bernoulli process arrivals. This method adds new cars to a stretch of road instead of using CAs with circular boundary conditions. We address this issue by generating a road with the desired flux with a fixed density of cars. For instance, when we make a road with density 0.15, the flux is maximal (which is about 0.52). If we make a road with density 0.07, then the influx of each lane is 0.34. We randomly generate roads by iterating simulations 1000 times to stabilize with the fixed density of cars. In Fig. 6, we confirm that the traffic jam does not show the periodic trends anymore and the lengths of traffic jams increase linearly.

4.2   The Traffic Flux with a Junction

Because the construction of new roads or traffic facilities costs a lot of money, it is better to predict a possible traffic flow before the construction. Especially, if we attach a new road to an existing road without a traffic jam, it may cause some traffic jams because of the new junction. The proposed model can estimate a traffic flow when building a new road and creating a junction.

The traffic flux is the number of cars passing through the lane in a time step. This is an important factor for adding new roads or building traffic facilities since high flux implies the efficient traffic flow. Thus it is better to maximize the traffic flux. In the simulation conducted by Clarridge and Salomaa [2], the flux is maximal when the density of the traffic is around 0.15.

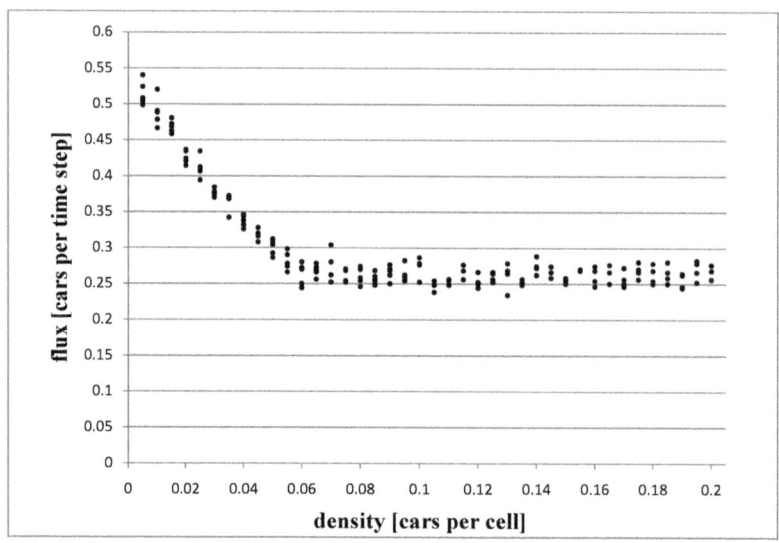

Fig. 7. The fundamental diagram of our model. We simulate with two lanes, where a lane L_1 has a constant density 0.15 and the other lane L_2 has a varying density. In simulation, we increase the density of L_2 as 0.005 in each step.

We examine the case when a lane L_1 is at maximal flux and the density of the other lane L_2 varies. We fix the density of L_1 as 0.15 and vary the density of L_2. The fundamental diagram of this simulation depicted in Fig. 7 shows a linear decrease until the density of L_2 becomes 0.06. After 0.06 the flux of L_1 is stabilized around 0.27. This is because the length of the queue of static cars in front of the traffic junction does not affect the minimal flux of L_1 anymore. Based on this observation, we can estimate the worst-case time complexity to pass the junction.

4.3 The Length of Traffic Jam

Since the traffic jams in our simulation often occur in front of a traffic junction, it is possible to compute the length of traffic jams and make use of them more easily than general start-stop waves. In Fig. 5(b), for instance, we can estimate the rough length of the jams. With the fixed densities of the traffic, it always has similar length of traffic jams in our simulation. If it occurs in the real-world traffic flow in a similar way, then we can use these simulation results for predicting the length of actual traffic jams in front of a junction. However, since there is no general method of measurement for the traffic jams, we design a new rule for the simulation. When we decide whether or not a traffic is jammed, we focus on the partial density of the traffic. If a partial traffic is denser than the other part, then we consider the part as a traffic jam. For the part from the starting point of the traffic to a junction, we measure the average density of the traffic. If the density is lower than the standard, then the starting point moves to the right.

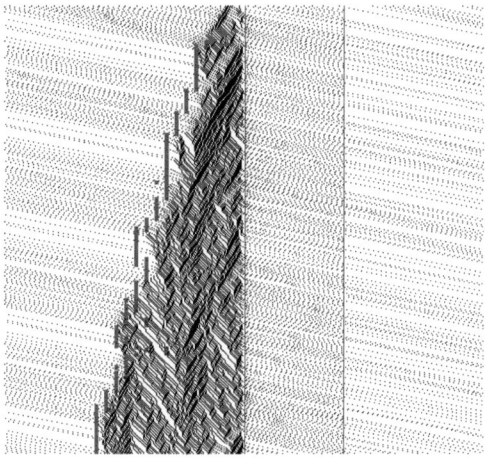

Fig. 8. Thick dots denote the point where the traffic jam begins. The standard density is 0.4.

It stops at the point where we have the average density for the part higher than the standard density.

We check the starting point of traffic jam as thick dots in Fig. 8. We use the standard density 0.4 since the maximal density is approximately 0.4 when a car is in a complete jam in our model. We observe that the length of traffic jam increases linearly.

4.4   Applications of the Proposed Model

When we use the simulation results for the real-world traffic flow, the quantitative comparison is needed. Since the unit in our simulation is abstract, we need some scaling between our model and the real-world traffic. One approach of scaling is to use the maximal velocity. Assume that the maximal velocity of the real-world traffic is 100km/h. The maximal velocity in our simulation is 5 cells per time step and, therefore, we can scale two values based on the following proportional equation:

$$\frac{5 \text{ cells}}{1 \text{ time step}} = \frac{100 \text{km}}{1 \text{ hour}} = \frac{100000 \text{m}}{3600 \text{ seconds}}$$

If we regard the size of a cell as 7m, then the time scale is 1.26 time steps to one second. Based on this scaling, we can calculate either the time to escape from the jam or the estimated arrival time. If a car in our simulation takes n time steps to escape from the jam and enters the single-lane, we can simply convert n time steps to $\frac{n}{1.26}$ seconds in real time.

In a similar way, we can predict the length of traffic jams in the real-world traffic. The simplest way is to use the size of one cell. Nagel and Schreckenberg [9] claimed that in a complete jam each car occupies about 7.5m of place, which becomes the length of one cell. The second method is the comparative method. We can simulate a traffic when there is a junction with various parameters using our model. When the traffic is in the standard condition and has a traffic jam whose length is 10km in reality, we can simulate the condition with our model and obtain the traffic jam whose length is 50 cells. Then, we can simulate with other parameters such as various amounts of flux and obtain the expected length of traffic jam. If the length of traffic jam is, say, 100 cells in simulation, it becomes 20km.

5   Conclusions

When two lanes join, there is always a junction. We have proposed a CA-based traffic simulation model with a 'form-one-lane' rule for simulating traffic flow with a junction. We have considered five possible cases near a junction based on the BJH model [1] and the 'slow-to-stop' model [2], and demonstrated that the proposed model can predict the traffic flow with a junction accurately.

With some diagrams and examples, we have analyzed the experimental results and examined the length of traffic jams when the traffic density varies and the flux of the traffic with a junction. Remark that the flux of the lane becomes stable when the density of a lane is greater than a certain threshold. Based on this observation, we can predict the worst-case arrival time when the density of the other lane varies. We have suggested two approaches of using the proposed model to the real-world traffic.

In future, we need to compare the empirical traffic data with our simulation result. From the real-world data and the simulation results, we can adjust parameters and adopt new scale ratio to establish more precise prediction.

Acknowledgements

We thank Adam Clarridge for providing his source code of the car traffic model with a slow-to-stop rule. We wish to thank the referees for providing us with constructive comments and suggestions.

This research was supported by the Basic Science Research Program through NRF funded by MEST (2010-0009168).

References

1. Benjamin, S.C., Johnson, N.F., Hui, P.M.: Cellular automata models of traffic flow along a highway containing a junction. Journal of Physics A: Mathematical and General 29(12), 3119–3127 (1996)
2. Clarridge, A., Salomaa, K.: Analysis of a cellular automaton model for car traffic with a slow-to-stop rule. Theoretical Computer Science 411(38-39), 3507–3515 (2010)
3. Emmerich, H., Rank, E.: An improved cellular automaton model for traffic flow simulation. Physica A: Statistical and Theoretical Physics 234(3-4), 676–686 (1997)

4. Knospe, W., Santen, L., Schadschneider, A., Schreckenberg, M.: Towards a realistic microscopic description of highway traffic. Journal of Physics A: Mathematical and General 33, 477–485 (2000)
5. Makowiec, D., Miklaszewski, W.: Nagel-Schreckenberg model of traffic - study of diversity of car rules. In: International Conference on Computational Science, vol. 3, pp. 256–263 (2006)
6. Nagel, K., Schreckenberg, M., Latour, A., Rickert, M.: Two lane traffic simulations using cellular automata. Physica A: Statistical and Theoretical Physics 231(4), 534–550 (1996)
7. Nagel, K., Wolf, D.E., Wagner, P., Simon, P.: Two-lane traffic rules for cellular automata: A systematic approach. Physical Review 58(2), 1425–1437 (1998)
8. Nagel, K.: Particle hopping models and traffic flow theory. Physical Review E 53(5), 4655–4672 (1996)
9. Nagel, K., Schreckenberg, M.: A cellular automaton model for freeway traffic. Journal de Physique I 2(12), 2221–2229 (1992)
10. Sasvári, M., Kertész, J.: Cellular automata models of single-lane traffic. Physical Review E 56(4), 4104–4110 (1997)
11. Schadschneider, A., Schreckenberg, M.: Cellular automaton models and traffic flow. Journal of Physics A: Mathematical and General 26, 679–683 (1993)
12. Wagner, P.: Traffic simulations using cellular automata: Comparison with reality. In: Traffic and Granular Flow. World Scientific, Singapore (1996)
13. Wagner, P., Nagel, K., Wolf, D.E.: Realistic multi-lane traffic rules for cellular automata. Physica A: Statistical and Theoretical Physics 234(3-4), 687–698 (1997)
14. Website, T.S.M.S.: Road rule changes (June, 2009),
    `http://www.seniorsmovingsafely.org.au/road_rules.html`
15. Wolfram, S.: Theory and Applications of Cellular Automata. World Scientific, Singapore (1986)
16. Wolfram, S.: A New Kind of Science. Wolfram Media, Champaign (2002)
17. Xiao, S., Kong, L., Liu, M.: A cellular automaton model for a bridge traffic bottleneck. Acta Mechanica Sinica 21, 305–309 (2005)

Loops and Overloops
for Tree Walking Automata

Pierre-Cyrille Héam*, Vincent Hugot**, and Olga Kouchnarenko

LIFC, Université de Franche-Comté & INRIA CASSIS, Besançon, France
{pcheam,vhugot,okouchnarenko}@lifc.univ-fcomte.fr

Abstract. Tree Walking Automata (TWA) have lately received renewed interest thanks to their tight connection to XML. This paper introduces the notion of tree overloops, which is closely related to tree loops, and investigates the use of both for the following common operations on TWA: testing membership, transformation into a Bottom-Up Tree Automaton (BUTA), and testing emptiness. Notably, we argue that transformation into a BUTA is slightly less straightforward than was assumed, show that using overloops yields much smaller BUTA in the deterministic case, and provide a polynomial over-approximation of this construction which detects emptiness with surprising accuracy against randomly generated TWA.

Keywords: Tree Walking Automata, loops, overloops, membership, emptiness, approximation.

1 Introduction

Tree Walking Automata (TWA for short) are a well-established sequential model for recognising tree languages which was introduced in 1969 by Aho and Ullman [1]. While they originally received far less attention than the better known branching model of tree automata, they have been steadily gathering interest in the last few years. Notably, important questions which had remained open for decades have recently been closed. This renewed interest is owed in great part to the ever-growing popularity of XML, with which they and their variants are tightly connected, in particular through Core XPath [6] and streaming [13].

In this context, it becomes helpful to have reasonably efficient algorithms for essential operations on TWA such as deciding membership and emptiness, as well as transformation into a BUTA. Until now, research has been mainly focused on closing fundamental open problems concerning the expressiveness of TWA [5,2,4]. While algorithms for the above operations are known, they appear in print mostly as proof sketches, and there has been no focus on finding tighter complexity bounds. In contrast, this paper provides explicit algorithms for these

---

* This author is supported by the project ANR 2010 BLAN 0202 02 FREC.
** This author is supported by the French DGA (Direction Générale de l'Armement).

B. Bouchou-Markhoff et al. (Eds.): CIAA 2011, LNCS 6807, pp. 166–177, 2011.

tasks and deals with complexity issues. The common thread of our contributions is the notion of *tree loop*, which is pervasive to the algorithms we give. This notion may be related to Knuth's construction for testing circularity of attribute grammars [11]. The contributions are organised as follows:

⋄ Section 2 gives a thorough introduction to tree loops – which are more or less folklore – and introduces a new notion of *tree overloop*. Simple algorithms for testing membership follow naturally from this work. To the best of our knowledge, no such algorithm exists in the literature.

⋄ Section 3 treats the transformation from TWA to BUTA, based on the proof sketches in [3] and [12, p143]. Two variants are given: one using loops and one using overloops. The latter yields slightly smaller automata in general. Then we show that, in the deterministic case, the overloops-based construction admits a much smaller upper bound on the number of generated states.

⋄ The emptiness problem is known to be ExpTime-complete for TWA, and is traditionally tested by first transforming the TWA into a BUTA. Section 4 provides a polynomial algorithm which computes an "over-approximation" of this BUTA, and thus can – with luck – decide emptiness positively. This approach is tested against randomly generated TWA, and turns out to be astonishingly accurate. Should it prove inefficient against some families of TWA, then the approximation can be refined as much as needed.

Notations.   Let $R \subseteq Q^2$ be a binary relation on a set Q; we denote by R^+ and R^* its transitive and reflexive-transitive closure, respectively. The notation $[\![n, m]\!]$ denotes the integer interval $[n, m] \cap \mathbb{Z}$.

We denote by $\mathbb{N}^*$ the set of words over $\mathbb{N}$; if $v, w \in \mathbb{N}^*$, then $v.w$ stands for the concatenation of the words v and w. A *ranked alphabet* is a finite set of symbols, equipped with an arity function $arity : \Sigma \to \mathbb{N}$. The subset of symbols of Σ with arity k is denoted by Σ_k. The set $\mathcal{T}(\Sigma)$ of trees over Σ is defined inductively as the smallest set such that $\Sigma_0 \subseteq \mathcal{T}(\Sigma)$ and, if $k \geqslant 1$, $f \in \Sigma_k$ and $u_1, \ldots, u_k \in \mathcal{T}(\Sigma)$, then $f(u_1, \ldots, u_k) \in \mathcal{T}(\Sigma)$. If $t \in \mathcal{T}(\Sigma)$ is a tree, then the set of *positions* (or *nodes*) $\mathcal{P}os(t) \subseteq \mathbb{N}^*$ is defined inductively by $\mathcal{P}os(t) = \{\varepsilon\}$ if t is a constant – that is to say, $t \in \Sigma_0$ – and $\mathcal{P}os(f(u_1, \ldots, u_n)) = \{\varepsilon\} \cup \{k.\alpha_k \mid k \in [\![0, n-1]\!]$ and $\alpha_k \in \mathcal{P}os(u_{k+1})\}$ otherwise, where n is the arity of f. We see a tree t as a function $t : \mathcal{P}os(t) \to \Sigma$ which maps a position to the symbol at that position in t. In this paper we consider only binary trees, that is to say we assume that $k \notin \{0, 2\} \implies \Sigma_k = \varnothing$. Positions are equipped with a non-strict (resp. strict) partial order $\unlhd$ (resp. $\lhd$), such that $\alpha \unlhd \beta$ iff β is a prefix of α (resp. $\alpha \unlhd \beta$ and $\alpha \neq \beta$). The *size* of a tree t is denoted by $\|t\|$ and defined by $\|t\| = |\mathcal{P}os(t)|$.

The *parent function* $\mathfrak{p}(\cdot) : \mathcal{P}os(t) \setminus \{\varepsilon\} \to \mathcal{P}os(t)$ maps any (non-root) *child* node $\alpha.k$ (where $k \in \{0, 1\}$) to its *father* α. We denote by $t|_\alpha$ the subtree of t under α. The reader is assumed to be well-acquainted with the bottom-up variety of branching tree automata (see for instance [7]). A Tree-Walking Automaton (TWA) is a tuple $\mathcal{A} = \langle \Sigma, Q, I, F, \Delta \rangle$ where Q is a finite set of states, Σ a ranked alphabet, $I \subseteq Q$ is the set of initial states, $F \subseteq Q$ the subset of final – or accepting – states, and

$$\Delta \subseteq \Sigma \times Q \times \underbrace{\{\star, \mathbf{0}, \mathbf{1}\}}_{\mathbb{T} \,:\, \text{types}} \times \underbrace{\{\uparrow, \circlearrowleft, \swarrow, \searrow\}}_{M \,:\, \text{moves}} \times Q$$

is the set of transitions. In this paper the tuple $\langle \Sigma, Q, I, F, \Delta \rangle$ will be assumed whenever we speak of a TWA $\mathcal{A}$. Each node α of a tree t has a *type* in $\mathbb{T}$, denoted by $\natural \alpha$, such that $\natural \varepsilon = \star$ (root), $\natural(\beta.0) = \mathbf{0}$ (left son), $\natural(\beta.1) = \mathbf{1}$ (right son). As we will seldom deal with the root in practice, we define for short the *sons* $\mathbb{S} = \{\mathbf{0}, \mathbf{1}\} \subset \mathbb{T}$. We will also put in relation types and moves through the function $\chi(\cdot) : \mathbb{S} \to \{\swarrow, \searrow\}$ such that $\chi(\mathbf{0}) = \swarrow$ and $\chi(\mathbf{1}) = \searrow$. For our convenience, we will take the special notation $\langle f, p, \tau \to \mu, q \rangle$ for the tuple $(f, p, \tau, \mu, q) \in \Delta$. Using this notation, some of the parameters can be replaced by sets, with the obvious meaning that we consider the set of all transitions thus described. For instance $\langle \Sigma_2, p, \mathbb{T} \to \circlearrowleft, q \rangle = \{(\sigma, p, \tau, \circlearrowleft, q) \mid \sigma \in \Sigma_2, \tau \in \mathbb{T}\}$. Note that all the transitions from $\langle \Sigma_0, Q, \mathbb{T} \to \{\swarrow, \searrow\}, Q \rangle \cup \langle \Sigma, Q, \star \to \uparrow, Q \rangle$ are invalid.

A *configuration* of $\mathcal{A}$ on a tree t is a pair $c = (\beta, q) \in \mathit{Pos}(t) \times Q$; it is *initial* if $c \in \{\varepsilon\} \times I$ and *final* (or *accepting*) if $c \in \{\varepsilon\} \times F$. It is a *successor* of a configuration (α, p) if $\langle t(\alpha), p, \natural \alpha \to \mu, q \rangle \in \Delta$, where μ is $\uparrow$ if $\beta = \mathfrak{p}(\alpha)$, $\circlearrowleft$ if $\beta = \alpha$, $\swarrow$ if $\beta = \alpha.0$ and $\searrow$ if $\beta = \alpha.1$. We write $c_1 \twoheadrightarrow_{\mathcal{A}} c_2$ (or simply $c_1 \twoheadrightarrow c_2$ whenever $\mathcal{A}$ is clear from the context) if the configuration c_2 is a successor of c_1. A *run* is a (not necessarily finite) sequence of successive configurations $c_1 \twoheadrightarrow c_2 \twoheadrightarrow \ldots c_n \twoheadrightarrow \ldots$. A run is *accepting* (or *successful*) if it starts with an initial configuration and reaches a final configuration. A tree t is *accepted* or *recognised* by $\mathcal{A}$ if there exists an accepting run of $\mathcal{A}$ on t. The set of all accepted trees is the *language* of $\mathcal{A}$, denoted by $\mathcal{L}ng(\mathcal{A})$.

Example: Let $\mathcal{X}$ be a TWA such that $\Sigma_0 = \{a, b, c\}$ and $\Sigma_2 = \{f, g, h\}$, $Q = \{q_\ell, q_u\}$, $I = \{q_\ell\}$, $F = \{q_u\}$, and $\Delta = \langle a, q_\ell, \{\star, \mathbf{0}\} \to \circlearrowleft, q_u \rangle \cup \langle \Sigma, q_u, \mathbf{0} \to \uparrow, q_u \rangle \cup \langle \Sigma_2, q_\ell, \{\star, \mathbf{0}\} \to \swarrow, q_\ell \rangle$. Then $\mathcal{X}$ accepts exactly all trees whose leftmost leaf is labelled by a. We shall use this (trivial) example throughout the paper.

2 Loops, Overloops and the Membership Problem

The notion of *loop* turned out to be very useful to deal with TWA. Informally, loops arise naturally as a generalisation of the definition of an accepting run, where the automaton enters the root in a given initial state p_{in}, moves along the tree, and then comes back to the root in a certain final state p_{out}. In practice, the details of the moves which form the loop itself are largely irrelevant and are discarded: the most useful information is the pair of states $(p_{\mathrm{in}}, p_{\mathrm{out}})$.

Definition 1 (Tree Loops). Let $\mathcal{A}$ be a TWA, t a tree and $\alpha \in \mathit{Pos}(t)$. A pair of states $(p, q) \in Q^2$ is a *loop* of $\mathcal{A}$ on the subtree $t|_\alpha$ if there exist $n \geqslant 0$ and a run $(\alpha, p), (\beta_1, s_1), \ldots, (\beta_n, s_n), (\alpha, q)$ such that for all $k \in [\![1, n]\!], \beta_k \trianglelefteq \alpha$. Such a run is a *looping run*, and we say that it *forms* the loop (p, q).

Data: A TWA $\mathcal{A} = \langle \Sigma, Q, I, F, \Delta \rangle$
Result: A BUTA $\mathcal{B}$ such that $\mathcal{L}\mathrm{ng}\,(\mathcal{B}) = \mathcal{L}\mathrm{ng}\,(\mathcal{A})$

initialise States and Rules to $\varnothing$
foreach $a \in \Sigma_0, \tau \in \mathbb{T}$ **do**

A $\quad\lfloor$ let $P = (a, \tau, \mathcal{H}_a^{\tau*})$; add $a \to P$ to Rules and P to States

repeat
$\quad$ **foreach** $f \in \Sigma_2, \tau \in \mathbb{T}$ **do**

B $\qquad$ add every $f(P_0, P_1) \to P$ to Rules and P to States
$\qquad$ **where** $P_0, P_1 \in$ States such that $P_0 = (\sigma_0, \mathbf{0}, S_0)$ and $P_1 = (\sigma_1, \mathbf{1}, S_1)$
$\qquad$ and $P = (f, \tau, (\mathcal{H}_f^{\tau} \cup S)^*)$, with

$$S = \left\{ (p, q) \;\middle|\; \exists \theta \in \mathbb{S}, (p_\theta, q_\theta) \in S_\theta : \begin{array}{l} \langle f, p, \tau \to \chi(\theta), p_\theta \rangle \in \Delta \text{ and} \\ \langle \sigma_\theta, q_\theta, \theta \to \uparrow, q \rangle \in \Delta \end{array} \right\}$$

until Rules *remains unchanged*
return $\mathcal{B} = \langle \Sigma, \mathsf{States}, \{ (\sigma, \star, L) \in \mathsf{States} \mid L \cap (I \times F) \neq \varnothing \}, \mathsf{Rules} \rangle$

Algorithm 1. Tranformation into BUTA, with loops

Data: An escaped TWA $\mathcal{A} = \langle \Sigma, Q, I, F, \Delta \rangle$ (see Def. 13)
Result: A BUTA $\mathcal{B}$ such that $\mathcal{L}\mathrm{ng}\,(\mathcal{B}) = \mathcal{L}\mathrm{ng}\,(\mathcal{A})$

initialise States and Rules to $\varnothing$
foreach $a \in \Sigma_0, \tau \in \mathbb{T}$ **do**

C $\quad\lfloor$ let $P = (\tau, \mathcal{U}_a^{\tau}[\mathcal{H}_a^{\tau*}])$; add $a \to P$ to Rules and P to States

repeat
$\quad$ **foreach** $f \in \Sigma_2, \tau \in \mathbb{T}$ **do**

D $\qquad$ add every $f(P_0, P_1) \to P$ to Rules and P to States
$\qquad$ **where** $P_0, P_1 \in$ States such that $P_0 = (\mathbf{0}, S_0)$ and $P_1 = (\mathbf{1}, S_1)$ and
$\qquad$ $P = (\tau, \mathcal{U}_f^{\tau}[(\mathcal{H}_f^{\tau} \cup S)^*])$, with

$$S = \left\{ (p, q_\theta) \;\middle|\; \exists \theta \in \mathbb{S}, p_\theta \in Q : \begin{array}{l} \langle f, p, \tau \to \chi(\theta), p_\theta \rangle \in \Delta \\ \text{and } (p_\theta, q_\theta) \in S_\theta \end{array} \right\}$$

until Rules *remains unchanged*
return $\mathcal{B} = \langle \Sigma, \mathsf{States}, \{ (\star, O) \in \mathsf{States} \mid O \cap (I \times \{\checkmark\}) \neq \varnothing \}, \mathsf{Rules} \rangle$

Algorithm 2. Tranformation into BUTA, with overloops

Example: The looping run $(0, q_\ell), (0.0, q_\ell), (0.0, q_u), (0, q_u)$ of $\mathcal{X}$ on the subtree $g(f(a, b), c)|_0 = f(a, b)$ forms the loop (q_ℓ, q_u).

Notice that loops are not only defined on whole trees, but on subtrees as well with the restriction that the automaton cannot leave the subtree during the looping run. It is in fact this restriction which grants loops their usefulness. TWA, unlike their branching cousins, whose runs are defined inductively, do not naturally lend themselves to inductive reasoning; and yet, thanks to the above restriction, loops are easily computed by induction. Thus loops and their variants can be thought of as convenient devices which hide the sequential, stateful aspect of TWA runs beneath a much more "user-friendly" layer of induction.

Data: An escaped TWA $\mathcal{A} = \langle \Sigma, Q, I, F, \Delta \rangle$ (see Def. 13)
Result: Empty (only if $\mathcal{L}ng\,(\mathcal{A}) = \varnothing$) or Unknown

initialise $\mathcal{L}_0, \mathcal{L}_1, \mathcal{L}_\star$ to $\varnothing$; **foreach** $a \in \Sigma_0, \tau \in \mathbb{T}$ **do** $\mathcal{L}_\tau \leftarrow \mathcal{L}_\tau \cup \mathcal{U}_a^\tau[\mathcal{H}_a^{\tau *}]$
repeat

$\quad$ **foreach** $f \in \Sigma_2, \tau \in \mathbb{T}$ **do** $\mathcal{L}_\tau \leftarrow \mathcal{L}_\tau \cup \mathcal{U}_f^\tau\big[(\mathcal{H}_f^\tau \cup S)^*\big]$

$\quad$ **where** $S = \left\{ (p, q_\theta) \;\middle|\; \exists \theta \in \mathbb{S}, p_\theta \in Q : \begin{array}{l} \langle f, p, \tau \to \chi(\theta), p_\theta \rangle \in \Delta \\ \text{and } (p_\theta, q_\theta) \in \mathcal{L}_\theta \end{array} \right\}$

until $\mathcal{L}_0, \mathcal{L}_1, \mathcal{L}_\star$ *remain unchanged*
return Empty *if* $\mathcal{L}_\star \cap (I \times \{\checkmark\}) = \varnothing$, *else* Unknown

Algorithm 3. Approximation for emptiness, with overloops

In the next few paragraphs we compute the loops of a TWA $\mathcal{A}$ on a subtree $t|_\alpha$.

Definition 2 (Kinds of Loops). Clearly for all $p \in Q$, (p, p) is a loop; we call them *trivial loops*. A looping run of $\mathcal{A}$ on $t|_\alpha$ is *simple* if it reaches α exactly twice. It is *non-trivial* if it reaches α at least twice. A loop is *simple* (resp. *non-trivial*) if there exists a *simple* (resp. *non-trivial*) looping run forming it.

Example: The loop (q_ℓ, q_u) in the above example is simple, because $(0, q_\ell)$, $(0.0, q_\ell)$, $(0.0, q_u), (0, q_u)$ only reaches $\alpha = 0$ twice, on the first and last configuration. The TWA $\mathcal{X}$ forms only trivial and simple loops, but suppose that we alter it so that it also checks that the *right-most* leaf is a. During an accepting run it would go down and left, back up to the root, down and right, and back up to the root again, in a final state. Thus all accepting runs would be non-trivial and non-simple, reaching the root exactly three times.

Fortunately, we only ever need to compute simple loops, as we can deduce the rest from them thanks to the following lemma:

Lemma 3 (*Loop Decomposition*). *If $S \subseteq Q^2$ is the set of all simple loops of $\mathcal{A}$ on a given subtree $u = t|_\alpha$, then S^* is the set of all loops of $\mathcal{A}$ on u.*

Proof. Every looping run is either trivial or non-trivial. All trivial loops are in S^* by reflexive closure. Furthermore, every non-trivial looping run can easily be decomposed into one or more simple runs. Indeed, any non-trivial looping run ℓ has the following general form, where $\beta_i^k \lhd \alpha$ for all k, i, and the notation $[x_k]^{k \in [\![1,m]\!]}$ designates the run obtained by concatenating the runs $x_1, \dots, x_m$:

$$ \ell = (\alpha, p^0), \big[(\beta_1^k, s_1^k), \dots, (\beta_{n_k}^k, s_{n_k}^k), (\alpha, p^k) \big]^{k \in [\![1,m]\!]} \;. $$

This can be seen as the composition of m simple looping runs ℓ_k, for $k \in [\![1, m]\!]$, where $\ell_k = (\alpha, p^{k-1}), (\beta_1^k, s_1^k), \dots, (\beta_{n_k}^k, s_{n_k}^k), (\alpha, p^k)$. Let us compute the loops formed by the looping run ℓ: for every $k, l \in [\![1, m]\!]$, $k \leqslant l$, we can build a looping run $\ell_k, \ell_{k+1}, \dots, \ell_l$, and it follows that (p^{k-1}, p^l) is a loop. Since only the states p^k appear at position α, ℓ forms no other loops. But we have

$$\left\{\,(p^{k-1},p^k)\mid k\in[\![1,m]\!]\,\right\}^+ = \left\{\,(p^{k-1},p^l)\mid k,l\in[\![1,m]\!]:k\leqslant l\,\right\}\;.$$

Note that each loop (p^{k-1},p^k) is formed by ℓ_k. Therefore the loops formed by the non-trivial looping run ℓ are the transitive closure of the loops formed by the simple looping runs of which it is composed. □

Let us denote $\mho^\tau(u)$ the set of all loops of $\mathcal{A}$ on a subtree u, where τ is the type of the root of u. Concretely, if u is the whole tree, then $\tau = \star$ and, more generally, if u is a subtree, say, $u = t|_\alpha$, then $\tau = \natural\alpha$. Note that thanks to the above-mentioned restriction in the definition of loops, the type of the subtree's root is the only information which is actually needed from the context.

Let $a \in \Sigma_0$ be a leaf of type τ. We compute the loops on a. By definition of a looping run, $\mathcal{A}$ cannot move up; nor can it move down since leaves have no children. So the only transitions which can be activated are $\circlearrowleft$-transitions. As we are only interested in *simple* loops, we can only activate one of these transitions *once*, thus creating runs of the form $(\alpha,p) \twoheadrightarrow (\alpha,q)$, and the corresponding loops (p,q). Let us have a general notation for this:

Definition 4 (Simple Here-Loops). $\mathcal{H}_\sigma^\tau \overset{\text{def}}{=} \{\,(p,q)\mid \langle\sigma,p,\tau\to\circlearrowleft,q\rangle\in\Delta\,\}$.

Thus the simple loops on a are $\mathcal{H}_a^\tau$. By Lemma 3 we have $\mho^\tau(a) = (\mathcal{H}_a^\tau)^*$. We now deal with inner nodes. Let $f \in \Sigma_2$, and $u = f(u_0,u_1)$; again, τ denotes the type of the root of u. Clearly the elements of $\mathcal{H}_f^\tau$ are loops on u, as above, but this time $\mathcal{A}$ can move down as well. It cannot move up on the first move (that would mean leaving the subtree), but it will obviously *need* to move up to rejoin the root if it ever moves down. To clarify all that, let us reason on what the first move of a simple looping run can be. It cannot be $\uparrow$ and all simple loops whose first move is $\circlearrowleft$ are already computed in $\mathcal{H}_f^\tau$. Say the first move is $\swarrow$: then the run can do whatever it wants in the left subtree u_0, after which it has to move back up to the root to complete the loop. Again, we only consider *simple* loops, so no move can be made past this point, as the root has been reached twice already. Thus the general form of such a run is $(\varepsilon,p),(0,p_0),(\beta_1,s_1),\ldots,(\beta_n,s_n),(0,q_0),(\varepsilon,q)$, with all $\beta_k \trianglelefteq 0$. But by definition, this means that (p_0,q_0) is a loop on u_0, ie. $(p_0,q_0) \in \mho^0(u_0)$. Needless to say, the same applies (with *1* instead of *0*) if the first move is $\searrow$. It follows that to determine whether (p,q) forms a simple loop on u, we need only check three things: 1. $\mathcal{A}$ can move down (left or right) from state p into a state p_0, 2. there is a loop (p_0,q_0) on this subtree and 3. in state q_0, $\mathcal{A}$ can move up from this subtree and into the state q. Formally:

$$\mho^\tau(u) = \left(\mathcal{H}_f^\tau \cup \left\{(p,q)\;\middle|\; \begin{array}{l}\exists\theta\in\mathbb{S}: \\ \exists(p_\theta,q_\theta)\in\mho^\theta(u_\theta)\end{array}\;\text{st.}\;\begin{array}{l}\langle f,p,\tau\to\chi(\theta),p_\theta\rangle\in\Delta \\ \langle u_\theta(\varepsilon),q_\theta,\theta\to\uparrow,q\rangle\in\Delta\end{array}\right\}\right)^*.$$

Theorem 5 (*Loops*). Let $\mathcal{A}$ be a TWA and $t \in \mathcal{T}(\Sigma)$. Then for all $\alpha \in \mathcal{P}os(t)$, $\mho^{\natural\alpha}(t|_\alpha)$, as defined above, is the set of all loops of $\mathcal{A}$ on $t|_\alpha$.

Example: For the TWA $\mathcal{X}$, $\mho^0(a) = \{\,(q_\ell,q_u)\,\}^* = \{\,(q_\ell,q_\ell),(q_u,q_u),(q_\ell,q_u)\,\}$, and $\mho^\star(f(a,b)) = (\varnothing \cup \{(q_\ell,q_u)\})^*$ (no simple here-loop, and one loop built on the left child). On the other hand, $\mho^\star(f(b,a)) = \varnothing^*$, because $\mho^1(a) = \mho^0(b) = \varnothing^*$.

Note that a reasonably efficient algorithm for testing membership is straightfor-
wardly derived from the above computation of loops:

Corollary 6 (*TWA Membership*). *Let $\mathcal{A}$ be a TWA and $t \in \mathcal{T}(\Sigma)$. Then we
have $t \in \mathcal{L}ng\,(\mathcal{A})$ if and only if $\mho^\star(t) \cap (I \times F) \neq \varnothing$.*

Corollary 7. *The complexity of TWA membership is $O\left(\|t\| \cdot (|Q|^3 + |\Delta|)\right)$.*

We now introduce a new notion related to tree loops: *tree overloops*.

Definition 8 (Over-Root, Extended Positions and Transitions). The *extended
positions* $\overline{Pos}(t)$ of a tree $t \in \mathcal{T}(\Sigma)$ are the set $Pos(t) \cup \{\overline{\varepsilon}\}$, where $\overline{\varepsilon}$ is called
the *overroot*. The parent function $\mathfrak{p}(\cdot)$ is extended over $\overline{Pos}(t)$ into the *extended
parent function* $\overline{\mathfrak{p}}(\cdot)$, such that $\overline{\mathfrak{p}}(\varepsilon) = \overline{\varepsilon}$ and $\varepsilon \lhd \overline{\varepsilon}$. The notion of configuration is
extended as well, so that the transitions of $\langle \Sigma, Q, \star \to \uparrow, Q \rangle$ become valid. Their
application yields configurations of the form $(\overline{\varepsilon}, q)$.

Definition 9 (Tree Over-Loops). Let $\mathcal{A}$ be a TWA and t a tree. A pair of states
$(p, q) \in Q^2$ forms an *overloop* of $\mathcal{A}$ on $t|_\alpha$ if there exists a run $(\alpha, p), (\beta_1, s_1), \ldots,$
$(\beta_n, s_n), (\overline{\mathfrak{p}}(\alpha), q)$ such that for all $k \in [\![1, n]\!], \beta_k \trianglelefteq \alpha$.

A way to compute overloops is to compute loops, then check for $\uparrow$-transitions:

Definition 10 (Up-Closure). Let $L \subseteq Q^2, \tau \in \mathbb{T}$ and $\sigma \in \Sigma$:
$$\mho_\sigma^\tau[L] \stackrel{\text{def}}{=} \left\{ (p, q) \in Q^2 \mid \exists p' \in Q : (p, p') \in L \text{ and } \langle \sigma, p', \tau \to \uparrow, q \rangle \in \Delta \right\} \ .$$

Lemma 11 (*Up-Closure*). *Let $\mathcal{A}$ be a TWA. If L is the set of all loops of $\mathcal{A}$ on
a subtree $u = t|_\alpha$, then $\mho_{t(\alpha)}^{\natural\alpha}[L]$ is the set of all overloops of $\mathcal{A}$ on u.*

Similarly to loops, we denote $\mathbb{O}^\tau(u)$ the set of all overloops of $\mathcal{A}$ on a subtree u,
where τ is the type of the root of u. By Lem. 11 we have $\mathbb{O}^\tau(u) = \mho_{u(\varepsilon)}^\tau[\mho^\tau(u)]$,
and in the case of leaves this yields $\mathbb{O}^\tau(a) = \mho_a^\tau[(\mathcal{H}_a^\tau)^*]$. However, in the case
of inner nodes (say $u = f(u_0, u_1)$), in order to have an inductive computation
of overloops instead of one based on loops, we need to compute the overloops
of the father, knowing the overloops of the children. The simplest way is to
compute the loops of the father and take the up-closure. We only need to check
whether 1. the automaton can go down and left (resp. right) from p to a state
p_0 and 2. there is a left (resp. right) overloop (p_0, q_0): this forms a loop (p, q_0).
Formally:

$$\mathbb{O}^\tau(u) = \mho_f^\tau\left[\left(\mathcal{H}_f^\tau \cup \left\{ (p, q_\theta) \ \middle| \ \begin{array}{c} \exists \theta \in \mathbb{S} : \\ \exists p_\theta \in Q \end{array} \text{ st.} \ \begin{array}{l} \langle f, p, \tau \to \chi(\theta), p_\theta \rangle \in \Delta \\ \text{and } (p_\theta, q_\theta) \in \mathbb{O}^\theta(u_\theta) \end{array} \right\} \right)^*\right] \ .$$

Theorem 12 (*Overloops*). *Let $\mathcal{A}$ be a TWA and $t \in \mathcal{T}(\Sigma)$. Then for all $\alpha \in
Pos(t)$, $\mathbb{O}^{\natural\alpha}(t|_\alpha)$, as defined above, is the set of all overloops of $\mathcal{A}$ on $t|_\alpha$.*

Example: For the TWA $\mathcal{X}$, $\mathbb{O}^0(a) = \mho_a^0[\mho^0(a)] = \{(q_u, q_u), (q_\ell, q_u)\}$. However
$\mho^\star(f(a, b))$ is the empty set. Thus a small adjustment is needed to test member-
ship using overloops, as standard TWA – such as $\mathcal{X}$ – never admit any overloop
at the root of a tree, for lack of $\uparrow$-transitions.

Definition 13 (Overfinal State & Escaped TWA). *Let* $\mathcal{A} = \langle \Sigma, Q, I, F, \Delta \rangle$ *be a TWA; it can be transformed into an* escaped TWA

$$\mathcal{A}' = \langle\, \Sigma,\ Q \uplus \{\checkmark\},\ I, F,\ \Delta \uplus \langle \Sigma, F, \star \to \uparrow, \checkmark \rangle\, \rangle\ ,$$

where $\checkmark \notin Q$ *is a fresh state, called* overfinal state. *[Clearly* $\mathcal{L}ng\,(\mathcal{A}) = \mathcal{L}ng\,(\mathcal{A}')$.]

Example: Once $\mathcal{X}$ is escaped, we have $\mho^*(f(a,b)) = \{\,(q_{\mathrm{u}}, \checkmark), (q_{\ell}, \checkmark)\,\}$.

Corollary 14 (*TWA Membership Redux*). *Let* $\mathcal{A}$ *be an escaped TWA and* $t \in \mathcal{T}(\Sigma)$. *Then* $t \in \mathcal{L}ng\,(\mathcal{A})$ *if and only if* $\mho^*(t) \cap (I \times \{\checkmark\}) \neq \varnothing$.

3 Transforming TWA into Equivalent BUTA

It is well-known that every TWA is equivalent to a BUTA; a more general version of this result has been proven in [8] – using game-theoretic arguments – and the main idea of a loop-based transformation from TWA into BUTA is outlined in [3] and [12, p143]. In this section we present two versions of it: the classical, loop-based one (Algo. 1[p169]) and an overloop-based variant (Algo. 2[p169]). We go on to show that, in the case of deterministic TWA, the overloop-based construction results in much smaller equivalent BUTA than the classical one.

3.1 Two Variants: Loops and Overloops

Lemma 15 (*Loop-Based Algorithm*). *Let* $\mathcal{A}$ *be a TWA,* $\mathcal{B}$ *its equivalent BUTA by Algorithm 1,* $t \in \mathcal{T}(\Sigma)$ *and a subtree* $u = t|_\alpha$. *Then for every type* $\tau \in \mathbb{T}$, *there is one unique run* ρ *of* $\mathcal{B}$ *on* u *such that* $\rho(\varepsilon) = (u(\varepsilon), \tau, L)$. *Furthermore,* L *is the set of all loops of* $\mathcal{A}$ *on* u, *provided that* $\natural\alpha = \tau$.

Proof. Both claims are shown by structural induction on u. *First claim:* If $u = a \in \Sigma_0$, then by line A in Algorithm 1, $\rho(\varepsilon) = P = (a, \tau, L) = (u(\varepsilon), \tau, L)$. It is unique, as only one transition $a \to P$ is generated for each couple a, τ. If $u = f(u_0, u_1), f \in \Sigma_2$, then by induction hypothesis there exists one run ρ_0 on u_0 such that $\rho_0(\varepsilon) = P_0 = (u_0(\varepsilon), \mathbf{0}, S_0)$, and one run ρ_1 on u_1 such that $\rho_1(\varepsilon) = P_1 = (u_1(\varepsilon), \mathbf{1}, S_1)$. Thus by line B in Algo. 1 we use the rule $f(P_0, P_1) \to P$ to build the run ρ such that $\rho(\varepsilon) = P = (f, \tau, L) = (u(\varepsilon), \tau, L)$, $\rho|_0 = \rho_0$ and $\rho|_1 = \rho_1$. Since ρ_0 and ρ_1 are unique, so is ρ. *Second claim:* If $u = a \in \Sigma_0$, then $\rho(\varepsilon) = (a, \tau, \mathcal{H}_a^{\tau*})$, and by Theorem 5 we have $\mathcal{H}_a^{\tau*} = \mho^\tau(a)$. If $u = f(u_0, u_1)$, then $\rho(\varepsilon) = (f, \tau, (\mathcal{H}_f^\tau \cup S)^*)$ and by induction hypothesis $S_\theta = \mho^\theta(u_\theta)$ and $\sigma_\theta = u_\theta(\varepsilon)$, for all $\theta \in \mathbb{S}$. Thus by Theorem 5, $(\mathcal{H}_f^\tau \cup S)^* = \mho^\tau(u)$. $\square$

Theorem 16. *Algorithm 1 is correct; that is,* $\mathcal{L}ng\,(\mathcal{A}) = \mathcal{L}ng\,(\mathcal{B})$.

Proof. If $t \in \mathcal{L}ng\,(\mathcal{A})$, then there is a loop $(q_{\mathrm{i}}, q_{\mathrm{f}}) \in I \times F$ of $\mathcal{A}$ on t. Therefore there is a run ρ of $\mathcal{B}$ on t such that $\rho(\varepsilon) = (t(\varepsilon), \star, L)$, with $(q_{\mathrm{i}}, q_{\mathrm{f}}) \in L$. Thus $\rho(\varepsilon)$ is a final state and $t \in \mathcal{L}ng\,(\mathcal{B})$. Conversely, if $t \in \mathcal{L}ng\,(\mathcal{B})$ then there is an accepting run ρ of $\mathcal{B}$ on t, that is to say such that $\rho(\varepsilon) = (t(\varepsilon), \star, \mho^*(t))$ and there exists $(q_{\mathrm{i}}, q_{\mathrm{f}}) \in (I \times F) \cap \mho^*(t)$. Thus by Cor. 6 we have $t \in \mathcal{L}ng\,(\mathcal{A})$. $\square$

Two short but important remarks are in order. First: it might seem strange that our states are in $\Sigma \times \mathbb{T} \times 2^{Q^2}$, and not more simply in $\mathbb{T} \times 2^{Q^2}$, as suggested in [12]. In [3] a similar construction – albeit deterministic, see the second remark – is proposed, which does not include Σ either. However, it is not clear how loops could be considered independently from the root symbol of the subtree that bears them. Consider for instance $a, b \in \Sigma_0$ with only the transitions $\langle \{a, b\}, p, \tau \to \circlearrowleft, q\rangle$ and $\langle b, q, \tau \to \uparrow, s'\rangle \in \Delta$. Then the loops on a and b are exactly the same – $\{(p, q)\}^*$ – and yet, from their father's point of view, they behave very differently. If $\mathcal{A}$ can go down from a state s to p, it can form a loop (s, s') if the child is b, but not if it is a. In contrast to the loop-based construction, the overloop-based algorithm (Algo. 2) suppresses this problem completely.

Second: the observation made in Lemma 15 that the run of $\mathcal{B}$ is unique, given a subtree and a type, makes it easy to adapt the algorithm to yield a deterministic BUTA. Indeed, every tree in $\mathcal{T}(\Sigma)$ is non-deterministically evaluated by $\mathcal{B}$ into exactly three possible states (one per type); the correct one is chosen according to the context during the run. Recall that rules $f(P_0, P_1) \to P$ are built such that the "type" component of P_θ is θ, and final states bear the root type $\star$. Hence, it suffices to group those three possible states into one element of $\Sigma \times (2^{Q^2})^{|\mathbb{T}|}$ to achieve determinism, which brings us back to the states suggested in [3].

Lemma 17 (*Overloop-Based Algorithm*). *Let $\mathcal{A}$ be a TWA, $\mathcal{B}$ its equivalent BUTA by Algorithm 2, $t \in \mathcal{T}(\Sigma)$ and a subtree $u = t|_\alpha$. Then for all $\tau \in \mathbb{T}$, there is one unique run ρ of $\mathcal{B}$ on u such that $\rho(\varepsilon) = (\tau, O)$. Furthermore, O is the set of all overloops of $\mathcal{A}$ on u, provided that $\natural\alpha = \tau$.*

Theorem 18. *Algorithm 2 is correct; that is, $\mathcal{L}ng(\mathcal{A}) = \mathcal{L}ng(\mathcal{B})$.*

Note that this construction can be adapted to yield deterministic BUTA in exactly the same way as for Algo. 1.

3.2   Overloops and the Deterministic Case

Definition 19 (Deterministic TWA). A TWA $\mathcal{A} = \langle \Sigma, Q, I, F, \Delta\rangle$ is *deterministic* (ie. a *DTWA*) if[a] for all $\sigma \in \Sigma, p \in Q, \tau \in \mathbb{T}$, $|\langle\sigma, p, \tau \to \mathbb{M}, Q\rangle \cap \Delta| \leqslant 1$.

Definition 20 (Functional Relation). A relation $R \subseteq Q^2$ is *functional* (or *right-unique*, or *a partial function*) if, for all $p, q, q' \in Q$, pRq and $pRq' \implies q = q'$.

Remark 21. There are $2^{|Q|^2}$ binary relations on Q, of which $|Q + 1|^{|Q|}$ are partial functions, of which $|Q|^{|Q|}$ are total functions.

Remark 22. If a relation R is functional, then so is R^k, for any $k \in \mathbb{N}$.

By construction, a BUTA built by Algo. 1 (loop-based) has at most $|\Sigma| \cdot |\mathbb{T}| \cdot 2^{|Q|^2}$ states, while one built by Algo. 2 (overloop-based) has at most $|\mathbb{T}| \cdot 2^{|Q|^2}$. We will see in this section that, in the deterministic case, this upper bound is in fact much lower for the overloop-based algorithm than for the traditional loop-based one. More specifically, we will show that the following holds:

---

[a] In this paper we do not need the usual, stronger definition, where I is a singleton.

Theorem 23 (*Deterministic Upper-Bound*). *Let $\mathcal{A}$ be a deterministic TWA and $\mathcal{B}$ its equivalent BUTA built by application of Algorithm 2. Then $\mathcal{B}$ has at most $|\mathbb{T}| \cdot 2^{|Q| \log_2(|Q|+1)}$ states.*

The idea is that every state which we build corresponds exactly to the set L of all loops (resp. overloops) of the automaton $\mathcal{A}$ on a certain subtree u. Since $L \subseteq Q^2$, we can see it as a binary relation on the states. The intuition here is that, if $\mathcal{A}$ is deterministic, and enters the root of u in one given state p, then there "should be" only one possible outcome. More formally:

Lemma 24. *If $\mathcal{A}$ is a deterministic TWA, then $\twoheadrightarrow_\mathcal{A}$ is functional.*

Proof. In a given configuration (α, p), over a tree t, $|\langle t(\alpha), p, \natural\alpha \rightarrow \mathbb{M}, Q \rangle \cap \Delta|$ $\leqslant 1$. Therefore, (α, p) has at most one successor. $\square$

However, in the case of loops, this does not suffice to make L functional because, determinism notwithstanding, a single (non-trivial) loop may reach the root several times, and in different states, before exiting the subtree. Thus there is nothing to prevent us from having both pLq and pLq', for $q \neq q'$; we show next that in that case, one of these loops is simply an extension of the other.

Lemma 25 (*Hidden Loops*). *Let $p, q, q' \in Q$, $q \neq q'$ such that (p, q) and (p, q') are loops of the TWA $\mathcal{A}$ on a given subtree $t|_\alpha$. Then if $\mathcal{A}$ is deterministic, either (q, q') or (q', q) must be a loop of $\mathcal{A}$ on $t|_\alpha$.*

Proof. By Definition 1, there exist two runs $c_0, \ldots, c_n$ and $d_0, \ldots, d_m$ such that $c_0 = d_0 = (\alpha, p)$, $c_n = (\alpha, q)$ and $d_m = (\alpha, q')$. If $n = m$ then $c_0 \twoheadrightarrow^n c_n$ and $c_0 \twoheadrightarrow^n d_n$ and by Lemma 24 and Remark 22, it follows that $c_n = d_m$. But this contradicts $q \neq q'$, so we must have $n \neq m$. Say that $n < m$. Then $c_n = d_n$, and $(\alpha, q) = d_n, \ldots, d_m = (\alpha, q')$ forms a run. Therefore (q, q') is a loop. Similarly, if $n > m$, then by the same arguments (q', q) is a loop. $\square$

Contrariwise, two overloops cannot be combined to form another overloop on the same subtree, which satisfies the above intuition of a "single outcome":

Lemma 26. *Let $p, q, q' \in Q$, such that (p, q) and (p, q') are overloops of the TWA $\mathcal{A}$ on a given subtree $t|_\alpha$. Then if $\mathcal{A}$ is deterministic, $q = q'$.*

Proof. By Def. 9, there exist $s, s' \in Q$ such that $(\alpha, p), \ldots, (\alpha, s), (\overline{\mathsf{p}}(\alpha), q)$ and $(\alpha, p), \ldots, (\alpha, s'), (\overline{\mathsf{p}}(\alpha), q')$ are runs; thus (p, s) and (p, s') are loops. If $s \neq s'$, then by Lem. 25, say, (s, s'), is a loop. So there exist $s_1, \ldots, s_n \in Q, \beta_1 \trianglelefteq \alpha$ $, \ldots, \beta_n \trianglelefteq \alpha$ such that $(\alpha, s), (\beta_1, s_1), \ldots, (\beta_n, s_n), (\alpha, s')$ is a run. Thus we have in particular $(\alpha, s) \twoheadrightarrow (\overline{\mathsf{p}}(\alpha), q)$ and $(\alpha, s) \twoheadrightarrow (\beta_1, s_1)$. It follows that $\overline{\mathsf{p}}(\alpha) = \beta_1 \trianglelefteq \alpha$, which is contradictory. Hence $s = s'$. We have both $(\alpha, s) \twoheadrightarrow (\overline{\mathsf{p}}(\alpha), q)$ and $(\alpha, s) \twoheadrightarrow (\overline{\mathsf{p}}(\alpha), q')$. Since $\twoheadrightarrow$ is functional (Lem. 24), we have finally $q = q'$. $\square$

With this, we can conclude the proof of Theorem 23.

Proof of Theorem 23. By construction, for every state $P = (\tau, L)$ generated for $\mathcal{B}$ by Algorithm 2, there exists at least a subtree t such that L is the set of overloops of $\mathcal{A}$ on t. Thus, by Lemma 26, L is functional. Therefore, by Remark 21, there are at most $|\mathbb{T}| \cdot |Q + 1|^{|Q|}$ states (or, equivalently, $|\mathbb{T}| \cdot 2^{|Q| \log_2(|Q|+1)}$). $\square$

4   The Emptiness Problem and Experimental Results

Polynomial Over-Approximation for the Emptiness Problem. Testing emptiness of a TWA $\mathcal{A}$ is an ExpTime-complete problem [3]. This is rather unfortunate, as there are practical questions – such as satisfiability of some XPath fragments – which reduce to the emptiness of the language of a TWA. We present in this section a crude but fairly accurate and very expeditious overloops-based algorithm capable of detecting emptiness in a number of cases. Algorithm 3[p170] is a variant of Algorithm 2 with the following properties:

Lemma 27 (*Overloops Over-Approximation*). *Let $\mathcal{A}$ be a TWA, then when the execution of Algorithm 3 ends, for any $\tau \in \mathbb{T}$, $\mathcal{L}_\tau \supseteq \bigcup_{t \in \mathcal{T}(\Sigma)} \Phi^\tau(t)$.*

Theorem 28. *Algorithm 3 is correct; that is, it yields Empty only if $\mathcal{L}ng(\mathcal{A}) = \varnothing$.*

Corollary 29 (*Complexity of the Approximation*). *The execution of Algorithm 3 is done in a time polynomial in the size of $\mathcal{A}$ – more precisely: $O(|\Sigma| \cdot |\mathbb{T}|^2 \cdot |Q|^4 \cdot |\Delta|)$.*

Note that Algorithm 3 can easily be made just as coarse or as fine as the need dictates. At the coarse end of that gamut we have a variant of Algorithm 3 which forgoes type information, thus hoarding up all overloops in a single set $\mathcal{L}$ instead of three, and at the fine end we find something equivalent to Algorithm 2.

Experimental Results. APPROXIMATION. The approximation has yielded astonishingly good results with randomly generated TWA: out of the – roughly – ten thousands of automata of various sizes ($2 \leqslant |Q| \leqslant 20$) on which it was tested, 75% of which had empty languages, only two of them yielded Unknown instead of Empty. Those results are – unfortunately – probably much better than what can be expected in practice, as our generation scheme is, for now, very simplistic. It is therefore likely that the generated instances are in some sense trivial wrt. emptiness. Two approaches which we plan on taking to obtain more meaningful results are a study similar to that of [9] to identify interesting instances, and the use of statistically-exploitable generation schemes as in [10]. GENERAL RESULTS. Comparing the output of Algos. 1 & 2, we noted that the latter generates smaller automata – the cardinality of each state being ignored – by a factor two or more, depending on the size of the input TWA. The same caveat as above applies concerning the random TWA. DEMONSTRATION SOFTWARE. Readers interested in experimenting with this paper's algorithms will find online [(b)] a proof of concept (binaries and OCaml source code), as well as instructions for use.

[(b)] On `http://lifc.univ-fcomte.fr/~vhugot/TWA`

5   Conclusion

In this paper we have introduced tree overloops, and applied both loops and overloops to common operations on TWA: deciding membership, transforming a TWA into a BUTA, and inexpensively testing emptiness. We have shown that the use of overloops simplifies transformation into BUTA, and substantially lowers the upper bound in the deterministic case. We intend to pursue this further by using overloops to characterise useful classes of TWA and perform significant simplifications on the automata, hopefully leading to applications to XPath.

Acknowledgements. The authors would like to thank the members of the INRIA ARC ACCESS for interesting discussions on this topic. Our thanks go as well to the anonymous reviewer who provided a tighter complexity bound for Cor. 7, and whose careful proofreading improved the readability of this paper.

References

1. Aho, A., Ullman, J.: Translations on a context free grammar. Information and Control 19(5), 439–475 (1969)
2. Bojańczyk, M.: 1-bounded TWA cannot be determinized. In: Pandya, P.K., Radhakrishnan, J. (eds.) FSTTCS 2003. LNCS, vol. 2914, pp. 62–73. Springer, Heidelberg (2003)
3. Bojańczyk, M.: Tree-walking automata. In: Martín-Vide, C., Otto, F., Fernau, H. (eds.) LATA 2008. LNCS, vol. 5196, pp. 1–2. Springer, Heidelberg (2008), http://www.mimuw.edu.pl/~bojan/papers/twasurvey.pdf
4. Bojańczyk, M., Colcombet, T.: Tree-walking automata cannot be determinized. Theoretical Computer Science 350(2-3), 164–173 (2006)
5. Bojańczyk, M., Colcombet, T.: Tree-walking automata do not recognize all regular languages. In: STOC 2005, pp. 234–243. ACM Press, New York (2005)
6. ten Cate, B., Segoufin, L.: Transitive closure logic, nested tree walking automata, and XPath. J. ACM 57(3), 251–260 (2010)
7. Comon, H., Dauchet, M., Gilleron, R., Löding, C., Jacquemard, F., Lugiez, D., Tison, S., Tommasi, M.: Tree automata techniques and applications (2007)
8. Cosmadakis, S., Gaifman, H., Kanellakis, P., Vardi, M.: Decidable optimization problems for database logic programs. In: STOC 1988, pp. 477–490. ACM Press, New York (1988)
9. Heam, P., Hugot, V., Kouchnarenko, O.: Random Generation of Positive TAGEDs wrt. the Emptiness Problem. Tech. Rep. RR-7441, INRIA (November 2010)
10. Héam, P.-C., Nicaud, C., Schmitz, S.: Random generation of deterministic tree (Walking) automata. In: Maneth, S. (ed.) CIAA 2009. LNCS, vol. 5642, pp. 115–124. Springer, Heidelberg (2009)
11. Knuth, D.: Semantics of context-free languages: Correction. Theory of Computing Systems 5(2), 95–96 (1971)
12. Samuelides, M.: Automates d'arbres à jetons. Ph.D. thesis, Université Paris-Diderot - Paris VII (December 2007), http://tel.archives-ouvertes.fr/tel-00255024/en/
13. Segoufin, L., Vianu, V.: Validating Streaming XML Documents. In: Popa, L. (ed.) PODS, pp. 53–64. ACM, New York (2002)

Nondeterministic State Complexity of Star-Free Languages

Markus Holzer, Martin Kutrib, and Katja Meckel

Institut für Informatik, Universität Giessen,
Arndtstraße 2, 35392 Giessen, Germany
{holzer,kutrib,meckel}@informatik.uni-giessen.de

Abstract. We investigate the nondeterministic state complexity of several operations on finite automata accepting star-free languages. It turns out that in most cases exactly the same tight bounds as for general regular languages are reached. This nicely complements the results recently obtained in [8] for the operation problem of star-free languages accepted by deterministic finite automata.

1 Introduction

The operation problem on a language family is the question of cost (in terms of states) of operations on languages from this family with respect to their representations. More than a decade ago the operation problem for regular languages represented by deterministic finite automata (DFAs) as studied in [32,33] renewed the interest in descriptional complexity issues of finite automata in general. Although the research area of finite automata dates back to the beginning of the 1950s, their (descriptional) complexity with respect to the operation problem had attracted surprisingly less attention in the early days. This lack of interest may be one reason for the prevailing view on regular languages during the late seventies [32]:

> Since the late seventies, many believed that everything of interest about regular languages is known except for a few very hard problems, [...] It appeared that not much further work could be done on regular languages.

Nowadays descriptional complexity of finite automata and related structures is a vivid area of research, for which the (recent) surveys on this area give evidence [16,17,18,32].

It is well known that nondeterministic and deterministic finite automata are computationally equivalent. More precisely, given some n-state NFA one can always construct a language equivalent DFA with at most 2^n states [27] and, therefore, NFAs can offer exponential savings in space compared with DFAs. In fact, later it was shown independently in [23,25,26] that this exponential upper bound is best possible, that is, for every n there is an n-state NFA which cannot be simulated by any DFA with strictly less than 2^n states. Recently, in [2] it was shown that this exponential tight bound for the determinization

B. Bouchou-Markhoff et al. (Eds.): CIAA 2011, LNCS 6807, pp. 178–189, 2011.

of NFAs also holds when restricting the NFAs to accept only subregular language families such as star languages [3], (two-sided) comet languages [5], ordered languages [30], star-free languages [24], power-separating languages [31], prefix-closed languages, etc. On the other hand, there are also subregular language families known, where this exponential bound is not met. Prominent examples are the family of unary regular languages, where an asymptotic bound of $e^{\Theta(\sqrt{n \cdot \ln n})}$ states for determinization has been shown in [9,10], and the family of finite languages with a tight bound of $O(k^{\frac{n}{\log_2(k)+1}})$, where k is the size of the alphabet [28]. The significant different behavior with respect to the relative succinctness of NFAs compared to DFAs is also reflected in the operation problem for these devices. The operation problem for NFAs was first investigated in [15]. It turned out that in most cases when an operation is cheap for DFAs it is costly for NFAs and *vice versa*. We give two examples: (i) the complementation operation applied to a language accepted by an n-state DFA results in a DFA of exactly the same number of states, while complementing NFAs gives an exponential tight bound of 2^n states [19], and conversely (ii) for two languages accepted by m- and n-state DFAs we have a tight bound of $m \cdot 2^n - t \cdot 2^{n-1}$ states for concatenation [32,33], where t is the number of accepting states of the "left" automaton, and $m + n + 1$ states when considering NFAs [15]. All these results are for general regular languages. So, the question arises what happens to these bounds if the operation problem is restricted to subregular language families.

In fact, for some subregular language families this question was recently studied in the literature [6,7,8,12,13,20,21,22] mostly for DFAs. To this end, the notion of quotient complexity [4] which has been studied in a series of papers [6,7,8] is a useful tool for exploring the deterministic state complexity. An example for a subregular language family whose DFA operation problems meet the general bounds for most operations is the family of star-free languages [8], while prefix-, infix-, and suffix-closed languages [7], bifix-, factor-, and subword-free languages [6] show a diverse behavior mostly not reaching the general bounds. For a few language families, in particular prefix- and suffix-free regular languages, also the operation problem for NFAs was considered [12,13,20,22], but for the exhaustively studied family of star-free languages it is still open. The family of star-free (or regular non-counting) languages is an important subfamily of the regular languages, which can be obtained from the elementary languages $\{a\}$, for $a \in \Sigma$, and the empty set $\emptyset$ by applying the Boolean operations union, complementation, and concatenation finitely often. They obey nice characterizations in terms of aperiodic monoids and permutation-free DFAs [24]. Here we investigate their operation problem for NFAs with respect to the basic operations union, intersection, complementation, concatenation, Kleene star, and reversal. It turns out that in most cases exactly the same tight bounds as in the general case are reached. This nicely complements the results recently obtained for the operation problem of star-free languages accepted by DFAs [8]. We summarize our results in Table 1, where we also list the results for DFAs accepting star-free languages [8], for comparison reasons.

Table 1. Deterministic and nondeterministic state complexities for the operation problem on star-free languages summarized. The results for DFAs are from [8].

Operation	Star-free language accepted by ...	
	DFA	NFA
$\cup$	mn	$m + n + 1$
$\cap$	mn	mn
$\sim$		2^n
$\cdot$	$(m-1)2^n + 2^{n-1}$	$m + n$
$*$	$2^{n-1} + 2^{n-2}$	$n + 1$
R	$2^n - 1$	$n + 1$

2 Preliminaries

For $n \geq 0$ we write $\Sigma^{\leq n}$ for the set of all words whose lengths are at most n and Σ^n for the set of all words of length n. The empty word is denoted by λ. The reversal of a word w is denoted by w^R, and for the length of w we write $|w|$. Set inclusion is denoted by $\subseteq$ and strict set inclusion by $\subset$. We write 2^S for the power set and $|S|$ for the cardinality of a set S.

A *nondeterministic finite automaton* (NFA) is a 5-tuple $\mathcal{A} = (S, \Sigma, \delta, s_0, F)$, where S is the finite set of *states*, Σ is the finite set of *input symbols*, $s_0 \in S$ is the *initial state*, $F \subseteq S$ is the set of *accepting states*, and $\delta : S \times \Sigma \to 2^S$ is the *transition function*. As usual the transition function is extended to $\delta : S \times \Sigma^* \to 2^S$ reflecting sequences of inputs: $\delta(s, \lambda) = \{s\}$ and $\delta(s, aw) = \bigcup_{s' \in \delta(s,a)} \delta(s', w)$, for $s \in S$, $a \in \Sigma$, and $w \in \Sigma^*$. A word $w \in \Sigma^*$ is *accepted* by $\mathcal{A}$ if $\delta(s_0, w) \cap F \neq \emptyset$. The *language accepted* by $\mathcal{A}$ is $L(\mathcal{A}) = \{w \in \Sigma^* \mid w$ is accepted by $\mathcal{A}\}$.

A finite automaton is *deterministic* (DFA) if and only if $|\delta(s, a)| = 1$, for all $s \in S$ and $a \in \Sigma$. In this case we simply write $\delta(s, a) = s'$ for $\delta(s, a) = \{s'\}$ assuming that the transition function is a mapping $\delta : S \times \Sigma \to S$. So, any DFA is complete, that is, the transition function is total, whereas for NFAs it is possible that δ maps to the empty set. A state s is *reachable* in $\mathcal{A}$ if there is an input word w with $s \in \delta(s_0, w)$. Without loss of generality we assume that any state of a nondeterministic finite automaton is reachable. A finite automaton is said to be *minimal* if there is no finite automaton of the same type with fewer states, accepting the same language. Note that a sink state is counted for DFAs, since they are always complete, whereas it is not counted for NFAs, since their transition function may map to the empty set.

Next, we briefly recall the so-called (extended) *fooling set* technique (see, for example, [1,11,16]) that is widely used for proving lower bounds on the number of states necessary for an NFA to accept a given language.

Theorem 1 (Extended Fooling Set Technique). *Let $L \subseteq \Sigma^*$ be a regular language and suppose there exists a set of pairs $S = \{(x_i, y_i) \mid 1 \leq i \leq n\}$ such that (1) $x_i y_i \in L$, for $1 \leq i \leq n$, and (2) $i \neq j$ implies $x_i y_j \notin L$ or $x_j y_i \notin L$, for $1 \leq i, j \leq n$. Then any nondeterministic finite automaton accepting L has at least n states. Here S is called an (extended) fooling set for L.*

Now we turn to the subregular language family of interest. A language $L \subseteq \Sigma^*$ is *star-free* (or regular *non-counting*) if and only if it can be obtained from the elementary languages $\{a\}$, for $a \in \Sigma$, and the empty set $\emptyset$ by applying the Boolean operations union, complementation, and concatenation finitely often. These languages are exhaustively studied in, for example, [24] and [29]. Since regular languages are closed under Boolean operations and concatenation, every star-free language is regular. On the other hand, not every regular language is star free. Here we sometimes utilize an alternative characterization of star-free languages by so called permutation-free automata [24]: A regular language $L \subseteq \Sigma^*$ is starfree if and only if the minimal DFA accepting L is *permutation-free*, that is, there is *no* word $w \in \Sigma^*$ that induces a non-trivial permutation on any subset of the set of states. Here a trivial permutation is simply the identity permutation. Note that word uw induces a non-trivial permutation $\{s_1, s_2, \ldots, s_n\} \subseteq S$ in a DFA with state set S and transition function δ if and only if wu induces a non-trivial permutation $\{\delta(s_1, u), \delta(s_2, u), \ldots, \delta(s_n, u)\}$ in the same automaton.

3   Results on the Operation Problem

We start our investigations with Boolean operations. For deterministic finite automata it was recently shown that in the worst case the Boolean operations union, intersection, and complementation have state complexity $m \cdot n$, $m \cdot n$, and n not only for general regular languages, but also for star-free languages. However, the state complexity of NFA operations for general regular languages is essentially different [15]. Namely, union, intersection, and complementation have nondeterministic state complexity $m + n + 1$, $m \cdot n$, and 2^n. It is worth mentioning that the exponential bound of 2^n states for complementation was shown to be tight in [19]. Here we prove that this is also the case for star-free languages. Note, that all the upper bounds are from [15]. Thus, we only have to give star-free witness languages meeting these bounds. At first we consider the union operation.

Theorem 2. *For any integers $m, n \geq 2$ let $\mathcal{A}$ be an m-state and $\mathcal{B}$ be an n-state NFA that accept star-free languages. Then $m + n + 1$ states are sufficient for an NFA to accept the language $L(\mathcal{A}) \cup L(\mathcal{B})$. The bound is tight for binary alphabets.*

Proof. As already mentioned, the upper bound of $m+n+1$ states is that for arbitrary regular languages shown in [15]. For the lower bound we argue as follows: Consider the NFA $\mathcal{A} = (S, \{a, b\}, \delta, s_0, F)$ with state set $S = \{0, 1, \ldots, m-1\}$, for $m \geq 2$. State 0 is the initial state s_0, and state $m-1$ is the only final state. The transition function is given by (cf. Figure 1):

- $\delta(i, a) = \{i+1\}$, for $0 \leq i < m-1$, and
- $\delta(m-1, b) = \{0\}$.

The language accepted by $\mathcal{A}$ is $a^{m-1}(ba^{m-1})^*$. Observe, that the automaton $\mathcal{A}$ is actually a *partial* DFA, where the sink state is missing. The corresponding

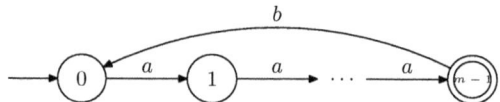

Fig. 1. The m-state NFA $\mathcal{A}$, $m \geq 2$, accepting a star-free language. The automaton $\mathcal{B}$ has n states and letters a and b are interchanged.

complete DFA is minimal and does not obey a non-trivial permutation on the state set. Therefore, the language $L(\mathcal{A})$ is a star-free language. Similarly, we define the automaton $\mathcal{B}$ by taking $\mathcal{A}$ with n states and interchanging the letters a and b. Hence, we obtain the star-free language $L(\mathcal{B}) = b^{n-1}(ab^{n-1})^*$.

It remains to be shown that $m+n+1$ states are needed by any NFA to accept $L(\mathcal{A}) \cup L(\mathcal{B})$. To this end, we construct the following set of pairs

$$S = \{\, (a^{m-1}ba^i, a^{m-1-i}ba^{m-1}) \mid 0 \leq i \leq m-1 \,\}$$
$$\cup \{\, (b^{n-1}ab^i, b^{n-1-i}ab^{n-1}) \mid 0 \leq i \leq n-1 \,\}.$$

First consider the pairs of the form $(a^{m-1}ba^i, a^{m-1-i}ba^{m-1})$ in S. Clearly, the word $a^{m-1}ba^i \cdot a^{m-1-j}ba^{m-1}$, for $0 \leq i,j \leq m-1$, is in the union of $L(\mathcal{A})$ and $L(\mathcal{B})$ if and only if $i = j$. Thus, the pair $(a^{m-1}ba^i, a^{m-1-i}ba^{m-1})$ induces a word that belongs to the union under consideration, but any word induced by crossing different pairs of the above form results in two words not in $L(\mathcal{A}) \cup L(\mathcal{B})$. Symmetrically we can argue for the pairs of the form $(b^{n-1}ab^i, b^{n-1-i}ab^{n-1})$. Finally, we have to compare pairs $(a^{m-1}ba^i, a^{m-1-i}ba^{m-1})$, for $0 \leq i \leq m-1$, with pairs $(b^{n-1}ab^j, b^{n-1-j}ab^{n-1})$, for $0 \leq j \leq n-1$. In this case we obtain the words $a^{m-1}ba^i \cdot b^{n-1-j}ab^{n-1}$ and $b^{n-1}ab^j \cdot a^{m-1-i}ba^{m-1}$, where the start and end blocks of a's and b's of the words do not correspond. Thus, both words do not belong to the union of $L(\mathcal{A})$ and $L(\mathcal{B})$. Hence, S is a fooling set for the language $L(\mathcal{A}) \cup L(\mathcal{B})$ of size $m+n$. To the upper bound one state is missing. We argue that the initial state of the automaton that accepts the language $L(\mathcal{A}) \cup L(\mathcal{B})$ is not one of the states induced by S.

Assume to the contrary that the initial state of the automaton accepting the language $L(\mathcal{A}) \cup L(\mathcal{B})$ is one of the states induced by S. If the initial state is equal to the state referenced by the pair $(a^{m-1}ba^i, a^{m-1-i}ba^{m-1})$, $0 \leq i \leq m-1$, then the word $a^{m-1}ba^i \cdot b^{n-1}$ is also accepted, because b^{n-1} is in $L(\mathcal{A}) \cup L(\mathcal{B})$ and must be accepted from the initial state. This contradicts the definition of $L(\mathcal{A}) \cup L(\mathcal{B})$. Symmetrically, we argue for the pairs $(b^{n-1}ab^i, b^{n-1-i}ab^{n-1})$, for $0 \leq i \leq n-1$. In all cases we obtain a contradiction to our assumption. This shows that an additional state is needed, which gives the $m+n+1$ lower bound for the language $L(\mathcal{A}) \cup L(\mathcal{B})$. □

Now we turn to the intersection of NFAs. Again, we make use of the upper bound already proven.

Theorem 3. *For any integers $m, n \geq 2$ let $\mathcal{A}$ be an m-state and $\mathcal{B}$ be an n-state NFA that accept star-free languages. Then $m \cdot n$ states are sufficient for an NFA to accept the language $L(\mathcal{A}) \cap L(\mathcal{B})$. This bound is tight for binary alphabets.*

Proof. The upper bound of $m \cdot n$ states follows from the construction presented in [15] for the intersection of general regular languages accepted by NFAs. In order to show a matching lower bound we apply the DFA used in [8] to prove the corresponding result for deterministic finite automata. Clearly, the DFA is also an NFA. However, here we have to show that the resulting NFA is minimal. So, let $\mathcal{A} = (S, \{a, b\}, \delta, s_0, F)$ with state set $S = \{0, 1, \ldots, m - 1\}$, $m \geq 2$. State 0 is the initial state s_0, and state $m - 1$ is the only final state. The transition function is given by (cf. Figure 2):

- $\delta(i, a) = \{i + 1\}$, for $0 \leq i < m - 1$, and
- $\delta(i, b) = \{i\}$, for $0 \leq i \leq m - 1$.

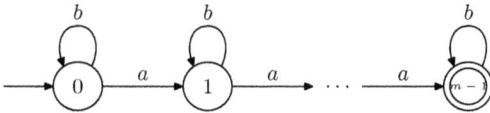

Fig. 2. The m-state NFA $\mathcal{A}$, $m \geq 2$, accepting the star-free language $b^*(ab^*)^{m-1}$. The automaton $\mathcal{B}$ has n states and letters a and b are interchanged.

The language accepted by $\mathcal{A}$ is $b^*(ab^*)^{m-1}$, which can easily be shown to be star free. Similarly, we define the automaton $\mathcal{B}$ by taking the automaton defined above but now with n states, and interchange the letters a and b. Hence we obtain the language $L(\mathcal{B}) = a^*(ba^*)^{n-1}$, which is star-free, too.

It is not hard to verify that

$$L(\mathcal{A}) \cap L(\mathcal{B}) = \{\, u \in \{a, b\}^* \mid |u|_a = m - 1 \text{ and } |u|_b = n - 1 \,\}.$$

It remains to be shown that this language needs at least mn states if accepted by an NFA. To this end, consider the following set of pairs

$$S = \{\, (a^i b^j, a^{m-1-i} b^{n-1-j}) \mid 0 \leq i \leq m - 1 \text{ and } 0 \leq j \leq n - 1 \,\}.$$

For each pair $(a^i b^j, a^{m-1-i} b^{n-1-j})$ in S the word $a^i b^j a^{m-1-i} b^{n-1-j}$ has $m - 1$ symbols a and $n - 1$ symbols b. So, it belongs to $L(\mathcal{A}) \cap L(\mathcal{B})$. Next, consider different pairs $(a^i b^j, a^{m-1-i} b^{n-1-j})$ and $(a^{i'} b^{j'}, a^{m-1-i'} b^{n-1-j'})$ from S with $i \neq i'$ or $j \neq j'$. At least one of the words induced by crossed pairs is not in the intersection of the languages accepted by $\mathcal{A}$ and $\mathcal{B}$. Thus, the set S is a fooling set for $L(\mathcal{A}) \cap L(\mathcal{B})$ of size $m \cdot n$, which proves the stated claim. □

Next we come to the complementation operation. Here the situation is a little bit more involved to come up with an NFA accepting a star-free language that meets the tight exponential bound of 2^n states [15,19]. The following easy example already gives an exponential lower bound. Consider the languages $L_n = \{a, b\}^* a \{a, b\}^n b \{a, b\}^*$, for $n \geq 0$, accepted by $(n + 3)$-state NFAs. Obviously, these languages are star free. Moreover, from [15] it is known that any NFA that accepts the complement of L_n needs at least 2^{n-2} states. Thus, we have proven a tight bound in order of magnitude. The question arises, whether one can do better. We answer the question in the affirmative by showing that the language used in [19] (accepted by the NFA depicted in Figure 3) is in fact star-free.

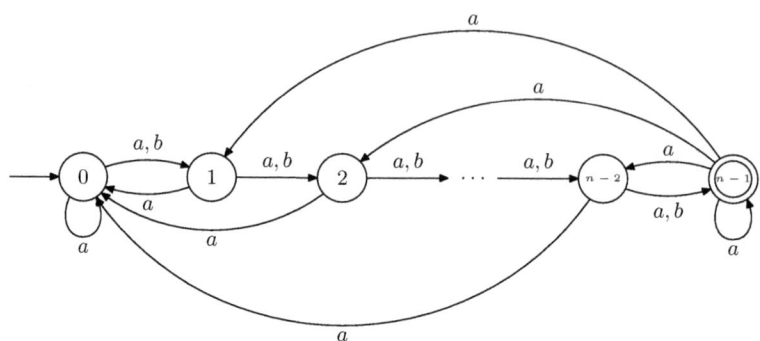

Fig. 3. The n-state NFA $\mathcal{A}$, $n \geq 3$, used for the lower bound on the complementation problem for star-free languages. Any NFA accepting the complement of $L(\mathcal{A})$ needs at least 2^n states.

Before we can start with our investigation we need some additional notation that gives some insights on permutation-free automata that are built by the powerset construction from NFAs [14].

Lemma 4. *Let $\mathcal{A}$ be an NFA with state set S over alphabet Σ, and assume that $\mathcal{A}'$ is the equivalent minimal DFA obtained by the powerset construction, which is non-permutation-free. If the word w in Σ^* induces a non-trivial permutation on the state set $\{P_1, P_2, \ldots, P_k\} \subseteq 2^S$ of $\mathcal{A}'$ such that $\delta'(P_i, w) = P_{i+1}$, for $1 \leq i < k$, and $\delta'(P_k, w) = P_1$, then there are no two states P_i and P_j with $i \neq j$ such that $P_i \subseteq P_j$.*

Now we are prepared for the next theorem.

Theorem 5. *For any integer $n \geq 2$ let $\mathcal{A}$ be an n-state NFA that accepts a star-free language. Then 2^n states are sufficient for an NFA to accept the complement of the language $L(\mathcal{A})$. The bound is tight for binary alphabets.*

Proof. It suffices to prove that the NFA $\mathcal{A} = (S, \Sigma, \delta, 0, F)$, where $\Sigma = \{a, b\}$, $S = \{0, 1, \ldots, n-1\}$, $F = \{n-1\}$, and

$$\delta(i, x) = \begin{cases} \{i+1\}, & \text{if } i < n-1 \text{ and } x = b \\ \{0, i+1\}, & \text{if } i < n-1 \text{ and } x = a \\ \{1, 2, \ldots, n-1\}, & \text{if } i = n-1 \text{ and } x = a \end{cases}$$

that is depicted in Figure 3 accepts a star-free language. To this end, we consider the equivalent minimal DFA $\mathcal{A}' = (S', \Sigma, \delta', \{0\}, F')$ obtained by the powerset construction where $S' \subseteq 2^S$.

The cardinality of the symmetric difference of two states R and T of $\mathcal{A}'$ is denoted by $\langle R, T \rangle = |R \setminus T| + |T \setminus R|$. The outline of the proof is as follows. We assume contrarily that the language accepted by $\mathcal{A}'$ is not star free. Then there exists a word $w \in \{a,b\}^*$ that induces a non-trivial permutation on a subset $P = \{P_1, P_2 \ldots, P_k\}$ of the states of $\mathcal{A}'$ such that $\delta'(P_i, w) = P_{i+1}$, $1 \le i < k$, and $\delta'(P_k, w) = P_1$. Next we consider arbitrary pairs $P_i \neq P_j$ and distinguish whether the state $n-1$ of $\mathcal{A}$ belongs to none of P_i and P_j, to both, or to exactly one of them. In all cases we will derive either a contradiction or a decrease of the cardinality of the symmetric difference. In particular, this shows that the cardinality can never increase. By $\langle P_i, P_j \rangle = \langle \delta'(P_i, w^k), \delta'(P_j, w^k) \rangle = \langle \delta'(P_i, w^{k \cdot m}), \delta'(P_j, w^{k \cdot m}) \rangle$, for $m \ge 0$, this implies a contradiction also in the case of decreasing cardinality.

We first show that w must be at least two letters long in order to induce a non-trivial permutation. If w would be equal to a, then at most $n-1$ applications of δ' to any P_i give a state P_i' which includes $n-1$. At most two more applications of δ' to P_i' give the set $\{0, 1, \ldots, n-1\}$ which includes any state from P. This contradicts Lemma 4. If w would be equal to b, then at most $n-1$ applications of δ' to any P_i give the emptyset, which is a rejecting sink state that can never be part of P.

Now we turn to distinguish the three cases for the occurrence of the state $n-1$. Let $P_i = \{i_1, i_2, \ldots, i_{\ell_i}\}$ and $P_j = \{j_1, j_2, \ldots, j_{\ell_j}\}$ be two arbitrary but different states from P, where $i_1 < i_2 < \cdots < i_{\ell_i}$ and $j_1 < j_2 < \cdots < j_{\ell_j}$.

First we consider the case that the state $n-1$ belongs to exactly one of P_i or P_j. Without loss of generality we assume $n-1 \in P_i \setminus P_j$. If, in this case, the first letter of w is an a we have

$$\delta'(P_i, a) = \begin{cases} \{1, 2, \ldots, n-1\} & \text{if } P_i = \{n-1\} \\ \{0, 1, \ldots, n-1\} & \text{otherwise.} \end{cases}$$

So, a contradiction to Lemma 4 follows if either $P_i \neq \{n-1\}$ or the prefixes of multiples of w start with a and include two a's that are not separated by exactly $n-2$ letters b, because in both cases P_i is transformed into $\{0, 1, \ldots, n-1\}$ which, in turn, includes any state from P. Therefore, we conclude $P_i = \{n-1\}$ which is transformed into $\{1, 2, \ldots, n-1\}$, and in order to avoid a contradiction, P_j must include at least one state from $\{0, 1, \ldots, n-2\}$. If 0 belongs to P_j, then $\delta'(P_i, ab^{n-2}) = \{n-1\}$ and $\delta'(P_j, ab^{n-2}) = \{n-2, n-1\}$, which is a contradiction to Lemma 4. If 0 does not belong to P_j, we consider the evolutions of P_i and P_j under input word ab^{n-2}. Then $\delta'(\{n-1\}, ab^{n-2}) = \{n-1\}$ and $\delta'(\{j_1, \ldots, j_{\ell_j}\}, ab^{n-2}) = \{n-2\}$ and $\delta'(\{n-2\}, ab^{n-2}) = \{n-2\}$. So, the evolution runs into cycles, and there is no way to reach state P_j from P_i and vice versa. Thus, they cannot be part of a non-trivial permutation.

We conclude that in the present case the first letter of w is a b and have

$$\delta'(P_i, b) = \begin{cases} \emptyset, & \text{if } P_i = \{n-1\} \\ \{i_1+1, \ldots, i_{\ell_i-1}+1\} & \text{otherwise} \end{cases}$$

as well as $\delta'(P_j, b) = \{j_1+1, \ldots, j_{\ell_j}+1\}$. Furthermore, P_i must be different from $\{n-1\}$ since otherwise it would be transformed into the emptyset. Together, this implies $\langle \delta'(P_i, b), \delta'(P_j, b) \rangle = \langle P_i, P_j \rangle - 1$. Thus, the cardinality of the symmetric difference is properly decreased. This concludes the case $n-1 \in P_i \setminus P_j$.

For the next case, we assume that state $n-1$ is not included in P_i and P_j, that is, $n-1 \notin P_i \cup P_j$. If the first letter of w is an a, we obtain $\delta'(P_i, a) = \{0, i_1+1, \ldots, i_{\ell_i}+1\}$ and $\delta'(P_j, a) = \{0, j_1+1, \ldots, j_{\ell_j}+1\}$. If the first letter of w is a b, we obtain $\delta'(P_i, b) = \{i_1+1, \ldots, i_{\ell_i}+1\}$ and $\delta'(P_j, b) = \{j_1+1, \ldots, j_{\ell_j}+1\}$. So, the single states belonging to P_i and P_j are "shifted" towards $n-1$. Since P_i and P_j are different, whatever the input is, applications of δ' evolve to a situation where one or both of the new states include $n-1$. Thus, to the case $n-1 \in P_i \setminus P_j$ covered before or to the following case $n-1 \in P_i \cap P_j$.

For the final case, we assume that state $n-1$ belongs to both P_i and P_j, that is, $n-1 \in P_i \cap P_j$. Clearly, now P_i as well as P_j must be different from $\{n-1\}$. Otherwise, one of both would be a subset of the other, which is a contradiction to Lemma 4. If, in the present case, the first letter of w is an a, then P_i and P_j are immediately transformed into $\{0, 1, 2, \ldots, n-1\}$ which causes again a contradiction to Lemma 4. So, we know that the first letter of w is a b. After consuming the letter b, that is, after one transition we have $\delta'(P_i, b) = \{i_1+1, \ldots, i_{\ell_i-1}+1\}$ and $\delta'(P_j, b) = \{j_1+1, \ldots, j_{\ell_j-1}+1\}$. If both new states $\delta'(P_i, b)$ and $\delta'(P_j, b)$ contain $n-1$, we repeat the argumentation of the present case. This means that we will be concerned with another application of δ' on input b, resulting in $\delta'(P_i, bb) = \{i_1+2, \ldots, i_{\ell_i-2}+2\}$ and $\delta'(P_j, bb) = \{j_1+2, \ldots, j_{\ell_j-2}+2\}$. Since P_i and P_j are different and the single states belonging to it are shifted towards $n-1$ during an application of δ' on input b, the argumentation can be repeated until we end up with either both new states do not contain $n-1$, or $n-1$ belongs to exactly one of them. The latter situation has completely been covered by the case $n-1 \in P_i \setminus P_j$ before. The former situation brings us to the case $n-1 \notin P_i \cup P_j$ which, in turn, may end up in the present case again. However, in every possible step of this cycle between both cases, the single states are shifted towards $n-1$. Moreover, only a 0 may additionally be included. So, since P_i and P_j are different, the cycle appears finitely often only. This concludes the case $n-1 \in P_i \cap P_j$ and, hence, the proof. □

In the remainder of this section we investigate the concatenation operation, and its iteration, the Kleene star, as well as the reversal operation. In general, these operations have deterministic state complexity $m \cdot 2^n - 2^{n-1}$, $2^{n-1} + 2^{n-2}$, and 2^n in the worst case, which is also met for star-free languages [8], except for reversal, which is one state less. For NFAs all these operations are cheap, in the sense that $m+n$, $n+1$, and $n+1$ states are sufficient and necessary in the worst case. We show that for star-free languages exactly the same bounds apply. For concatenation we find the following situation.

Theorem 6. *For any integers $m, n \geq 2$ let $\mathcal{A}$ be an m-state and $\mathcal{B}$ be an n-state NFA that accept star-free languages. Then $m+n$ states are sufficient for an NFA to accept the language $L(\mathcal{A}) \cdot L(\mathcal{B})$. The bound is tight for binary alphabets.*

Proof. Again, the upper bound is that for general regular languages [15]. For the lower bound we use the NFAs $\mathcal{A}$ and $\mathcal{B}$ introduced in the proof of Theorem 2. Recall that $L(\mathcal{A}) = a^{m-1}(ba^{m-1})^*$ and $L(\mathcal{B}) = b^{n-1}(ab^{n-1})^*$. In order to show that $m + n$ states are necessary for any NFA to accept the language $L(\mathcal{A}) \cdot L(\mathcal{B})$ we construct the set

$$S = \{\, (a^i, a^{m-1-i}ba^{m-1}b^{n-1}) \mid 0 \leq i \leq m - 1 \,\}$$
$$\cup \{\, (a^{m-1}b^{n-1}ab^j, b^{n-1-j}) \mid 0 \leq j \leq n - 1 \,\},$$

whose fooling set property is verified as follows: Consider pairs of the form $(a^i, a^{m-1-i}ba^{m-1}b^{n-1})$. The word $a^i \cdot a^{m-1-j}ba^{m-1}b^{n-1}$, for $0 \leq i, j \leq m - 1$, is in $L(\mathcal{A}) \cdot L(\mathcal{B})$ if and only if $i = j$. Thus, the pair $(a^i, a^{m-1-i}ba^{m-1}b^{n-1})$ induces a word that belongs to the concatenation of the languages under consideration, but the crossing of different pairs of this form gives two words that are not in $L(\mathcal{A}) \cdot L(\mathcal{B})$. Similarly, we can argue for the pairs $(a^{m-1}b^{n-1}ab^j, b^{n-1-j})$, for $0 \leq j \leq n-1$. Finally, we have to compare pairs of the form $(a^i, a^{m-1-i}ba^{m-1}b^{n-1})$, for $0 \leq i \leq m - 1$, with pairs $(a^{m-1}b^{n-1}ab^j, b^{n-1-j})$, for $0 \leq j \leq n - 1$. Since $a^i \cdot b^{n-1-j}$ belongs to $L(\mathcal{A}) \cdot L(\mathcal{B})$ if and only if $i = m - 1$ and $j = 0$, for the cases $0 \leq i < m - 1$ and $0 < j \leq n - 1$ at least one word induced by crossing the corresponding pairs is not in the concatenation of the languages accepted by the automata $\mathcal{A}$ and $\mathcal{B}$. For the remaining case $i = m - 1$ and $j = 0$ we find that the other induced word $a^{m-1}b^{n-1}ab^j \cdot a^{m-1-i}ba^{m-1}b^{n-1}$ does not belong to $L(\mathcal{A}) \cdot L(\mathcal{B})$. Therefore, S is a fooling set for the language $L(\mathcal{A}) \cdot L(\mathcal{B})$ of size $m + n$. Hence, the stated claim follows.   □

The star-free languages are not closed under Kleene star, which is seen by the finite language a^2. Since the minimal DFA accepting $(a^2)^*$ reads as $\mathcal{A} = (\{0, 1\}, \{a\}, \delta, 0, \{0\})$ with $\delta(0, a) = 1$ and $\delta(1, a) = 0$ and contains a non-trivial permutaton on the state set $\{0, 1\}$ by reading the word a, this language is *not* star-free. Nevertheless, one can consider the corresponding operation problem (leaving the family of star-free languages).

Theorem 7. *For any integer $n \geq 2$ let $\mathcal{A}$ be an n-state NFA that accepts a star-free language. Then $n + 1$ states are sufficient for an NFA to accept the Kleene star of the language of $\mathcal{A}$. This bound is tight for binary alphabets.*

Proof. The upper bound can be found in [15]. For the lower bound we again use the automaton $\mathcal{A}$ introduced in the proof of Theorem 2. Recall, that $L(\mathcal{A}) = a^{n-1}(ba^{n-1})^*$. We claim that

$$S = \{\, (a^{n-1}ba^i, a^{n-1-i}ba^{n-1}) \mid 0 \leq i \leq n - 1 \,\} \cup \{(\lambda, \lambda)\}$$

is a fooling set for $L(\mathcal{A})^*$. In fact, the word $a^{n-1}ba^i \cdot a^{n-1-j}ba^{n-1}$, for $0 \leq i, j \leq n - 1$, is in $L(\mathcal{A})^*$ if and only if $i = j$. Hence, the pair $(a^{n-1}ba^i, a^{n-1-i}ba^{n-1})$

gives a word that belongs to $L(\mathcal{A})^*$, but for different pairs of this form none of the induced words (by crossing) is a member of $L(\mathcal{A})^*$. Obviously, the remaining pair (λ, λ) gives the empty word that is a member of $L(\mathcal{A})^*$ by definition. Finally, for the words $a^{n-1}ba^i \cdot \lambda$ and $\lambda \cdot a^{n-1-i}ba^{n-1}$, for $0 \leq i \leq n-1$, at least one is not in $L(\mathcal{A})^*$. Thus, the pairs $(a^{n-1}ba^i, a^{n-1-i}ba^{n-1})$ from S with the pair (λ, λ) obey the properties required for being a fooling set. Therefore, the claim follows since S is of size $n+1$. $\qquad\square$

Our last result on the reversal operation already follows from the literature [19]. A slight modification of the automaton depicted in Figure 1 was used to show a tight bound of $n+1$ states for the reversal operation on languages accepted by NFAs. The modification simply is to make all states accepting. Since this does not effect the existence of non-trivial permutations on the state set of the minimal DFA which accepts this language, we may conclude that it is star-free. Thus, we obtain the following result.

Theorem 8. *For any integer $n \geq 2$ let $\mathcal{A}$ be an n-state NFA that accepts a star-free language. Then $n+1$ states are sufficient for an NFA to accept the reversal of the language $L(\mathcal{A})$. This bound is tight for binary alphabets.* $\qquad\square$

References

1. Birget, J.C.: Intersection and union of regular languages and state complexity. Inform. Process. Lett. 43, 185–190 (1992)
2. Bordihn, H., Holzer, M., Kutrib, M.: Determinization of finite automata accepting subregular languages. Theoret. Comput. Sci. 410, 3209–3222 (2009)
3. Brzozowski, J.A.: Roots of star events. J. ACM 14, 466–477 (1967)
4. Brzozowski, J.A.: Quotient complexity of regular languages. J. Autom. Lang. Comb. 15, 71–89 (2010)
5. Brzozowski, J.A., Cohen, R.S.: On decompositions of regular events. J. ACM 16, 132–144 (1969)
6. Brzozowski, J.A., Jirásková, G., Li, B., Smith, J.: Quotient complexity of bifix, factor, and subword-free languages. In: International Conference on Automata and Formal Languages (AFL 2011) (to appear)
7. Brzozowski, J.A., Jirásková, G., Zou, C.: Quotient complexity of closed languages. In: Ablayev, F., Mayr, E.W. (eds.) CSR 2010. LNCS, vol. 6072, pp. 84–95. Springer, Heidelberg (2010)
8. Brzozowski, J.A., Liu, B.: Quotient complexity of star-free languages. In: International Conference on Automata and Formal Languages (AFL 2011) (to appear)
9. Chrobak, M.: Finite automata and unary languages. Theoret. Comput. Sci. 47, 149–158 (1986)
10. Chrobak, M.: Errata to finite automata and unary languages. Theoret. Comput. Sci. 302, 497–498 (2003)
11. Glaister, I., Shallit, J.: A lower bound technique for the size of nondeterministic finite automata. Inform. Process. Lett. 59, 75–77 (1996)
12. Han, Y.S., Salomaa, K.: Nondeterministic state complexity for suffix-free regular languages. In: Descriptional Complexity of Formal Systems (DCFS 2010), EPTCS, vol. 31, pp. 189–204 (2010)
13. Han, Y.S., Salomaa, K., Wood, D.: Nondeterministic state complexity of basic operations for prefix-free regular languages. Fund. Inform. 90, 93–106 (2009)

14. Holzer, M., Jakobi, S., Kutrib, M.: The magic number problem for subregular language families. In: Descriptional Complexity of Formal Systems (DCFS 2010), EPTCS, vol. 31, pp. 110–119 (2010)
15. Holzer, M., Kutrib, M.: Nondeterministic descriptional complexity of regular languages. Int. J. Found. Comput. Sci. 14, 1087–1102 (2003)
16. Holzer, M., Kutrib, M.: Nondeterministic finite automata – Recent results on the descriptional and computational complexity. Int. J. Found. Comput. Sci. 20, 563–580 (2009)
17. Holzer, M., Kutrib, M.: Descriptional complexity – An introductory survey. In: Scientific Applications of Language Methods, pp. 1–58. Imperial College Press, London (2010)
18. Holzer, M., Kutrib, M.: Descriptional and computational complexity of finite automata – A survey. Inform. Comput. 209, 456–470 (2011)
19. Jirásková, G.: State complexity of some operations on binary regular languages. Theoret. Comput. Sci. 330, 287–298 (2005)
20. Jirásková, G., Krausová, M.: Complexity in prefix-free regular languages. In: Descriptional Complexity of Formal Systems (DCFS 2010). EPTCS, vol. 31, pp. 197–204 (2010)
21. Jirásková, G., Masopust, T.: Complexity in union-free regular languages. In: Gao, Y., Lu, H., Seki, S., Yu, S. (eds.) DLT 2010. LNCS, vol. 6224, pp. 255–266. Springer, Heidelberg (2010)
22. Jirásková, G., Olejár, P.: State complexity of intersection and union of suffix-free languages and descriptional complexity. In: Non-Classical Models of Automata and Applications (NCMA 2009). books@ocg.at, vol. 256, pp. 151–166. Austrian Computer Society (2009)
23. Lupanov, O.B.: A comparison of two types of finite sources. Problemy Kybernetiki 9, 321–326 (1963) (in Russian); German translation: Über den Vergleich zweier Typen endlicher Quellen. Probleme der Kybernetik 6, 328–335 (1966)
24. McNaughton, R., Papert, S.: Counter-Free Automata, Research Monographs, vol. 65. MIT Press, Cambridge (1971)
25. Meyer, A.R., Fischer, M.J.: Economy of description by automata, grammars, and formal systems. In: Symposium on Switching and Automata Theory (SWAT 1971), pp. 188–191. IEEE, Los Alamitos (1971)
26. Moore, F.R.: On the bounds for state-set size in the proofs of equivalence between deterministic, nondeterministic, and two-way finite automata. IEEE Trans. Comput. 20, 1211–1214 (1971)
27. Rabin, M.O., Scott, D.: Finite automata and their decision problems. IBM J. Res. Dev. 3, 114–125 (1959)
28. Salomaa, K., Yu, S.: NFA to DFA transformation for finite languages over arbitrary alphabets. J. Autom. Lang. Comb. 2, 177–186 (1997)
29. Schützenberger, M.P.: On finite monoids having only trivial subgroups. Inform. Control 8, 190–194 (1965)
30. Shyr, H.J., Thierrin, G.: Ordered automata and associated languages. Tamkang J. Math. 5, 9–20 (1974)
31. Shyr, H.J., Thierrin, G.: Power-separating regular languages. Math. Systems Theory 8, 90–95 (1974)
32. Yu, S.: Regular languages. In: Handbook of Formal Languages, vol. 1, pp. 41–110. Springer, Heidelberg (1997)
33. Yu, S.: State complexity of regular languages. J. Autom. Lang. Comb. 6, 221–234 (2001)

On the Containment and Equivalence Problems for GSMs, Transducers, and Linear CFGs

Oscar H. Ibarra

Department of Computer Science
University of California
Santa Barbara, CA 93106, USA
ibarra@cs.ucsb.edu

Abstract. We explore the boundaries between decidability and undecidability of the containment and equivalence problems for restricted
classes of nondeterministic generalized sequential machines (NGSMs),
nondeterministic finite transducers (NFTs), nondeterministic pushdown
transducers (NPDTs), and linear context-free grammars (LCFGs). We
believe that our results are the sharpest known to date concerning these
devices.

Keywords: generalized sequential machine, finite transducer, pushdown
transducer, linear context-free grammar, containment problem, equivalence problem.

1 Introduction

It is known that it is undecidable to determine, given two NGSMs A_1 and A_2,
whether $R(A_1) \subseteq R(A_2)$ (where $(R(A_i)$ is the input/output relation defined by
A_i) [3]. In fact, the undecidability holds even when the NGSMs have unary output (resp., input) alphabet, since the equivalence problem (is $R(A_1) = R(A_2)$?)
is undecidable even for this special case [6].

We strengthen the undecidability of containment and equivalence. In particular, we show that there is a *fixed* NGSM A with input alphabet Σ and unary
output alphabet $\Delta = \{1\}$ such that it is undecidable to determine, given a pos-
itive integer d, whether $\{(x, 1^{|x|-min\{d,|x|\}}) \mid x \text{ in } \Sigma^*\} \subseteq R(A)$. Note that the
relation on the left is realized by the trivial deterministic generalized sequential
machine (DGSM) with states $q_0, \ldots, q_d$, start state q_0, and transitions: For all a
in Σ, $\delta(q_i, a) = (q_{i+1}, \varepsilon)$ for $0 \leq i \leq d-1$ and $\delta(q_d, a) = (q_d, 1)$. This DGSM has
a "tail" of length d and this length is the only "input" to the decision problem.
Obviously, if d can only come from a finite set of positive integers, the problem
would not be undecidable, since there will only be a finite number of instances.

The result above shows that it is undecidable to determine, given a DGSM A_1
and an NGSM A_2, whether $R(A_1) \subseteq R(A_2)$. However, when A_1 is an NGSM and
A_2 is a DGSM, containment is decidable. In fact, we prove something stronger.

A nondeterministic finite transducer (NFT) is a generalization of an NGSM,
where the machine now has accepting states and can have ε-moves. DFT is

B. Bouchou-Markhoff et al. (Eds.): CIAA 2011, LNCS 6807, pp. 190–202, 2011.

the deterministic version. An NFT (DFT) augmented with a pushdown stack is called an NPDT (DPDT).

An NFT A is output finite-valued if there is a $k \geq 1$ such that for every x, there are most k distinct strings y such that (x, y) is in $R(A)$. Similarly, an NFT A is input finite-valued if there is a $k \geq 1$ such that for every y, there are at most k distinct strings x such that (x, y) is in $R(A)$. A is output/input finite-valued if it can effectively be decomposed into output finite-valued NFT A_1 and input finite-valued NFT A_2 such that $R(A) = R(A_1) \cup R(A_2)$. (Note that A_1 or A_2 may be the trivial NFT realizing the empty relation.) As an example, the relation over input/output alphabet $\{a, b, c\}$, $R = \{(xcy, x) \mid x, y \text{ in } \{a, b\}^+\} \cup \{(x, xcy) \mid x, y \text{ in } \{a, b\}^+\}$, can be realized by an output/input finite-valued NFT (with $k = 1$, i.e., single-valued).

Output finite-valued NFTs have been investigated before, where they were simply called finite-valued NFTs. It is decidable to determine, given an NFT, whether it is output finite-valued [8](resp., output k-valued for a given k [4]). The containment and equivalence problems for output finite-valued NFTs are decidable [2,9]. We show that the following problems are decidable:

1. Given an NFT A, is it input finite-valued (resp., input k-valued for a given k)?
2. Given an NPDT A and an output/input finite-valued NFT B, is $R(A) \subseteq R(B)$?
3. Given a DPDT A and an output/input finite-valued NFT B, is $R(A) = R(B)$?
4. Given a context-free language L and output/input finite-valued NFTs A and B, is $R(A) \subseteq R(B)$ on L? (i.e., is $\{y \mid (x, y) \text{ in } R(A)\} \subseteq \{y \mid (x, y) \text{ in } R(B)\}$ for all x in L?).

Next we look at restricted classes of linear context-free grammars (LCFGs). Let $\$$ be a special symbol. For any alphabet Σ not containing $\$$, let $\Sigma_\$ = \Sigma \cup \{\$\}$. Let $G = \langle V, \Sigma_\$, S, P \rangle$ be an LCFG, where V and $\Sigma_\$$ are the sets of nonterminals and terminals, respectively, S is the start nonterminal, and P is the set of rules. We investigate "marked" LCFGs where the rules in P are of the form: $A \to xBy$ or $A \to \$$ (called a $\$$-rule), where A, B are nonterminals, and x, y are in Σ^*. Thus in any derivation of a string in $L(G)$, the final step is an application of a $\$$-rule. Hence, $L(G) \subseteq \Sigma^* \$ \Sigma^*$.

Note: Throughout the paper, unless otherwise specified, we will refer to "marked" LCFG simply as LCFG.

If x is in Σ^*, let $R_G^x = \{y \mid y \text{ in } \Sigma^*, x\$y \text{ in } L(G)\}$. Similarly, let $L_G^x = \{y \mid y \text{ in } \Sigma^*, y\$x \text{ in } L(G)\}$. G is right-bounded (resp., left-bounded) if there is a positive integer k such that for every x in Σ^*, $|R_G^x| \leq k$ (resp., $|L_G^x| \leq k$). G is right/left-bounded bounded if it can effectively be decomposed into right-bounded LCFG G_1 and left-bounded LCFG G_2 such that $L(G) = L(G_1) \cup L(G_2)$. (Note that G_1 or G_2 may be the trivial LCFG generating the empty language.) As an example, the language over $\Sigma_\$ = \{a, b, c, \$\}$, $L = \{xcy\$x^r \mid x, y \text{ in } \{a, b\}^+\} \cup \{x^r\$xcy \mid x, y \text{ in } \{a, b\}^+\}$, can be generated by a right/left-bounded LCFG G (with $k = 1$).

We show that there exists a *fixed* LCFG G over some alphabet $\Sigma_\$$ containing symbol 1 such that it is undecidable to determine, given a positive integer d, whether $\{x\$1^{|x|-min\{d,|x|\}} \mid x \text{ in } \Sigma^*\} \subseteq L(G)$. Note that the language on the left is generated by a 1-right-bounded LCFG with nonterminals $S_0, \ldots, S_d$, start nonterminal S_0, and the following rules: For all a in Σ, $S_i \to aS_{i+1} \mid \$$ for $0 \le i \le d-1$, $S_d \to aS_d1 \mid \$$.

We also prove that the following problems are decidable:

1. Given an LCFG G, is it right-bounded (resp., left-bounded)?
2. Given an LCFG G is it k-right-bounded (resp., k-left-bounded) for a given k?
3. Given a nondeterministic pushdown automaton M and a right/left-bounded LCFG G, is $L(M) \subseteq L(G)$?
4. Given right/left-bounded LCFGs G_1 and G_2, is $L(G_1) \subseteq L(G_2)$?

In fact, (4) can be made stronger: It is decidable to determine, given a CFL L and right/left-bounded LCFGs G_1 and G_2, whether for all x in L and y in Σ^*, $x\$y$ in $L(G_1)$ implies $x\$y$ in $L(G_2)$.

In contrast to (3), we show that there is a *fixed* 1-turn deterministic pushdown automaton M (i.e., once it pops it can longer push) such that it is undecidable to determine, given a 1-right-bounded LCFG G, whether $L(G) \subseteq L(M)$.

Suppose, we no longer have the special marker symbol $\$$, and we now require that terminal rules are of form $A \to \varepsilon$. Clearly, any LCFG can be converted to one which has only rules of the form $A \to xBy$ or $A \to \varepsilon$. Although, (1) and (2) remain valid for this class of LCFGs, it is open whether (3) and (4) also hold.

Note: Because of space limitation, some proofs are omitted in this version of the paper.

2   Undecidability of Containment and Equivalence

A nondeterministic generalized sequential machine (NGSM) A is a 5-tuple $\langle Q, \Sigma, \Delta, \delta, q_0 \rangle$, where Q is the state set, Σ is the input alphabet, Δ is the output alphabet, q_0 is the start state, and δ is a (transition) function from $Q \times \Sigma$ into the finite subsets of $Q \times \Delta^*$. A move (p, y) in $\delta(q, a)$ means that A in state q on input symbol a outputs y and enters state p. A defines a relation $R(A) = \{(x, y) \mid A$ when started in its start state on input x outputs y and enters some state after scanning all the symbols in x$\}$. A is a DGSM (i.e., deterministic) if $|\delta(q, a)| \le 1$ for all q and a. A has unary output if $\Delta = \{1\}$.

If Σ is an alphabet and d is a positive integer, define the relation $R_\Sigma^d = \{(x, 1^{|x|-min\{d,|x|\}}) \mid x \text{ in } \Sigma^*\}$.

Theorem 1. *There is a fixed NGSM A with input alphabet Σ and unary output alphabet $\Delta = \{1\}$ such that it is undecidable to determine, given a positive integer d, whether $R_\Sigma^d \subseteq R(A)$.*

Proof. (Idea) Let U a single-tape deterministic Turing machine (DTM) with a unary input alphabet that accepts a recursively enumerable set $L \subseteq a^*$ that is not recursive. Hence, it is undecidable to determine, given a unary string a^d on its tape, whether U will halt on a^d. Without loss of generality, we make the following assumptions about U (for technical reasons):

1. The undecidability of halting holds even if we assume that d is a positive odd integer.
2. If U halts on a^d, it halts after at least one move.
3. U can only expand on the right and that when it scans a blank, it must rewrite it by a non-blank symbol. Thus the lengths of the configurations U goes through in the computation are non-decreasing.

Let Q and Γ be the state set and worktape alphabet of U and q_0 be its start state. Note that a is in Γ. We assume that $Q \cap \Gamma = \varnothing$. Let $\Sigma = Q \cup \Gamma \cup \{\#\}$ (where $\#$ is a new symbol).

We construct an NGSM A such that for a given odd $d \geq 1$, $R_\Sigma^d \subseteq R(A)$ if and only if U does not halt on a^d. Given input w, A nondeterministically selects (I) or (II) below to process:

(I) (Case: $d \geq |w|$.) Here, A scans the input w outputting ε for every symbol it sees until it falls off the end of the string. Thus, when $d \geq |w|$, A outputs ε on input w.

(II) (Case: $|w| > d$.) On an input w with $|w| > d$, A operates in such a way that it outputs $1^{|w|-d}$ if and only if w is not of the form:

$$ID_1 \# a^d \# ID_2 \# a^d \# ... \# a^d \# ID_k \#$$

for some $k \geq 3$, $ID_1 = ID_2 = q_0 a^d$ (the initial configuration of U), ID_k is a halting configuration, and $(ID_2, ID_3, ..., ID_k)$ is a halting sequence of configurations of U on input a^d i.e., configuration ID_{i+1} is a valid successor of ID_i. To do this, A nondeterministically selects one of subcases (a) - (d) below to verify. Because of space limitation, we do not include the description of how each subcase is accomplished by A.

(a) (Subcase: w does not have the correct format, i.e, it is not the form:
$$ID_1 \# a^{d_1} \# ID_2 \# a^{d_2} \# ... \# a^{d_{k-1}} \# ID_k \#$$
for some $k \geq 3$ and configurations $ID_1, ..., ID_k$, where $ID_1 = q_0 a^r$ and $ID_2 = q_0 a^s$.)

(b) (Subcase: $d_i \neq d$ for some $1 \leq i \leq k-1$.) There are two situations, which A chooses nondeterministically to verify: $d_i > d$ or $d_i < d$.

(c) (Subcase: $ID_1 \neq q_0 a^d$ or $ID_2 \neq q_0 a^d$.) We assume from Subcase (a) that $ID_1 = q_0 a^r$ and $ID_2 = q_0 a^s$. Then, as in Subcase (b), A can check if $r \neq d$ or $s \neq d$.

(d) (Subcase: w has the correct format, i.e.,
$$w = ID_1 \# a^d \# ID_2 \# a^d \# ... \# a^d \# ID_k, \text{ where } k \geq 3,$$
$$ID_1 = ID_2 = q_0 a^d.)$$

That is, on input w, A must output $1^{|w|-d}$ if w is not a halting sequence of configurations of U. This is where we need the fact that $ID_1 = ID_2 = q_0 a^d$ in the string w. This makes the construction for this subcase a bit easier.

It can be verified that $R_\Sigma^d \subseteq R(A)$ if and only if U does not halt on a^d. $\qquad\square$

There is an equivalent formulation of Theorem 1. If A is a unary-output NGSM with input alphabet Σ and start state q_0, and d is a positive integer, let A^d be a unary-output NGSM obtained from A by adding the following transitions to A, where $s_0, \ldots, s_d$ are new states and s_0 is the start state: For all a in Σ, $\delta(s_0, a) = \{(q_0, 1^{d+1}), (s_1, 1)\}$ and $\delta(s_i, a) = \{(s_{i+1}, 1)\}$ for $1 \le i \le d-1$. Clearly, $R(A^d) = \{(x, 1^d y) \mid (x, y) \text{ in } R(A)\} \cup \{(w, 1^{|w|}) \mid w \text{ in } \Sigma^*, |w| \le d\}$.

Define $U_\Sigma = \{(w, 1^{|w|}) \mid w \text{ in } \Sigma^*\}$. Note that this relation is realized by the trivial 1-state DGSM with transition $\delta(q_0, a) = (q_0, 1)$ for all a in Σ.

Theorem 2. *There is a fixed unary-output NGSM A such that it is undecidable to determine, given a positive integer d, whether $U_\Sigma \subseteq R(A^d)$.*

From Theorems 1 and 2, it follows that it is undecidable to determine, given a unary-output DGSM A_1 and a unary-output NGSM A_2, whether $R(A_1) \subseteq R(A_2)$. However, the problem is decidable when A_1 is an NGSM and A_2 is a DGSM. In fact, we will prove stronger results in Section 3.

Let Σ be an alphabet containing (among other symbols) $a, b, s, \#$. For a positive integer d, let $L_d = \{b\#a^1b\#a^2b\# \cdots \#a^{d-1}b\#a^d s\#\}\Sigma^*$. Clearly, L_d is regular. Hence, $\overline{L}_d$ (the complement of L_d) is also regular and can be accepted by a simple DFA M_d.

Now define an NGSMA to be an NGSM with accepting states. If A is an NGSM over input alphabet Σ (hence it contains symbols $a, b, s, \#$) and unary output alphabet $\Delta = \{1\}$, let A^d be an NGSMA obtained from A where the inputs are constrained to come from $\overline{L}_d$. Clearly, A^d can be easily constructed from A using the DFA M_d.

Notation: For an alphabet Σ, let $S_\Sigma = \{(x, 1^n) \mid n \ge 1, \text{ for some } x_1, x_2, x_3 \text{ in } \Sigma^*, x = x_1 x_2 x_3 \ne \varepsilon \text{ and } n = |x_1| + 2|x_2| + 3|x_3|\}$. Clearly, S_Σ can be realized by a 1-state NGSMA with transition $\delta(q_0, a) = \{(q_0, 1^k) \mid k = 1, 2, 3\}$, and q_0 is accepting.

Theorem 3. *There is a fixed NGSM A over input alphabet Σ containing symbols $a, b, s, \#$ (and other symbols) and unary output alphabet $\Delta = \{1\}$, such that it is undecidable to determine, given a positive integer d, whether $R(A^d) = S_\Sigma$.*

3 Output Finite-Valued NFTs

A nondeterministic finite transducer (NFT) is a generalization of an NGSMA in that ε-moves on the input are allowed, i.e., the transition function δ is now from $Q \times (\Sigma \cup \{\varepsilon\})$ into the finite subsets of $Q \times \Delta^*$. Note that the machine has accepting states, and the relation realized by A is now $R(A) = \{(x,y) \mid A$ when started in the start state on input x outputs y and enters some accepting state after scanning all the symbols of x$\}$. A is a deterministic finite transducer (DFT) if $|\delta(q,a)| \leq 1$ for all q in Q and a in $\Sigma \cup \{\varepsilon\}$; moreover, if $\delta(a, \varepsilon) \neq \varnothing$, then $\delta(q,a) = \varnothing$ for all a in Σ. Note that any NFT (DFT) can be normalized in that at each step, the output is in $\Delta \cup \{\varepsilon\}$ (the input at each step is in $\Sigma \cup \{\varepsilon\}$.)

An NFT (DFT) augmented with a pushdown stack is called an NPDT (DPDT). An NFT (NPDT) A is *output k-valued* if for each x, there are at most k distinct strings y such that (x,y) is in $R(A)$. It is *output finite-valued* if it is output k-valued for some k. When $k = 1$ we use the term *output single-valued*. Note that every DFT (DPDT) is output single-valued, but the converse is not true in general.

We will also be using the following notations: DFA – deterministic finite automaton, NFA – nondeterministic finite automaton, DPDA – deterministic pushdown automaton, NPDA – nondeterministic pushdown automaton.

It is important to note that the machines above are allowed to make ε-moves. However, for the deterministic versions, we require that if for a given state q (and topmost stack symbol in the case when there is a stack), there is a move on ε, then there is no move for all input symbols.

We will need a result concerning NPDAs augmented with 1-reversal counters. A nondeterministic multicounter machine (NCM) M is an NFA augmented with multiple counters which are initially set to zero. At each step, every counter can be incremented by 1, decremented by 1, or left unchanged, and can be tested for zero. A zero counter cannot be decremented. M is 1-reversal if it has the property that once a counter is decremented, it can no longer be incremented. A 1-reversal NCM augmented with a pushdown stack is called a 1-reversal NPCM. The deterministic versions are called 1-reversal DCM and 1-reversal DPCM, respectively. The following result is known [5]:

Theorem 4. *The emptiness problem (given A, is $L(A) = \varnothing$?) for 1-reversal NPCMs (NCMs) is decidable.*

To illustrate the basic ideas, we first consider the output single-valued (i.e., 1-valued) case. Define $domain(R) = \{x \mid (x,y)$ is in R for some $y\}$.

Theorem 5

1. It is decidable to determine, given an NPDT A_1 and an output single-valued NFT A_2, whether $R(A_1) \subseteq R(A_2)$.
2. It is decidable to determine, given an NFT A_1 and a DPDT A_2, whether $R(A_1) \subseteq R(A_2)$.
3. It is decidable to determine, given a DPDT A_1 and an output single-valued NFT A_2, whether $R(A_1) = R(A_2)$.

Proof. Let M_1 be an NPDA such that $L(M_1) = domain(R(A_1))$ and M_2 be a DFA such that $L(M_2) = domain(R(A_2))$. Then we can effectively construct M_1 and M_2 and check if $L(M_1) \subseteq L(M_2)$, which is is decidable. If $L(M_1) \not\subseteq L(M_2)$, then $R(A_1) \not\subseteq R(A_2)$. Otherwise, we proceed as follows.

Clearly, $R(A_1) \nsubseteq R(A_2)$ if and only if for some x, there are distinct y and z such that (x, y) is in $R(A_1)$ and (x, z) is in $R(A_2)$. We construct an NCM M to check this condition.

M has two 1-reversal counters. Given input x, M simulates the computation of A_1 but suppresses and does not record the outputs of A_1. In parallel, M also simulates the computation of A_2 and also suppresses and does not record the outputs of A_2. Thus M simulates the computation of A_1 which yields (x, y) for some y (may not be unique) and the computation of A_2 which yields (x, z) for some unique z (since A_2 is single-valued). If A_1 and A_2 enter accepting states after processing x, M accepts provided a certain condition is met, which we now describe.

During the simulation of A_1 and A_2, M guesses a discrepancy in y and z. M uses counter c_1 to record a nondeterministically chosen location r in string y_1 and remembers in the state the symbol, say a, in that location. Similarly, M uses counter c_2 to record a nondeterministically chosen location s in string y_2 and remembers in the state that symbol, say b, in that location, making sure that $a \neq b$. When A_1 and A_2 accept (x, y_1) and (x, y_2) respectively, M checks that $r = s$ by decrementing the counters c_1 and c_2 simultaneously verifying that they become zero at the same time. If so, N accepts; otherwise M rejects.

It follows that $R(A_1) \nsubseteq R(M_2)$ if and only if $L(M) \neq \varnothing$. Note that M makes ε-moves when it is checking that $r = s$. The result follows, since the emptiness problem for 1-reversal NPCMs is decidable (Theorem 4).

The proof of Part 2 is similar to that of Part 1, noting that it is decidable to determine, given an NFA M_1 and DPDA M_2, whether $L(M_1) \subseteq L(M_2)$. Part 3 follows from parts (1) and (2). □

Part 3 in Theorem 5 is not true when A_1 is an NPDT. To see this, suppose M_1 is a nondeterministic counter machine acceptor which makes 1-reversal on its counter (i.e., once the counter decrements it can no longer increment; hence, M_1 is a simple NPDA), and Σ is its input alphabet. Let M_2 be a DFA accepting Σ^*. Clearly, we can construct from M_1 and M_2, NPDT A_1 and DFT A_2 such that $R(A_1) = L(M_1) \times \{\varepsilon\}$ and $R(A_2) = L(M_2) \times \{\varepsilon\}$. Then $R(A_1) = R(A_2)$ if and only if $L(M_1) = L(M_2) = \Sigma^*$. However, the universe problem for nondeterministic 1-reversal counter machines is undecidable. On the other hand, if we restrict the problem to only NPDT A_1 satisfying the property that $domain(A_1) = \Sigma^*$, then we can show that Part 3 holds.

Similarly, Part 2 is not true when A_1 is also a DPDT. To see this, let M_1 and M_2 be two DPDAs. We can construct from M_1 and M_2 DPDTs A_1 and A_2 such that $R(A_1) = L(M_1) \times \{\varepsilon\}$ and $R(A_2) = L(M_2) \times \{\varepsilon\}$. Then $R(A_1) \subseteq R(A_2)$ if and only if $L(M_1) \subseteq L(M_2)$, which is undecidable, since the containment problem for DPDAs is undecidable.

We will need the following result in [9]:

Theorem 6. *There is an algorithm which, when given an NFT A, decides if A is output finite-valued and if so, constructs n output single-valued NFTs $A^1, \ldots, A^n$ (for some n) such that $R(A) = R(A^1) \cup \ldots \cup R(A^n)$.*

We now generalize Parts 1 and 3 of Theorem 5.

Theorem 7

1. *It is decidable to determine, given an NPDT A_1 and an output finite-valued NFT A_2, whether $R(A_1) \subseteq R(A_2)$.*
2. *It is decidable to determine, given a DPDT A_1 and an output finite-valued NFT A_2, whether $R(A_1) = R(A_2)$.*

Proof. Clearly (2) follows from (1) and Part 2 of Theorem 5. We now prove (1). From Theorem 6, we can construct n output single-valued NFTs $A_2^1, \ldots, A_2^n$ (for some n) such that $R(A_2) = R(A_2^1) \cup \ldots \cup R(A_2^n)$. We can also construct a DFA M_2 such that $L(M_2) = domain(R(A_2))$, and DFA M_2^i such that $L(M_2^i) = domain(R(A_2^i))$ for $1 \le i \le n$.

Claim: $R(A_1) \not\subseteq R(A_2)$ if and only if there exists an x such that the following two conditions are satisfied:

(a) For some y, (x, y) is in $R(A_1)$.
(b) x is not in $domain(R(A_2))$, or
 x is in $domain(R(A_2))$, and for $1 \le i \le n$, if x is in $domain(R(A_2^i))$ (note that there would be at least one such i), then there is a z_i such that (x, z_i) is in $R(A_2^i)$ and $y \ne z_i$.

To prove the Claim, suppose there is an x such that for some y, (x, y) is in $R(A_1)$. We consider two cases:

Case 1: x is not in $domain(R(A_2))$. Then (x, y) is not in $R(A_2)$.
Case 2: x is in $domain(R(A_2))$, and for $1 \le i \le n$, if x is in $domain(R(A_2^i))$, then there is a z_i such that (x, z_i) is in $R(A_2^i)$ and $y \ne z_i$. Since $R(A_2^i)$ is output single-valued, (x, y) cannot be in $R(A_2^i)$. Hence, (x, y) is not in $R(A_2)$.

Hence, in both cases, $R(A_1) \not\subseteq R(A_2)$.
Conversely, suppose $R(A_1) \not\subseteq R(A_2)$. Then there is some (x, y) in $R(A_1)$ that is not in $R(A_2)$. Then either x is not in $domain(R(A_2))$, or if x is in $domain(R(A_2))$, for $1 \le i \le n$, if x is in $domain(R(A_2^i))$, then there is a z_i such that (x, z_i) is in $R(A_2^i)$ and $y \ne z_i$, since (x, y) is not in $R(A_2)$.
Finally, we construct from NPDT A_1 and the output single-valued NFTs $A_2^1, \ldots, A_2^n$ and DFAs $M_2, M_2^1, \ldots, M_2^n$ a 1-reversal NPCM M with $2n$ 1-reversal counters. Given an input x, M accepts x if conditions (a) and (b) of the Claim above are satisfied. Hence $L(M) \ne \varnothing$ if and only if $R(A_1) \not\subseteq R(A_2)$.
The construction of M generalizes the idea in the proof of Theorem 5. Since x may be in the domain of all the A_2^i's (i.e., accepted by all the DFAs M_2^i's), and M needs to check that y is different from all the z_i's, M may need to use $2n$ 1-reversal counters. Note that the DFA M_2^i is used to determine if there exists a z_i such that (x, z_i) is in $R(A_2^i)$. We omit the details. $\qquad\square$

4   Right-Bounded LCFGs

Throughout the paper, by LCFG, we mean "marked" LCFG. Recall that for any alphabet Σ not containing the special symbol \$, $\Sigma_\$$ denotes the alphabet $\Sigma \cup \{\$\}$. are over the alphabet $\Sigma_\$$. We begin with the following result.

Theorem 8. *It is decidable to determine, given an LCFG G, whether G is right-bounded (resp., k-right-bounded for a given k).*

Proof. Given an LCFG G, we can effectively construct an NFT A such that $R(A) = \{(x\$, y) \mid x\$y^r \text{ is in } L(G)\}$. A, when given input $x\$$ simulates a derivation $S \Rightarrow^* xBy^r \Rightarrow x\$y^r$ (where the last step is an application of a \$-rule, $B \to \$$) as follows: A keeps track of the current nonterminal in the derivation, starting with S. When in the derivation a rule of the form $C \to uDv$ is applied, A reads u and outputs v^r, and remembers D. When a rule $B \to \$$ is finally applied, A reads $\$$ on the input, outputs ε, and enters an accepting state. Clearly, A is finite-valued (resp., k-valued) if and only if G is right-bounded (resp., k-right-bounded). The result follows, since it is decidable to determine, given an arbitrary NFT, whether it is finite-valued [8] (resp. k-valued [4]).   □

The following result follows from Theorems 1 and 2 and the construction in the proof of Theorem 8.

Theorem 9. *There exists a fixed LCFG G over some terminal alphabet $\Sigma_\$$ containing 1 whose rules are of the form: $A \to aB1^k$ or $A \to \$$ where a is in Σ, $k \geq 0$, and A, B are nonterminals, such that it is undecidable to determine, given a positive integer d:*

1. *Whether $\{x\$1^{|x|-min\{d,|x|\}} \mid x \text{ in } \Sigma^*\} \subseteq L(G)$.*
   (Note that the language on the left is generated by a simple 1-right-bounded LCFG with nonterminals $S_0, \ldots, S_d$, start nonterminal S_0, and the following rules: For all a in Σ, $S_i \to aS_{i+1} \mid \$$ for $0 \leq i \leq d-1$ and $S_d \to aS_d1 \mid \$$.)
2. *Whether $\{x\$1^{|x|} \mid x \text{ is in } \Sigma^*\} \subseteq L(G^d)$, where G^d is an LCFG obtained from G with start nonterminal S by adding the following rules where $A_0, \ldots, A_d$ are new nonterminals with A_0 the new start nonterminal: For all a in Σ, $A_0 \to aS1^{d+1} \mid aA_11$, and $A_i \to aA_{i+1}1$ for $1 \leq i \leq d-1$.*
   (Note that language on the left is generated by the trivial 1-right-bounded LCFG with one nonterminal S with the following rules: $S \to aS1$, $S \to \$$.)

The result above is not true when G is right-bounded, since we can prove the following theorem. The proof (which is rather involved) is a modification of our constructions for NFTs in the previous section, using the fact that if G is an LCFG, we can effectively construct an NFT A such that $R(A) = \{(x\$, y) \mid x\$y^r \text{ is in } L(G)\}$ (see the proof of Theorem 8).

Theorem 10. *It is decidable to determine, given an NPDA M and a right-bounded LCFG G, whether $L(M) \subseteq L(G)$.*

As in the proof of Theorem 1, let U be a single-tape DTM with a unary input alphabet that accepts a recursively enumerable set $L \subseteq a^*$ that is not recursive. Without loss of generality, assume that if U halts on a^d, it halts after at least four moves. Let Q and Γ be the state set and worktape alphabet of U and q_0 be its start state. Note that a is in Γ. Assume that $Q \cap \Gamma = \varnothing$. Let $\Sigma = Q \cup \Gamma \cup \{\#\}$ (where $\#$ is a new symbol).

Let L_1 be the language consisting of strings of the form:

$$ID_1\#ID_3\#...\#ID_{k-1}\$ID_k^r\#ID_{k-2}^r\#...\#ID_4^r\#ID_2^r$$

for some even $k \geq 4$, $ID_1 = q_0a^e$ (where e is a positive integer and q_0 the initial state of U), $ID_2, ..., ID_k$ are configurations of U, ID_k is a halting configuration, and for odd i, ID_{i+1} is the successor of ID_i. (Note that r denotes reverse.)

Thus, L_1 is the set of halting computations of U on unary inputs a^e, where e is any positive odd integer, of a special form [1]. Similarly, let L_2 be the language consisting of strings of the above form, but for even i, ID_{i+1} is the successor of ID_i.

For a given odd integer d, denote by L_1^d the language L_1, where e in ID_1 is set to d (i.e., q_0a^e becomes q_0a^d). Note that L_1, L_1^d, and L_2 can be be generated by LCFGs G_1, G_1^d, G_2 (and they can be also be accepted by 1-turn DPDAs). In fact, G_1 and G_1^d can be constructed to be 1-right-bounded.

Let $L_3 = \overline{L}_2$ (complement of L_2). Since L_2 can be accepted by a 1-turn DPDA, L_3 can also be accepted by a 1-turn DPDA, M_3.

Theorem 11. *There is a fixed 1-turn DPDA M_3 such that it is undecidable to determine, given a positive odd integer d, whether $L(G_1^d) \subseteq L(M)$.*

Proof. Clearly, for a given positive integer d, $L(G_1^d) \subseteq L(M_3)$ if and only if $L(G_1^d) \cap L(G_2) = \varnothing$, which is undecidable, since otherwise, we can decide if the single-tape DTM U halts on input a^d. $\square$

Note that not only is the 1-turn DPDA M fixed, but the 1-right-bounded LCFG G_1^d has also a fixed "template", and the only parameter is d.

From the above theorem we see that it is undecidable to determine, given a 1-right-bounded LCFG G and a 1-turn DPDA M, whether $L(G) \subseteq L(M)$. On the other hand, from Theorem 10, it is decidable to determine, given a 1-turn DPDA M and a 1-right-bounded LCFG G, whether $L(M) \subseteq L(G)$.

Open Question: Is it decidable to determine, given a 1-right-bounded LCFG G and a 1-turn DPDA M, whether $L(G) = L(M)$?

However, we have:

Theorem 12. *It is decidable to determine, given an NPDA M_1 and a 1-reversal DCM (i.e., a DFA with a finite number of 1-reversal counters) M_2, whether $L(M_1) \subseteq L(M_2)$.*

Proof. Construct from M_2 a 1-reversal DCM M_3 such that $L(M_3) = \overline{L}(M_2)$. This is possible, since 1-reversal DCM languages are effectively closed under complementation [5]. We then construct, from M_1 and M_3, a 1-reversal NPCM M (i.e., an NPDA with a finite number of 1-reversal counters) M_2, which when given an input string x simulates M_1 and M_3 (in parallel) and accepts if M_1 and M_3 accept. Then $L(M_1) \nsubseteq L(M_2)$ if and only if $L(M) \neq \varnothing$, which is decidable, since the emptiness problem for 1-reversal NPCMs is decidable [5]. $\square$

Corollary 1. *The containment and equivalence of right-bounded LCFGs are decidable.*

Proof. It is sufficient to show that containment is decidable. Given right-bounded LCFGs G_1 and G_2, we construct a (1-turn) NPDA M_1 accepting $L(G_1)$. Then by Theorem 10, we can check if $L(M_1) \subseteq L(G_2)$. $\square$

The above corollary can be made stronger. Let L be a CFL and G_1 and G_2 be LCFGs. We say that $L(G_1) \subseteq L(G_2)$ on L if for all x in L, if $x\$y$ is in $L(G_1)$, then $x\$y$ is also in $L(G_2)$. $L(G_1) = L(G_2)$ on L if $L(G_1) \subseteq L(G_2)$ on L and $L(G_2) \subseteq L(G_1)$ on L.

Corollary 2. *It is decidable to determine, given a CFL L and right-bounded LCFGs G_1 and G_2, whether $L(G_1) \subseteq L(G_2)$ on L (resp., $L(G_1) = L(G_2)$ on L).*

5 Output/Input Finite-Valued NFTs

Let A be an NFT with start state q_0, which can be decomposed into two (state-) disjoint NFTs A_1 and A_2 with start states q_{01} and q_{02}, respectively. There are only two transitions from q_0: transitions on ε with output ε to q_{01} and q_{02}. Thus $R(A) = R(A_1) \cup R(A_2)$. (Note that A_1 or A_2 may be the trivial NFT defining the empty relation.) If A_1 is output finite-valued (resp., output k-valued) and A_2 is input finite-valued (resp., input k-valued), then was say that A is output/input finite-valued (resp., output/inputt k-valued). First we note:

Theorem 13. *It is decidable to determine, given an NFT A, if it is input finite-valued (resp., input k-valued for a given k).*

Proof. If A is an NFT, then the set of tuples $\{(y,x) \mid (x,y) \text{ is in } R(A)\}$ can effectively be realized by an NFT A', and A' is output finite-valued (resp., output k-valued for a given k) if and only if A is input finite-valued (resp., input k-valued). The former is decidable [8] (resp., [4]). $\square$

We can prove:

Theorem 14. *It is decidable to determine, given an NPDT A and an output/input finite-valued NFT B, whether $R(A) \subseteq R(B)$.*

Corollary 3. *The following problems are decidable:*

1. *Given a DPDT A_1 and an output/input finite-valued NFT A_2, is $R(A_1) = R(A_2)$?*
2. *Given a context-free language (CFL) L and output/input finite-valued NFTs A_1 and A_2, is $R(A_1) \subseteq R(A_2)$ on L? (resp., is $R(A_1) = R(A_2)$ on L?)*

Proof. Part 1 follows from Theorem 14 and Part 2 of Theorem 5.

For Part 2, let M_L be an NPDA accepting L. Clearly, we can construct from M and NFT A_1 an NPDT A_L such that $R(A_L) = \{(x,y) \mid (x,y) \text{ in } R(A_1) \text{ and } x \text{ in } L\}$. Then $R(A_1) \subseteq R(A_2)$ on L if and only if $R(A_L) \subseteq R(A_2)$, which is decidable by Theorem 14. $\square$

6   Right/Left-Bounded LCFGs

Recall that if G is an LCFG and x in Σ^*, $L_G^x = \{y \mid y$ in $\Sigma^*, y\$x$ in $L(G)\}$. G is left-bounded if there is a positive integer k such that for every x in Σ^*, $|L_G^x| \le k$.

Theorem 15. *It is decidable, given an LCFG G, whether it is left-bounded (resp., k-left-bounded for a given k).*

Proof. This follows from the results in Section 4 and the following observation: Given an LCFG G, denote by G^r, the LCFG obtained from G by "reversing" the right-hand sides of the rules in G, i.e., a rule of the form $A \to \alpha$ in G becomes a rule $A \to \alpha^r$ in G^r. Then, $L(G^r) = (L(G))^r$, and G^r is right-bounded if and only if G is left-bounded.   □

We can prove:

Theorem 16.   1. *It is decidable to determine, given an NPDA M and a right/left-bounded LCFG G, whether $L(M) \subseteq L(G)$.*
2. *It is decidable to determine, given a CFL L and right/left-bounded LCFGs G_1 and G_2, whether $L(G_1) \subseteq L(G_2)$ on L? (resp., is $L(G_1) = L(G_2)$ on L?).*

7   Generalizations

Consider the following more general definition of valuedness. An NFT A is fully k-valued if for every (x, y), either there at most k strings y' such that (x, y') is in $R(A)$, or there at most k strings x' such that (x', y) is in $R(A)$.

Theorem 17. *The following problems are decidable:*

1. *It is decidable to determine, given an NFT A and a positive integer k, whether A is fully k-valued.*
2. *Given an NPDT A_1 and a fully 1-valued NFT A_2, is $R(A_1) \subseteq R(A_2)$?*
3. *Given a CFL L and fully 1-valued NFTs A_1 and A_2, is $R(A_1) \not\subseteq R(A_2)$ on L? (resp., is $R(A_1) = R(A_2)$ on L?)*

Open Questions: (1)Is it decidable to determine, given an NFT, whether it is fully k-valued for some k? (2)Are the containment and equivalence problems for fully k-valued NFTs decidable for $k \ge 2$?

A LCFG G is fully k-bounded if for every x in Σ^*, $|R_G^x| \le k$ or $|L_G^x| \le k$.

Theorem 18. *The following problems are decidable:*

1. *It is decidable to determine, given a LCF G and a positive integer k, whether G is fully k-bounded for some k.*
2. *Given an NPDA M and a fully 1-bounded LCFG G, is $L(M) \subseteq L(G)$?*
3. *Given a CFL L and fully 1-bounded LCFGs G_1 and G_2, is $L(G_1) = L(G_2)$ on L? (resp., is $L(G_1) = L(G_2)$ on L?).*

Open Questions: (1)Is it decidable to determine, given an LCFG, whether it is fully k-bounded for some k? (2)Are the containment and equivalence problems for fully k-bounded LCFGs decidable for $k \ge 2$?

References

1. Baker, B., Book, R.: Reversal-bounded multipushdown machines. J. Comput. Syst. Sci. 8, 315–322 (1974)
2. Culik, K., Karhumaki, J.: The equivalence of finite valued transducers (on HDTOL languages) is decidable. Theoret. Comput. Sci. 47, 71–84 (1986)
3. Griffiths, T.: The unsolvability of the equivalence problem for Λ-free nondeterministic generalized sequential machines. J. Assoc. Comput. Mach. 15, 409–413 (1968)
4. Gurari, E., Ibarra, O.H.: A note on finite-valued and finitely ambiguous transducers. Math. Systems Theory 16, 61–66 (1983)
5. Ibarra, O.H.: Reversal-bounded multicounter machines and their decision problems. J. Assoc. Comput. Mach. 25, 116–133 (1978)
6. Ibarra, O.H.: The unsolvability of the equivalence problem for ε-free NGSM's with unary input (output) alphabet and applications. SIAM J. Computing 7, 524–532 (1978)
7. Minsky, M.: Recursive unsolvability of Post's problem of Tag and other topics in the theory of Turing machines, Ann. of Math. 74, 437–455 (1961)
8. Weber, A.: On the valuedness of finite transducers. Acta Inform. 27, 749–780 (1990)
9. Weber, A.: Decomposing finite-valued transducers and deciding their equivalence. SIAM. J. on Computing 22, 175–202 (1993)

Computing All ℓ-Cover Automata Fast[*]

Artur Jeż[1],[**] and Andreas Maletti[2],[***]

[1] Institute of Computer Science, University of Wrocław
ul. Joliot-Curie 15, 50–383 Wrocław, Poland
`aje@cs.uni.wroc.pl`
[2] Institute for Natural Language Processing, Universität Stuttgart
Azenbergstraße 12, 70174 Stuttgart, Germany
`andreas.maletti@ims.uni-stuttgart.de`

Abstract. Given a language L and a number ℓ, an ℓ-cover automaton for L is a DFA M such that its language coincides with L on all words of length at most ℓ. It is known that an equivalent minimal ℓ-cover automaton can be constructed in time $\mathcal{O}(n \log n)$, where n is the number of states of M. This is achieved by a clever and sophisticated variant of HOPCROFT's algorithm, which computes the ℓ-similarity inside the main algorithm. This contribution presents an alternative simple algorithm with running time $\mathcal{O}(n \log n)$, in which the computation is split into three phases. First, a compact representation of the gap table is created. Second, this representation is enriched with information about the length of a shortest word leading to the states. These two steps are independent of the parameter ℓ. Third, the ℓ-similarity is extracted by simple comparisons against ℓ. In particular, this approach allows the calculation of all the sizes of minimal ℓ-cover automata (for all valid ℓ) in the same time bound.

1 Introduction

Deterministic finite automata (DFA) are widely used in computer science due to their simplicity and flexibility. Their minimisation is one of the oldest problems that is motivated both theoretically and practically and almost every DFA toolkit implements it. More precisely, the DFA minimisation problem asks for a smallest DFA that recognises the same language as a given input DFA M. The asymptotically best solution is due to HOPCROFT [9,7], who presented an $\mathcal{O}(n \log n)$ algorithm where n is the number of states of M. Whether an asymptotically faster algorithm exists, remains one of the most challenging open questions in the area. In many applications the desired language L is finite. It was

---

[*] This work was done when A. Maletti was visiting Wrocław University thanks to the support of the "Visiting Professors" programme of the Municipality of Wrocław.
[**] Supported by the MNiSW grant N206 492638 2010–2012 and by the Young Researcher scholarship of University of Wrocław.
[***] Supported by the German Research Foundation (DFG) grant MA/4959/1-1.

B. Bouchou-Markhoff et al. (Eds.): CIAA 2011, LNCS 6807, pp. 203–214, 2011.

observed in [3] that membership of a word w in L can then be decided by:
(i) checking whether w is short (i.e., $|w| \leq \ell$ where $\ell = \max \{ |u| : u \in L \}$) and
(ii) checking it with a DFA M. This allows M to accept words that are longer
than ℓ, which yields that M need not recognise L. Thus, we arrive at the notion
of 'cover automata'. We say that a DFA M is a *deterministic finite cover automaton* (DFCA or *cover automaton*) for a finite language L if $L(M) \cap \Sigma^{\leq \ell} = L$,
where $\ell = \max\{ |u| : u \in L \}$ and $\Sigma^{\leq \ell}$ contains all words of length at most ℓ. It
is a minimal DFCA for L if no DFCA for L has (strictly) fewer states.

It is well-known that the minimal DFCA for L can be substantially smaller
than the minimal DFA for L. Already [3] presents a DFCA minimisation algorithm that runs in time $\mathcal{O}(n^2 \cdot \ell^2)$. It also allowed the input language to be
presented as a DFA M, which could potentially recognise an infinite language.
In that case, an explicit word length ℓ needs to be supplied. An ℓ-DFCA for M
is simply a DFCA for $L(M) \cap \Sigma^{\leq \ell}$. CÂMPEANU et al. [2] improved the minimisation algorithm for finite languages to $\mathcal{O}(n^2)$. Their algorithm can be trivially
extended to arbitrary DFA, but it then runs in time $\mathcal{O}(n^2 \cdot \ell^2)$. The currently
fastest algorithm for DFCA minimisation is due to KÖRNER [12], who developed
an algorithm that runs in time $\mathcal{O}(n \log n)$, and is a clever and refined modification of HOPCROFT's algorithm for DFA minimisation.

Minimal DFCA are theoretically characterised [3,12,4]. All known algorithms
for constructing a minimal ℓ-DFCA are based on a similarity relation $\sim_\ell$ on
states, which is defined such that a minimal ℓ-DFCA consists of pairwise dissim-
ilar states. The relation $\sim_\ell$ is defined using two very basic notions: (i) the level
of a state, which is the length of a shortest word leading to it, and (ii) the gap
between two states, which is the length of a shortest word on which they differ.

Lossy compression of DFA has received some attention recently, and DFCA
minimisation can be considered as an instance. *Hyper-minimisation* [1] is another
instance and aims to find a smallest DFA N for a given DFA M such that
$L(M)$ and $L(N)$ have finite symmetric difference. This notion was refined to
ℓ-*minimisation* [5], where the languages are allowed to differ only on words of
length at most ℓ. Yet another variant was proposed by SCHEWE [13].

It is noteworthy that ℓ-minimisation and ℓ-DFCA minimisation are dual. It was
already observed by BADR et al. [1] that there are languages L, which are best
represented by a pair consisting of an ℓ-minimal automaton (that makes errors on
words of length at most ℓ) and a minimal ℓ-DFCA. This combination can be substantially smaller than a single minimal DFA for L. An input word w is processed
by such a pair by selecting the authoritative DFA based on the word's length.

In principle, this approach works for all possible values of ℓ. Thus, it is desir-
able to construct an algorithm that decides for which value of ℓ the size of the
representation is minimal. For this, we need to have algorithms that for a given
DFA M return the size of an ℓ-minimal DFA and a minimal ℓ-DFCA for several
values ℓ. We note that such an algorithm is known for ℓ-minimal DFA [6], and
the current contribution adds the algorithm for minimal ℓ-DFCA.

In this paper, we give an alternative ℓ-DFCA minimisation algorithm, which
proceeds in three phases. First, we calculate the function 'gap' and represent it

compactly in a gap-tree. We show that its computation can be done by a slightly augmented version of HOPCROFT's algorithm, which means that it can be prepared in the DFA minimisation step. Second, we take the level of states into account and annotate the gap-tree. Up to this point, the computation is independent of the value of ℓ, and the obtained annotated gap-tree can be reused for all ℓ. In the third step, we identify the states that should be preserved in the minimal ℓ-DFCA (which naturally depends on ℓ) and determine its transition function. Our approach has several advantages. First, it is much easier to understand, verify, and implement. Its first phase closely resembles HOPCROFT's algorithm, which is well-known and understood. Second, since the first two phases are independent of ℓ, we can easily compute the size of all minimal ℓ-DFCA (for all valid ℓ) without overhead. In addition, we present an algorithm that constructs (a compact representation of) minimal ℓ-DFCA for consecutive values of ℓ in time $\mathcal{O}(n \log n)$.

We would like to point out that the minimisation algorithm presented in this paper shares the general outline with the ℓ-minimisation algorithm [6]: they both divide the computation of the minimal (with respect to the proper relation) DFA into phases, out of which only the last one depends on ℓ. Moreover, in both cases we present an ultrametric as an ultrametric tree and then annotate it. Due to differences in the similarity relations, the details vary significantly.

2   Preliminaries

In the following, let $M = \langle Q, \Sigma, \delta, q_0, F \rangle$ be a minimal DFA, and let $m = |Q \times \Sigma|$ and $n = |Q|$. As usual, we let $\min \emptyset = \infty$. For every state $q \in Q$, we let $\mathrm{level}(q) = \min \{ |w| : \delta(q_0, w) = q \}$ and call it the level of q. Given two states $p, q \in Q$, we define their *gap* by

$$ \mathrm{gap}(p, q) = \min \{ |w| : w \in L(p) \bigtriangleup L(q) \} \, , $$

where $\bigtriangleup$ is the symmetric difference operator. Note that $d(p, q) = 2^{-\,\mathrm{gap}(p,q)}$ with $2^{-\infty} = 0$ defines an ultrametric. We continue to work with $\mathrm{gap}(p, q)$ because it is used in the ℓ-similarity relation $\sim_\ell$, which is defined by

$$ p \sim_\ell q \iff \max(\mathrm{level}(p), \mathrm{level}(q)) + \mathrm{gap}(p, q) > \ell \, , $$

for all $p, q \in Q$. The currently fastest algorithm [12] for calculating minimal cover automata uses $\sim_\ell$, which in general is not an equivalence relation, but only a compatibility relation (i.e., reflexive and symmetric). Some additional, useful properties of $\sim_\ell$ are presented in [4]. In particular, they allow us to form an equivalence relation as follows.

Definition 1 (cf. [12, Definition 3]). *Let* $\pi\colon Q \to P$ *be a mapping for some* $P \subseteq Q$ *such that* $\pi(p) = p$ *for every* $p \in P$. *Then* π *is an* ℓ-similarity state decomposition *(ℓ-SSD) of* Q *if*

1. $\mathrm{level}(q) \geq \mathrm{level}(\pi(q))$ *for all* $q \in Q$,

2. $q \sim_\ell \pi(q)$ for every $q \in Q$, and
3. $p \not\sim_\ell p'$ for all $p, p' \in P$ with $p \neq p'$.

In other words, an ℓ-SSD is a partition of Q into $|P|$ blocks such that (1) each block has a representative with minimal level, (2) all elements in a block are ℓ-similar to their representative, and (3) the representatives of different blocks are pairwise ℓ-dissimilar. It is easy to observe that an ℓ-SSD $\pi \colon Q \to P$ contains a maximal (with respect to set inclusion) set P of pairwise ℓ-dissimilar states. Consequently, every ℓ-SSD π yields a minimal ℓ-DFCA by taking the quotient of M with respect to the equivalence relation π represents.

Theorem 2 (cf. [12, Theorem 1]). *For every ℓ-SSD $\pi \colon Q \to P$, the DFA $(M/\pi) = \langle P, \Sigma, \mu, \pi(q_0), F \cap P \rangle$ is a minimal ℓ-DFCA, where $\mu(p,a) = \pi(\delta(p,a))$ for every $p \in P$ and $a \in \Sigma$.*

KÖRNER's algorithm constructs an ℓ-SSD using a clever modification of HOPCROFT's algorithm [9]. It initially partitions the states into F and $Q \setminus F$ and then refines this partition while preserving Property 3 of Definition 1. Once the algorithm stops, also Property 2 of Definition 1 will be satisfied.

Part of the difficulty of KÖRNER's algorithm stems from the fact that it takes both 'gap' and 'level' into account when refining the partition. Our approach separates these two properties. We show that $\text{gap}(p,q)$ can be calculated by a standard run of HOPCROFT's algorithm. Moreover, the gap-matrix can be compactly represented as a gap-tree $\mathcal{G}$. With the help of $\mathcal{G}$, we can then compute an ℓ-SSD in a simpler manner by only taking 'level' into account.

3   Gap-Trees

The gap-matrix has size $\Theta(n^2)$, thus any algorithm that explicitly uses it is doomed to run in time $\Omega(n^2)$. To obtain a minimisation algorithm that runs in time $\mathcal{O}(m \log n)$ we need to represent it more compactly. This is achieved with the help of the gap-tree $\mathcal{G}$, which contains a leaf for each state of Q. The tree is organised such that each subtree contains only states whose pairwise gap exceeds a certain value. More precisely, for each subtree t' there exists $s \in \mathbb{N}$ such that

$$\text{gap}(p,q) \begin{cases} \geq s & \text{if } p \text{ and } q \text{ occur in } t' \\ < s & \text{otherwise.} \end{cases}$$

In the next section, it is shown that gap tree can be created during a standard run of a slightly augmented variant of HOPCROFT's algorithm.

Before we start with the formal definition, we recall some notions on trees. We generally use rooted trees, which are special undirected graphs with a dedicated vertex r (the root) such that there is exactly one path from each vertex to r. Moreover, we use weighted edges, where the edge weights are nonnegative integers. The sum of the edge weights along the unique path from a vertex v to the root r is denoted by $d(v)$ and called the *depth of v*. A leaf is a vertex with only one adjacent edge. A tree is an *ultrametric tree* [8,10,11] if the depth of all leaves is equal. Finally, for two vertices v and v', their join $v \vee v'$ is the lowest common ancestor (i.e., the deepest vertex such that both v and v' occur in its subtree).

Definition 3. *An ultrametric tree for* gap *(for short:* gap tree*) is an ultrametric tree with leaves Q such that* $\mathrm{gap}(p,q) = \mathrm{d}(p \vee q)$ *for all* $p, q \in Q$ *with* $p \neq q$.

Next, we want to determine representatives of similarity blocks. Since each vertex of the gap-tree determines a subtree and thus a block of states, which are the states that occur in the subtree, we assign a state to each vertex. To satisfy Property 1 of Definition 1, we select a state with minimal level among all states assigned to the direct subtrees. Formally, given a gap-tree $\mathcal{G}$ with vertices V, we let state: $V \to Q$ be a mapping such that (i) $\mathrm{state}(q) = q$ for all $q \in Q$, (ii) $\mathrm{state}(v) = \mathrm{state}(v')$ for all $v \in V \setminus Q$, where v' is some direct child vertex of v, and (iii) $\mathrm{level}(\mathrm{state}(v)) \leq \mathrm{level}(\mathrm{state}(v'))$ for all $v \in V$ and v' being a direct child vertex of v. Note there can be several mappings 'state' that fulfill the requirements (i)–(iii), which correspond to different choices of representatives. In the following, we assume that 'state' is any such mapping.

The selected mapping 'state' labels all vertices of $\mathcal{G}$ with a state of Q. Recall that $\mathrm{state}(q) = q$ for every $q \in Q$. Consequently, for every $q \in Q$ there exists a minimal (i.e., of minimal depth) vertex $v_q \neq q$ such that $\mathrm{state}(v) = q$ for all vertices besides v_q along the path (towards the root) starting in the leaf q to v_q. Note that the vertex v_q is unique, and called the *termination vertex* of q. The *termination state* of $q \in Q$ is $\mathrm{state}(v_q)$. Recall that r is the root vertex. Note that the termination state of q is always different from q unless $q = \mathrm{state}(r)$. Moreover, for all states $q \neq \mathrm{state}(r)$ we have $\mathrm{gap}(q, \mathrm{state}(v_q)) = \mathrm{d}(q \vee \mathrm{state}(v_q)) = \mathrm{d}(v_q)$, which motivates the following definitions.

Definition 4. *For every* $q \in Q$, *let*

- *the state-gap $g(q)$ be such that*

$$g(q) = \begin{cases} -\infty & \text{if } q = \mathrm{state}(r), \\ \mathrm{d}(v_q) & \text{otherwise.} \end{cases}$$

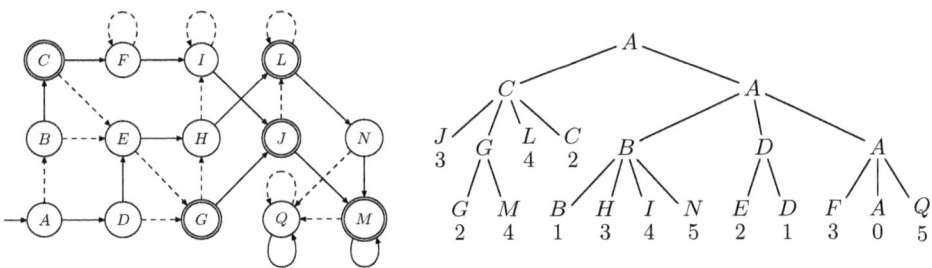

Fig. 1. Example DFA (left) and a gap-tree (right) for it

– $\text{value}(q) = \text{level}(q) + g(q)$.

Example 5. Let us consider the minimal DFA and the gap-tree for it that are displayed in Fig. 1. Below the leaves of the gap-tree we annotated the state's level. In addition, we already labelled the inner nodes with states. Now, we can determine the termination state for each state. For example, C is the termination state of G because it is the first label on the path from the leaf G towards the root that differs from G. Consequently, $g(G) = \text{d}(v_G) = 1$ and $\text{value}(G) = 2 + 1 = 3$. Overall, we obtain:

$$\text{value}(A) = -\infty \quad \text{value}(B) = 2 \quad \text{value}(C) = 2 \quad \text{value}(D) = 2 \quad \text{value}(E) = 4$$
$$\text{value}(F) = 5 \quad \text{value}(G) = 3 \quad \text{value}(H) = 5 \quad \text{value}(I) = 6 \quad \text{value}(J) = 4$$
$$\text{value}(L) = 5 \quad \text{value}(M) = 6 \quad \text{value}(N) = 7 \quad \text{value}(Q) = 7.$$

It is important to note that all the previous notions on the gap-tree are independent of the selection of ℓ. Nevertheless, we can use them to transform the gap-tree $\mathcal{G}$ into an ℓ-SSD. Let $P = \{q \in Q : \text{value}(q) \le \ell\}$. Note that $P \ne \emptyset$ because $\text{state}(r) \in P$. For every state $q \in Q$, its ℓ-state $\pi(q)$ is the label $\text{state}(v)$ of the first vertex v on the path from q to the root r such that $\text{value}(\text{state}(v)) \le \ell$.

Lemma 6. *The ℓ-state mapping π is an ℓ-SSD.*

Proof. We have to show the conditions of Definition 1. For every $p \in P$ we have $\text{value}(p) \le \ell$. Consequently, their ℓ-state $\pi(p)$ is p. Moreover, for every $q \in Q$ we have $\text{level}(q) \ge \text{level}(\pi(q))$ because $\pi(q)$ is the label of an ancestor of q. Let us continue with Condition 2 of Definition 1. It trivially holds for $q = \pi(q)$, so suppose that $q \in Q$ is such that $q \ne \pi(q)$. Consequently, $q \notin P$. Let p be the label of the last vertex v on the path from q to the root r such that $\text{value}(\text{state}(v)) > \ell$. Clearly, the next vertex along this path is the ℓ-vertex of q, which is labelled $\pi(q)$. Note that $p = q$ is possible. Then $q \vee \pi(q) = p \vee \pi(q)$ and thus $\text{gap}(q, \pi(q)) = g(p)$. In addition, $\text{level}(q) \ge \text{level}(p) \ge \text{level}(\pi(q))$. These two estimations together yield that

$$\max(\text{level}(q), \text{level}(\pi(q)) + \text{gap}(q, \pi(q))$$
$$= \text{level}(q) + g(p) \ge \text{level}(p) + g(p) = \text{value}(p) > \ell ,$$

which proves $q \sim_\ell \pi(q)$.

Finally, we have to show Condition 3 of Definition 1. Let $p, p' \in P$ be such that $p \ne p'$. Consequently, $\text{value}(p) \le \ell$ and $\text{value}(p') \le \ell$. Without loss of generality, suppose that (i) $\text{level}(p) \ge \text{level}(p')$, (ii) $p \vee p'$ is not labelled with p. If these conditions are not met for the pair (p, p'), then they are met for the pair (p', p). Since $\text{state}(p \vee p') \ne p$, the termination vertex of p is on the path from p to $p \vee p'$ and so $g(p) \ge \text{d}(p \vee p') = \text{gap}(p, p')$. Taking this and assumption (i) into account, we obtain

$$\max(\text{level}(p), \text{level}(p')) + \text{gap}(p, p')$$
$$\le \text{level}(p) + g(p) = \text{value}(p) \le \ell ,$$

which proves $p \nsim_\ell p'$. □

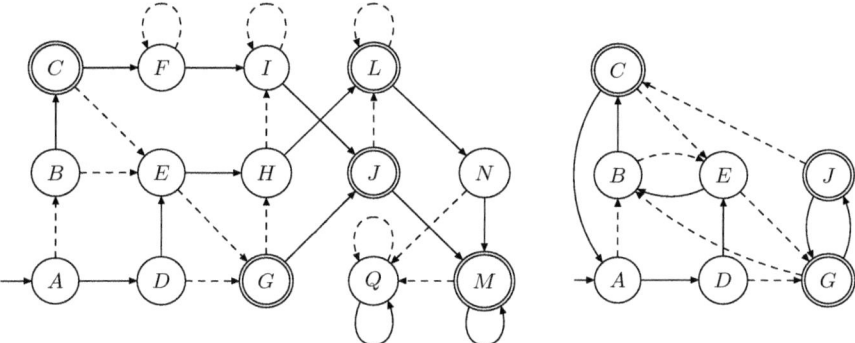

Fig. 2. Example DFA (left) and a minimal 4-DFCA (right) for it

Example 7 (Example 5 continued). Take $\ell = 4$, then $P = \{A, B, C, D, E, G, J\}$. Besides the obvious entries (identity on P) we have:

$$\pi(F) = \pi(Q) = A \qquad \pi(H) = \pi(I) = \pi(N) = B \qquad \pi(L) = C \qquad \pi(M) = G \ .$$

The resulting minimal 4-DFCA is displayed in Fig. 2.

Lemma 6 shows that the gap-tree with the help of 'value' indeed contains a characterisation of $\sim_\ell$ for all potential ℓ. This allows a fast construction of a minimal ℓ-DFCA for M whenever a gap-tree $\mathcal{G}$ is provided. We say that a gap-tree $\mathcal{G}$ with vertices V is *small* if $|V| \in \mathcal{O}(n)$.

Theorem 8. *Given a small gap-tree $\mathcal{G}$ with $\mathrm{d}(q) = s$ for all $q \in Q$, we can*

1. *calculate the sizes of all minimal ℓ-DFCA (for all valid ℓ) in time $\mathcal{O}(m + s)$,*
2. *construct a minimal ℓ-DFCA for a given ℓ in time $\mathcal{O}(m)$, and*
3. *iteratively construct (representations of) minimal ℓ-DFCA for all ℓ in time $\mathcal{O}(m \log n + s)$*

Proof. Let V be the set of vertices of $\mathcal{G}$. First, we compute a proper state labelling state: $V \to Q$ in the obvious manner. This can be done in time $\mathcal{O}(n)$ because $\mathcal{G}$ is small. Similarly, we can compute 'level' in time $\mathcal{O}(m)$ because every transition needs to be considered only once. A simple bottom-up procedure on $\mathcal{G}$ can calculate value(q) for every state q using 'level' and 'state'. Overall, we can complete these steps in time $\mathcal{O}(m)$.

For the first claim, we sort the elements of Q by their 'value' in time $\mathcal{O}(n + s)$ using, for example, COUNTING-SORT. We know that value(q) $\leq n + s$ for every $q \in Q$, hence we can obtain the mentioned time-bound for sorting. From this sorted list of states, we can now determine the sizes of all minimal ℓ-DFCA in time $\mathcal{O}(n + s)$ by iteration over the list because for a given ℓ the size of a minimal ℓ-DFCA is $|\{\, q \,:\, \text{value}(q) \leq \ell \}|$.

Next, let us move to the second claim. Theorem 2 and Lemma 6 show that given an efficient representation of an ℓ-state mapping π, we can construct a minimal ℓ-DFCA in time $\mathcal{O}(m)$. Consequently, it only remains to determine an ℓ-state mapping $\pi\colon Q \to P$. Clearly, the set $P = \{q \;:\; \text{value}(q) \le \ell\}$ can be constructed in time $\mathcal{O}(n)$ by a simple iteration over Q. Finally, we need to determine π. To this end, we traverse the gap-tree $\mathcal{G}$ top-down. Every time, we encounter a vertex v' such that $\text{value}(\text{state}(v')) > \ell$ but $\text{value}(\text{state}(v)) \le \ell$, where v is the direct ancestor of v', we set $\pi(q) = \text{state}(v)$ for all states $q \in Q$ that occur in the subtree of v'. Overall, this can be achieved in time $\mathcal{O}(|V|)$. Since $\mathcal{G}$ is small, we obtain the time bound $\mathcal{O}(m)$.

Finally, we have to show how to create minimal ℓ-DFCA sequentially, so that the total execution time is $\mathcal{O}(m \log n + s)$. Let $\ell_{\max+1} = \max_{q \in Q} \text{value}(q)$. In each step $\ell \in \{\ell_{\max}, \dots, 1, 0\}$ our algorithm keeps the states $P_\ell = \{q \;:\; \text{value}(q) \le \ell\}$. Consequently, it merges each state $q \in P_{\ell+1}$ such that $\text{value}(q) = \ell + 1$ into its terminating state p, which by construction satisfies

$$\text{value}(p) = \text{level}(p) + g(p) \le \text{level}(q) + (g(q) - 1) = \text{value}(q) - 1 \le \ell \;.$$

However, to obtain the stated running time, we need to organise the process properly. First, we note that there is a change in at most n steps because there can be at most n merges. Thus, we first filter out the steps, in which no changes occur. This can be done in time $\mathcal{O}(s)$. Second, we represent the DFA as a list of transitions. For each state q, we keep a list of all pairs (a, p) such that $\delta(q, a) = p$, where p is implemented as a pointer to a pointer to the actual state p, which allows a fast modification of all transitions leading to p by simply replacing the final pointer to p. In addition, for every state q, we keep a counter $c(q)$, which is initially 1 and counts how many states were merged into q. Now assume that we want to merge the state q into p. First, we assume that $c(q) \le c(p)$. In this case, we simply redirect each incoming transition of q to p (by a constant-time pointer replacement). However, if $c(q) > c(p)$, then we redirect each incoming transition of p to q (i.e., we do not merge q into p, but rather merge p into q). In addition, we replace the outgoing transitions of q by the outgoing transitions of p, which can be done in constant time by simply replacing the pointer to the list. We complete this case by renaming q to p. Finally, in both cases we update $c(p)$ by $c(p) \leftarrow c(p) + c(q)$. In this manner, every time the transition target $\delta(q, a)$ is modified due to a merge, the value $c(\delta(q, a))$ at least doubles. Since $c(q) \le n$ for each $q \in Q$, each transition can be modified at most $\log n$ times. Consequently, we obtain the overall running time $\mathcal{O}(m \log n + s)$. □

Note that the third statement of Theorem 8 only provides a (compact) representation of the minimal ℓ-DFCA in the presented running time $\mathcal{O}(m \log n + s)$. If we output the obtained DFCA for all ℓ, then we require time $\mathcal{O}(m^2 \log n)$ because we need $\mathcal{O}(m)$ steps for each output DFCA. The summand s disappears due to the fact that it can always be chosen such that $s \le n^2 \le m^2$.

4   Computing a Gap-Tree

We already showed that we can easily construct minimal ℓ-DFCA provided that we have access to a small gap-tree $\mathcal{G}$ for M. In this section, we show how to

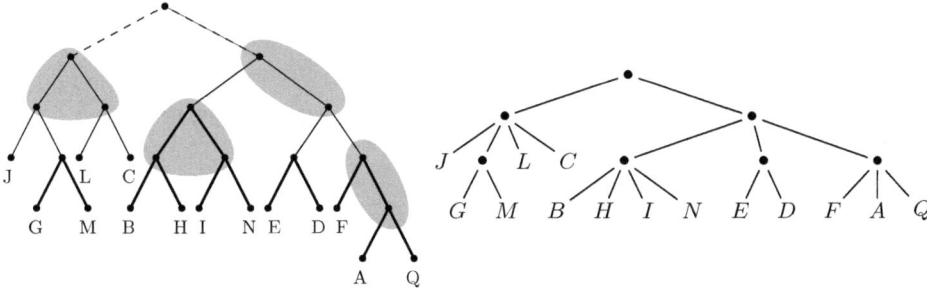

Fig. 3. The pre-gap tree (left) for the DFA of Fig. 1. The gap-tree (right) of Fig. 1 can be obtained from it by merging appropriate nodes, which are marked in grey. The edges labelled with different 'gap' were drawn in different styles (dashed = 0, normal = 1, thick = 2).

construct such a $\mathcal{G}$. Actually, a simple modification of HOPCROFT's algorithm [9] can perform the construction for us. Roughly speaking, we keep track of the length of words that cause a split of a set of states in the run of the algorithm. Already KÖRNER's algorithm [12] followed a similar strategy. Our modification is less drastic and yields a solution that is simpler and easier to understand.

Algorithm 1 presents the slightly modified version of HOPCROFT's algorithm that is suitable for our purposes. It creates a *pre-gap tree*, which keeps track of how the final partition was obtained. In particular, it stores the lengths of the used splitting words. The length of such a splitting word coincides with the gap between the affected states. The obtained pre-gap tree (see Fig. 3) is basically a binarisation of a gap-tree. It can easily be transformed into a gap-tree by merging appropriate nodes.

We marked the modifications (compared to a standard implementation of HOPCROFT's algorithm) by ▷. Clearly, all lines referring to gap calculations are new. However, they do neither affect the correctness of the overall algorithm nor the analysis of its run-time. In addition, the queue T is restricted to a FIFO-queue, which is essential for our purposes. Finally, although we split Q' into Q_{r-1} and Q_r in line 11, we do not replace Q' in T. When $v_{Q'}$ is extracted from T, we no longer have Q' as an element of P. However, we can recreate it by listing all the states that occur in the subtree of $v_{Q'}$. A similar approach was also used by KÖRNER [12], who proved that this does not affect the running time.

Next, we show that the pre-gap tree has the following properties:

1. It is a binary tree.
2. The states p and q point to the same node if and only if $p = q$.
3. If p and q with $p \neq q$ point, respectively, to v_p and v_q, then the edges to $v_p \vee v_q$ are labelled with $\mathrm{gap}(p, q)$.

The first statement follows clearly from HOPCROFT's strategy. Every time a leaf turns into an inner node in line 13, two children are created. Moreover, the previous line removed the corresponding set from P, which yields that the vertex is never split again. In particular, this statement yields that the pre-gap tree is

---

Algorithm 1. Modification of HOPCROFT's algorithm

1: $Q_1 \leftarrow F$, $Q_2 \leftarrow Q \setminus F$, $r \leftarrow 2$, $P \leftarrow \{Q \setminus F, F\}$
2: $T \leftarrow \{(v_F, 1)\}$ ▷ FIFO queue
3: create v_Q and its children $v_F, v_{Q \setminus F}$, $\mathrm{gap}(v_F, v_Q) \leftarrow \mathrm{gap}(v_{Q \setminus F}, v_Q) \leftarrow 0$ ▷
4: put two-way pointers $v_F \leftrightarrow F$ and $v_{Q \setminus F} \leftrightarrow Q \setminus F$ ▷
5: **while** $T \neq \emptyset$ **do**
6:     $(v_{Q_i}, k_i) \leftarrow$ first from T
7:     **for** $a \in \Sigma$ **do**
8:        $Q_a = \{q : \delta(q, a) \in Q_i\}$
9:        **for** $Q' \in P$ such that $Q' \not\subseteq Q_a$ and $Q' \cap Q_a \neq \emptyset$ **do**
10:           $r \leftarrow r + 2$
11:           $Q_{r-1} \leftarrow Q' \cap Q_a$, $Q_r \leftarrow Q' \setminus Q_a$
12:           $P \leftarrow (P \setminus \{Q'\}) \cup \{Q_{r-1}, Q_r\}$
13:           create nodes $v_{Q_{r-1}}$, v_{Q_r} and edges $(v_{Q_{r-1}}, v_{Q'})$, $(v_{Q_r}, v_{Q'})$ ▷
14:           $\mathrm{gap}(v_{Q_{r-1}}, v_{Q'}) \leftarrow \mathrm{gap}(v_{Q_r}, v_{Q'}) \leftarrow k_i$ ▷
15:           **if** $|Q_{r-1}| > |Q_r|$ **then**
16:              add $(v_{Q_r}, k_i + 1)$ to T ▷ Do not remove Q' from T
17:           **else**
18:              add $(v_{Q_{r-1}}, k_i + 1)$ to T
19:     **for** $Q' \in P$ **do**
20:        **for** $q \in Q'$ **do**
21:           add pointer from q to v, where Q' points to v ▷ Partition of states

---

small. The second statement is the correctness of HOPCROFT's algorithm, so we do not reprove it. Before, we can prove the third (and essential) statement, we first identify some properties of the maintained data structure.

Lemma 9 (cf. [12, Lemma 4]). *Let $(v_{Q_1}, k_1), \ldots, (v_{Q_s}, k_s)$ be the complete sequence of elements added to T during the run of Algorithm 1. Then $k_{i-1} \leq k_i$ for all $i \in \{2, \ldots, s\}$.*

Proof. We prove the statement by induction. For $i = 2$ it is obvious because $k_1 = 1$ and $k_2 = 2$. Now, let $i \geq 3$. The element (v_{Q_i}, k_i) was put into T while processing (v_{Q_j}, k_j) for some $j \leq i - 1$. Due to the FIFO strategy, its predecessor $(v_{Q_{i-1}}, k_{i-1})$ was put into T while processing $(v_{Q_{j'}}, k_{j'})$ for some $j' \leq j$. By the induction assumption, we have $k_{j'} \leq k_j$. Consequently, we obtain that $k_{i-1} = k_{j'} + 1 \leq k_j + 1 = k_i$.   □

Lemma 10. *For any two inequivalent states $p, q \in Q$ there is a set $Q' \in P$ at some point during the execution of Algorithm 1 that is split into Q_{r-1} and Q_r with $p \in Q_{r-1}$ and $q \in Q_r$. The corresponding edges $(v_{Q_{r-1}}, v_{Q'})$ and $(v_{Q_r}, v_{Q'})$ are labelled by $\mathrm{gap}(p, q)$.*

Proof. Since p and q are inequivalent, the states p and q will be split. Thus, the set Q' with the given properties exists. It remains to prove the property about the gap. Let $\mathrm{gap}'(p, q) = \mathrm{gap}(v_{Q_{r-1}}, v_{Q'})$. Next, we show that $\mathrm{gap}(p, q) = \mathrm{gap}'(p, q)$. To this end, we first show that $\mathrm{gap}(p, q) \leq \mathrm{gap}'(p, q)$ and then demonstrate that $\mathrm{gap}(p, q) \geq \mathrm{gap}'(p, q)$, which will conclude the proof.

The first part is shown by induction on the number i of elements of T considered by the algorithm. If $i = 0$, then $\text{gap}(p, q) = 0$ because exactly one of $\{p, q\}$ is in F. Since line 3 assigns the same gap, the claim holds. The inequivalent states p and q are eventually split by the algorithm. Let $Q' \in P$ be the element such that $\{p, q\} \subseteq Q'$ before they are split. Intuitively, the element $(v_{Q''}, k)$ of T that caused the split has the property that there exists a letter $a \in \Sigma$ such that exactly one of the states $p_a = \delta(p, a)$ and $q_a = \delta(q, a)$ is in Q''. Consequently, $p_a \neq q_a$ and $\text{gap}(p_a, q_a) \leq \text{gap}'(p_a, q_a)$ by the induction hypothesis (because p_a and q_a must have been split in a previous iteration). Then

$$\text{gap}(p, q) \leq \text{gap}(p_a, q_a) + 1 \leq \text{gap}'(p_a, q_a) + 1 = k = \text{gap}'(p, q) \ ,$$

which proves the induction step.

Finally, we show that $\text{gap}'(p, q) \leq \text{gap}(p, q)$ for all pairs (p, q) of states with $p \neq q$. Let $w = a_1 \cdots a_m$ be the shortest string such that exactly one of the states $\delta(p, w)$ and $\delta(q, w)$ is in F. Moreover, let (i) $p_0 = p$ and $q_0 = q$, and (ii) $p_i = \delta(p_{i-1}, a_i)$ and $q_i = \delta(q_{i-1}, a_i)$ for every $i \in \{1, \ldots, m\}$. Let us consider the maximal i such that $\text{gap}(p_i, q_i) < \text{gap}'(p_i, q_i)$. Trivially, we have $i < m$ because $\text{gap}(p_m, q_m) = \text{gap}'(p_m, q_m) = 0$, which follows because exactly one of $\{p_m, q_m\}$ is in F. By the first statement and the maximality of i, we have $\text{gap}(p_{i+1}, q_{i+1}) = \text{gap}'(p_{i+1}, q_{i+1})$. Due to the algorithm, there exists an element $(v_{Q''}, k)$ of T and $a \in \Sigma$ such that $\text{gap}'(p_i, q_i) = k$, where $p' = \delta(p_i, a)$, $q' = \delta(q_i, a)$, and exactly one of $\{p', q'\}$ is in S. The latest the split can happen is due to (p_{i+1}, q_{i+1}), but it can happen earlier, which allows us to conclude by Lemma 9 that

$$\begin{aligned}\text{gap}'(p_i, q_i) = k &\leq \text{gap}'(p_{i+1}, q_{i+1}) + 1 = \text{gap}(p_{i+1}, q_{i+1}) + 1 \\ &= \text{gap}(p_i, q_i) \ ,\end{aligned}$$

where the last equality follows from the fact that w is the shortest word. Consequently, $\text{gap}'(p_i, q_i) \leq \text{gap}(p_i, q_i)$, which contradicts the assumption and completes the proof. □

Now Property 3 of the pre-gap tree is an easy corollary of Lemma 10: consider any two inequivalent states p and q. The set Q' from Lemma 10 corresponds to the node $v_p \vee v_q$ in the pre-gap tree. Furthermore the lemma asserts that the edges to $v_p \vee v_q$ are labelled by $\text{gap}(p, q)$.

To obtain a gap-tree $\mathcal{G}$ from the pre-gap tree for the DFA M, it is enough to merge connected parts of the pre-gap tree with incoming edges labelled with the same value k into a single vertex v such that $d(v) = k$ (see Fig. 3). Moreover, Lemma 10 shows that $d(q) \leq n$ for every $q \in Q$, which allows us to state our main theorem.

Theorem 11. *For all DFA $M = \langle Q, \Sigma, \delta, q_0, F \rangle$ with $m = |Q \times \Sigma|$ and $n = |Q|$, we can perform the following in time $\mathcal{O}(m \log n)$:*

1. *Calculate the sizes of all minimal ℓ-DFCA (for all valid ℓ).*
2. *Construct a minimal ℓ-DFCA for a given ℓ.*
3. *Iteratively construct (representations of) minimal ℓ-DFCA for all ℓ.*

References

1. Badr, A., Geffert, V., Shipman, I.: Hyper-minimizing minimized deterministic finite state automata. RAIRO Theoret. Inform. Appl. 43(1), 69–94 (2009)
2. Câmpeanu, C., Paun, A., Yu, S.: An efficient algorithm for constructing minimal cover automata for finite languages. Int. J. Found. Comput. Sci. 13(1), 83–97 (2002)
3. Câmpeanu, C., Santean, N., Yu, S.: Minimal cover-automata for finite languages. Theor. Comput. Sci. 267(1-2), 3–16 (2001)
4. Champarnaud, J.-M., Guingne, F., Hansel, G.: Similarity relations and cover automata. RAIRO Theoret. Inform. Appl. 39(1), 115–123 (2005)
5. Gawrychowski, P., Jeż, A.: Hyper-minimisation made efficient. In: Královič, R., Niwiński, D. (eds.) MFCS 2009. LNCS, vol. 5734, pp. 356–368. Springer, Heidelberg (2009)
6. Gawrychowski, P., Jeż, A., Maletti, A.: On minimising automata with errors. Corr. abs/1102.5682 (2011)
7. Gries, D.: Describing an algorithm by Hopcroft. Acta Inf. 2(2), 97–109 (1973)
8. Hartigan, J.A.: Representation of similarity matrices by trees. J. Amer. Statist. Assoc. 62(320), 1140–1158 (1967)
9. Hopcroft, J.E.: An $n \log n$ algorithm for minimizing states in a finite automaton. In: Kohavi, Z. (ed.) Theory of Machines and Computations, pp. 189–196. Academic Press, London (1971)
10. Jardine, C.J., Jardine, N., Sibson, R.: The structure and construction of taxonomic hierarchies. Math. Biosci. 1(2), 173–179 (1967)
11. Johnson, S.C.: Hierarchical clustering schemes. Psychometrika 32(3), 241–254 (1967)
12. Körner, H.: A time and space efficient algorithm for minimizing cover automata for finite languages. Int. J. Found. Comput. Sci. 14(6), 1071–1086 (2003)
13. Schewe, S.: Beyond hyper-minimisation — Minimising DBAs and DPAs is NP-complete. In: Proc. Ann. Conf. Foundations of Software Technology and Theoretical Computer Science, LIPIcs, vol. 8, pp. 400–411. Schloss Dagstuhl (2010)

Preset and Adaptive Homing Experiments for Nondeterministic Finite State Machines

Natalia Kushik[1], Khaled El-Fakih[2], and Nina Yevtushenko[1]

[1] Tomsk State University
Tomsk, Russia
[2] American University of Sharjah, UAE
kushiknatalya@yahoo.com, kelfakih@aus.edu,
ninayevtushenko@yahoo.com

Abstract. In this paper, we present algorithms for preset and adaptive homing experiments for a given observable reduced nondeterministic finite state machine (NFSM). We show that the tight upper bound on a shortest preset homing sequence for a NFSM with n states and with two or more initial states is of order 2^{n^2}. The upper bound on a shortest adaptive homing sequence of a NFSM with m initial states, $m \leq n$, states is of order $\sum_{j=2}^{m} C_n^j$ and this upper bound is of order 2^n when m tends to n.

Keywords: Preset and adaptive homing experiments, nondeterministic finite state machines.

1 Introduction

Finite State Machines (FSMs) are widely used for modeling systems in many application domains. An FSM is a state transition system which has a finite number of inputs, outputs, states and a finite number of transitions each labeled by a pair of an input and output. In FSM-based testing, we have a machine or an implementation under test about which we lack some information, and we want to deduce this information by conducting experiments on this machine. An experiment consists of applying input sequences to the machine, observing corresponding output responses and drawing conclusion about the machine under test. An experiment is *preset* if input sequences are known before starting the experiment, and an experiment is *adaptive* if at each step of the experiment the next input is selected based on previously observed outputs.

Well-known types of experiments include distinguishing, homing, and synchronization experiments. Given an FSM, assuming that the initial state is unknown, a *distinguishing experiment* determines the initial state of the FSM. A *homing experiment* identifies the final state reached at the end of the experiment. An applied input sequence when performing such an experiment is called a *homing* sequence (HS). A *synchronization experiment* guarantees that the machine reaches a given state by the end of the experiment. A corresponding input sequence is called a *synchronizing* sequence (SS).

B. Bouchou-Markhoff et al. (Eds.): CIAA 2011, LNCS 6807, pp. 215–224, 2011.
© Springer-Verlag Berlin Heidelberg 2011

Since the seminal paper on "gedanken experiments" by Moore [1], there has been a lot of work on preset and adaptive homing experiments for deterministic FSMs. Homing experiments are typically used in FSM-based conformance testing when no reset, that takes the machine from any current state to a designated state, is assumed in the implementation under test. For information and surveys on FSM-based experiments and some related algorithms, a reader may refer to [2], [3-5]; in particular, the detailed survey given by Sandberg in [6] contains information about homing experiments for deterministic FSMs. In summary, based on the algorithm in [1], Ginsburg [7] presented an algorithm for preset homing experiment for a reduced deterministic FSM. Hibbard [8] showed that Moore's algorithm can be used for computing a homing sequence that is not longer than $n(n - 1)/2$, where n is the number of states of the given deterministic FSM. Hibbard also showed that machines possessing preset homing sequences with minimal length $n(n - 1)/2$ require adaptive homing sequences of the same length. Derivation of minimal length preset homing sequences can be done using the homing tree method introduced by Gill [9] and reported in details in Kohavi [2]. In addition, parallel algorithms for homing experiments for deterministic FSMs and for many other related problems are surveyed by Ravikumar in [5].

Some work has also been done on experiments for nondeterministic machines. Nondeterminism may occur due various reasons such as limited controllability, abstraction, modeling concurrency and real time systems, etc. [10]. For instance, Sandberg [6], based on some work in Rystsov [11], reports that the problem of finding a synchronizing sequence of a nondeterministic finite state automata (NFA) is PSPACE-complete when the NFA possesses proper features. That work can be applied for finding homing sequences for a special class of nondeterministic FSMs, namely FSMs with synchronizing sequences, as it is known that a synchronizing sequence is also a homing sequence; however, the converse is not true. We note that outputs are not needed for deriving SSs, and thus the application of NFA algorithms for deriving synchronizing experiments for such a class of nondeterministic FSMs can be carried out by ignoring the outputs of the machine. Burkhard [12] gave the sharp exponential upper bound $2^n - n - 1$ on minimum length of synchronization sequences of an n-state NFA. Imresh and Steinby [13] studied the same problem for a special class of NFAs and report lower and upper bounds in the paper [14].

In this paper, we present preset and adaptive algorithms for deriving a homing sequence for a given complete reduced observable nondeterministic FSM (denoted NFSM hereafter) with n states, when such a sequence exists. Differently from deterministic FSMs a HS may not exist even for a reduced NFSM. We show that the tight upper bound on a shortest preset homing sequence is of order 2^{n^2}. Further, we show that for an NFSM with proper features, a shortest preset HS has length $2^{(n^2/4-1)}$. In particular, this holds for NFSMs that have only a separating sequence as a HS. An input sequence is a *separating* sequence for two states of a NFSM if the sets of outputs produced by the NFSM at these states to the input sequence do not intersect. Separating sequences are introduced in [15] and have been studied in [16] and [17]. A preset algorithm is given for the case when a NFSM has two initial states and it is shown that the same algorithm can be used for an arbitrary number of initial states. The established upper bounds are shown to be of the same order for two or for more than two initial states. Finally, we show that the upper bound on a shortest adaptive HS for a set of $m \leq n$ (initial) states is of order $\sum_{j=2}^{m} C_n^j$ and thus, this upper bound is of order 2^n when m tends to n.

This paper is organized as follows. Preliminaries are introduced in Section 2. Algorithms for deriving preset and adaptive homing sequences are given in Sections 3 and 4, respectively. Section 5 concludes the paper.

2 Preliminaries

A *finite state machine (FSM)*, or simply a *machine* throughout this paper, is a 5-tuple $S = \langle S, I, O, h_S, S' \rangle$, where S is a finite nonempty set of states with a non-empty subset S' of initial states; I and O are finite input and output alphabets; and $h_S \subseteq S \times I \times O \times S$ is a behavior relation. A machine is *deterministic* if for each pair $(s, i) \in S \times I$ there exists at most one pair $(o, s') \in O \times S$ such that $(s, i, o, s') \in h_S$; otherwise, the machine is *nondeterministic*. If for each pair $(s, i) \in S \times I$ there exists $(o, s') \in O \times S$ such that $(s, i, o, s') \in h_S$ then FSM S is said to be *complete*; otherwise, the machine is called *partial*. A machine is *observable* if for each triple $(s, i, o) \in S \times I \times O$ there exists at most one state $s' \in S$ such that $(s, i, o, s') \in h_S$; otherwise, the machine is *nonobservable*. In this paper, we consider complete and observable nondeterministic machines, hereafter denoted as NFSMs.

In a usual way, the behavior relation is extended to input and output sequences. Given states $s, s' \in S$, an input sequence $\alpha = i_1 i_2 \ldots i_k \in I^*$ and an output sequence $\beta = o_1 o_2 \ldots o_k \in O^*$, there is a transition $(s, \alpha, \beta, s') \in h_S$ if there exist states $s_1 = s, s_2, \ldots, s_k, s_{k+1} = s'$ such that $(s_i, i_i, o_i, s_{i+1}) \in h_S$, $i = 1, \ldots, k$. Given states s and s', the input sequence α can *take* (or simply *takes*) the FSM S from state s to state s' if there exists an output sequence β such that $(s, \alpha, \beta, s') \in h_S$. The set $out(s, \alpha)$ denotes the set of all output sequences (responses) that the FSM S can produce at state s in response to the input sequence α, i.e. $out(s, \alpha) = \{\beta : \exists s' \in S \; [(s, \alpha, \beta, s') \in h_S]\}$. The pair α/β, $\beta \in out(s, \alpha)$, is an *Input/Output (I/O) sequence* (or a trace) at state s; if s is the initial state s_1 then the pair α/β is an *Input/Output (I/O) sequence* of the FSM S. Given states s and s', the *I/O* sequence α/β can *take* (or simply *takes*) the FSM S from state s to state s' if $(s, \alpha, \beta, s') \in h_S$. This property can be expressed by a function $next\_state(s, \alpha/\beta)$: $next\_state(s, \alpha/\beta) = s'$ if α/β takes the FSM S from state s to state s'. For observable FSMs, there is at most one state s' such that $next\_state(s, \alpha/\beta) = s'$.

Given two complete FSMs $S = \langle S, I, O, h_S, S' \rangle$ and $R = \langle R, I, O, h_R, R' \rangle$, two states s of S and r of R are *equivalent* [9, 15] if for each input sequence $\alpha \in I^*$ it holds that $out(s, \alpha) = out(r, \alpha)$. Otherwise, we say that states s and t are *distinguishable*. An FSM is said to be *reduced* if its states are pair-wise distinguishable. Given two complete FSMs $S = \langle S, I, O, h_S, S' \rangle$ and $R = \langle R, I, O, h_R, R' \rangle$, state r of R and state s of S are *non-separable* if for each input sequence $\alpha \in I^*$ it holds that $out(r, \alpha) \cap out(s, \alpha) \neq \varnothing$, i.e., the sets of output responses to each input sequence at state r and at state s intersect; otherwise, states r and s are *separable*. For separable states r and s, there exists an input sequence $\alpha \in I^*$ such that $out(r, \alpha) \cap out(s, \alpha) = \varnothing$, i.e., the sets of output responses of FSMs S and R at states r and s to the input sequence α are disjoint. In this case, α is a *separating* sequence of states r and s or simply α *separates* r and s. In this paper, when deriving preset and adaptive homing sequences for NFSMs we consider those NFSMs to be complete and observable.

3   Preset Homing Experiments

In this section, we propose an algorithm for deriving a preset homing sequence (PHS) for a given NFSM (if such a sequence exists). We start with a NFSM with two initial states and then show how the algorithm can be adapted for an arbitrary set of initial states.

3.1   Deriving a Preset Homing Sequence

3.1.1   Preliminaries

Given a complete NFSM $S = \langle S, I, O, h_S, \{s_1, s_2\} \rangle$, a sequence $\alpha \in I^*$ is a *homing sequence* (HS) for NFSM S if for the pair $\{s_1, s_2\}$ of initial states the following holds:

$$\forall \beta \in out(s_1, \alpha) \cap out(s_2, \alpha) \ [next\_state(s_1, \alpha/\beta) = next\_state(s_2, \alpha/\beta)] \tag{1}$$

We note that according to the well-known definitions, for a deterministic FSM $S = \langle S, I, O, h_S, \{s_1, s_2\} \rangle$, a sequence $\alpha \in I^*$ is a *homing sequence* if

$$next\_state(s_1, \alpha) \neq next\_state(s_2, \alpha) \Rightarrow out(s_1, \alpha) \neq out(s_2, \alpha) \tag{2}$$

and one can easily observe that (2) is a particular case of (1).

A homing sequence α is said to be *adaptive* (AHS) if the next input i_j is derived based on the output response $o_1 o_2 \dots o_{j-1}$ to the prefix $i_1 i_2 \dots i_{j-1}$. Otherwise, the homing sequence α is said to be *preset* (PHS).

A PHS not necessary exists even if a given complete observable nondeterministic machine is reduced and connected. As an example, consider the NFSM in Fig. 1 with the set $\{1, 2\}$ of initial states. If the machine outputs $o_1 \dots o_1$ to an input sequence $i \dots i$, we can never be sure which state is reached after this input sequence. However, a PHS always exists if there is an input sequence that separates the initial states s_1 and s_2 of a given NFSM.

Proposition 1. Given NFSM $S = \langle S, I, O, h_S, \{s_1, s_2\} \rangle$, if states s_1 and s_2 are separable then S has a PHS.

Proof. By definition, a separating sequence of states s_1 and s_2 is a PHS for NFSM S.

□

The condition of Proposition 1 is only sufficient but not necessary. However, there exist NFSMs with two initial states for which only a separating sequence can be a PHS. The following statement describes NFSMs with such features.

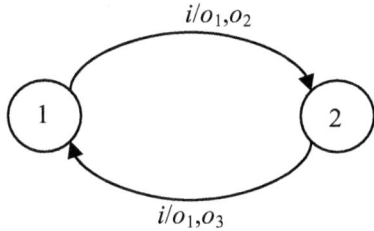

Fig. 1. A NFSM with no PHS

Proposition 2. Given NFSM $S = \langle S, I, O, h_S, \{s_1, s_2\}\rangle$, let S possess the following property:

$$\forall \alpha \in I^* \exists \beta \in out(s_1, \alpha) \cap out(s_2, \alpha) \; [next\_state(s_1, \alpha/\beta) \neq next\_state(s_2, \alpha/\beta)],$$

There exists a PHS for NFSM S if and only if states s_1 and s_2 are separable. Moreover, if states s_1 and s_2 are separable then each PHS is a separating sequence of states s_1 and s_2.

Proof. A PHS exists for NFSM S iff there exists an input sequence α such that for each $\beta \in out(s_1, \alpha) \cap out(s_2, \alpha)$ it holds that $[next\_state(s_1, \alpha/\beta) = next\_state(s_2, \alpha/\beta)]$. However, according to the proposition condition, for a NFSM S it holds that $\forall \alpha \in I^* \exists \beta \in out(s_1, \alpha) \cap out(s_2, \alpha) \; [next\_state(s_1, \alpha/\beta) \neq next\_state(s_2, \alpha/\beta)]$. The above two statements can be satisfied simultaneously if and only if $out(s_1, \alpha) \cap out(s_2, \alpha) = \varnothing$. The latter means that states s_1 and s_2 are separable and moreover, each PHS separates states s_1 and s_2.

$\square$

The conditions of Proposition 2 are only sufficient but not necessary. As an example, consider an FSM in Fig. 2 and an input sequence i. If the machine produces an output o_1 or o_2 to an input i then the machine enters state 2 independent of the initial state 1 or 2. If an output o_3 is produced to an input i then the machine enters state 1, since this output can be produced only when the machine is at state 2. The tight upper bound on the length of a separating sequence is known to be exponential [16]. According to Proposition 2, it can be expected that the upper bound on the length of a shortest PHS is also exponential. In order to confirm this, consider two NFSMs S and T with n states. In [17], it is shown that initial states of these machines can be separated only with a

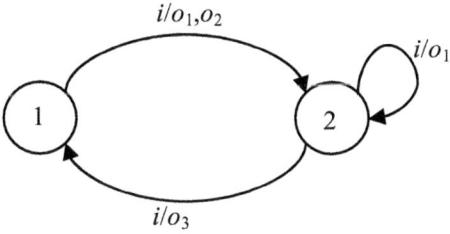

Fig. 2. FSM S that has a PHS despite of the fact that its states are not separable

sequence that has at least 2^{n^2-1} inputs. By direct inspection, one can assure that the conditions of Proposition 2 hold for the direct sum [9] of NFSMs S and T when the initial states s_0 and t_0 of S and T are two initial states of the direct sum. For this reason, only a separating sequence of s_0 and t_0 can be a PHS for such a direct sum and thus, since the direct sum has $m = 2n$ states, a shortest PHS for the direct sum has length $2^{(m/2)^2-1}$, i.e., is of the order 2^{m^2}.

3.1.2 Deriving a Preset Homing Sequence for a NFSM with Two Initial States

Given a NFSM $S = \langle S, I, O, h_S, \{s_1, s_2\}\rangle$ and the pair $\{s_1, s_2\}$ of initial states, in order to derive a PHS for S, we construct a truncated successor tree (TST) with proper termination rules given below.

The root of the tree is a node labeled with the pair $\{s_1, s_2\}$ while other tree nodes are labeled by sets of pairs $\{s, s'\}$, $s, s' \in S$, $s \neq s'$, or by the empty set because of observability of NFSM. Edges of the tree are labeled by inputs. Given an input i and an intermediate node labeled by a set P, there is an edge labeled with i to the node labeled with the following set P':

$\{s, s'\} \in P' \Leftrightarrow \exists \{s_i, s_j\} \in P \; \exists o \in O \; ((s_i, i, o, s) \in h_S \; \& \; (s_j, i, o, s') \in h_S)$ and $s \neq s'$.

Given a node labeled by a set P, the node is *terminal* if

Rule-1: Set P contains a subset that labels another node of the path from the root to the node labeled by the set P;

Rule-2: P is the empty set.

We note that termination Rule-1 allows truncating from TST branches that will never provide minimal length paths to a node labeled by the empty set. Correspondingly, we propose the following algorithm.

Algorithm 1 for deriving a minimal length PHS for a NFSM with two initial states.

Input: NFSM $S = \langle S, I, O, h_S, \{s_1, s_2\}\rangle$ with n states

Output: A minimal length PHS for S or a proper message when such a sequence does not exist

Derive a truncated successor tree with the root $\{s_1, s_2\}$.

If the successor tree has no nodes labeled with the empty set

Return the message "there is no PHS for NFSM S".

Otherwise,

Determine a path with minimal length to a node labeled with the empty set;

Return as a PHS the input sequence α that labels the selected path.

$\square$

By construction of the truncated successor tree, the following statements hold.

Theorem 1. Complete NFSM S has a PHS if and only if the TST derived by Algorithm 1 has a node labeled with the empty set. Moreover, a shortest PHS α of S labels a path in TST from the root to a node with the empty set.

Proof. Indeed, by definition of PHS, a PHS can only label a path to a node labeled with the empty set. Suppose there exists PHS $\alpha = \eta\gamma\sigma$ for S such that η takes the FSM

S from the pair of initial states to all pairs of the set P while $\eta\gamma$ takes the FSM S from the pair of initial states to all pairs of the set P' and $P \subseteq P'$. As $\eta\gamma\sigma$ is a HS for S, then σ is a homing sequence for each pair of the set P', i.e., σ is a homing sequence for each pair of the set P, and thus, $\eta\sigma$ is also a PHS for S.

□

Proposition 3. Algorithm 1 returns a PHS of length at most $2^{n(n-1)/2-1}$.

Proof. The maximal number of different pairs $\{s, s'\}$, $s \neq s'$, is $n(n-1)/2$, thus, the maximal number of different sets of such pairs equals $2^{n(n-1)/2}$.

According to the termination Rule-1, there is no path in the TST that has an intermediate node labeled by a set P containing the pair $\{s_1, s_2\}$. The maximal number of different sets of pairs of different states which do not include the pair $\{s_1, s_2\}$ equals $2^{n(n-1)/2-1}$.

□

Proposition 4. If observable NFSM S has a PHS then there exists a PHS with length at most $2^{n(n-1)/2-1}$.

Proof. Actually, the proposition holds since the maximal depth of the successor tree for NFSM S with two initial states is a most $2^{n(n-1)/2-1}$.

□

3.1.2 Deriving a Preset Homing Sequence for NFSM with Arbitrary Number of Initial States

Given NFSM $S = \langle S, I, O, h_S, S' \rangle$, $S' \subseteq S$, $| S' | = m \geq 2$, a sequence $\alpha \in I^*$ is a *homing sequence* (PHS) for S if for each subset $\{ S_{i_1}, ..., S_{i_j} \} \subseteq S'$, the following holds:

$$\forall \beta \in out(S_{i_1}, \alpha) \cap out(S_{i_2}, \alpha) \cap ... \cap out(S_{i_j}, \alpha) \ [next\_state(S_{i_1}, \alpha/\beta) =$$

$$next\_state(S_{i_2}, \alpha/\beta) = ... = next\_state(S_{i_j}, \alpha/\beta)], j \in \{1, ..., m\}, \qquad (3)$$

$$\{ S_{i_1}, S_{i_2}, ..., S_{i_j} \} \subseteq S'$$

Proposition 5. Given observable NFSM $S = \langle S, I, O, h_S, S' \rangle$, if there exists an input sequence $\alpha \in I^*$ such that for each pair $\{ S_{i_1}, S_{i_2} \} \subseteq S'$ it holds that $\forall \beta \in out(S_{i_1}, \alpha) \cap out(S_{i_2}, \alpha) \ [next\_state(S_{i_1}, \alpha/\beta) = next\_state(S_{i_2}, \alpha/\beta)]$, then α is a PHS for NFSM S.

Proof. Consider an input sequence $\alpha \in I^*$ such that for each pair $\{ S_{i_1}, S_{i_2} \} \subseteq S'$ it holds that $\forall \beta \in out(S_{i_1}, \alpha) \cap out(S_{i_2}, \alpha) \ [next\_state(S_{i_1}, \alpha/\beta) = next\_state(S_{i_2}, \alpha/\beta)]$ and a subset $\{ S_{i_1}, ..., S_{i_j} \} \subseteq S'$. Let $\beta \in out(S_{i_1}, \alpha) \cap out(S_{i_2}, \alpha) \cap ... \cap$

$out(S_{i_j}, \alpha)$. Since the FSM is observable, the set $next\_state(S_{i_1}, \alpha/\beta)$ is a singleton and moreover, for each two states S_{i_1} and S_{i_2}, $\{S_{i_1}, S_{i_2}\} \subseteq \{S_{i_1}, \ldots, S_{i_j}\}$, $next\_state(S_{i_1}, \alpha/\beta) = next\_state(S_{i_2}, \alpha/\beta)$, i.e., α is a HS for the FSM S.

$\square$

Proposition 5 shows how a PSH can be derived for a NFSM $S = \langle S, I, O, h_S, S' \rangle$ with the set S' of initial states where $|S'| > 2$. A TST is constructed as in Algorithm 1; however, the root of the tree is labeled by the set P of all pairs $\{S_{i_1}, S_{i_2}\} \subseteq S'$, $S_{i_1} \neq S_{i_2}$. As before, the TST is trimmed according to the above rules, i.e., Rule-1 and Rule-2. Thus, according to Proposition 4, the maximal length of a PHS for a given NFSM S with m initial states, $m = |S'| > 2$, is at most $2^{n(n-1)/2}$.

4 Deriving Adaptive Homing Sequences

In the previous section it has been shown that a method for deriving a PHS for two initial states of a given NFSM can be used for deriving a PHS for more than two initial states. For this reason, in this section, deriving an AHS we start directly with an arbitrary set of two or more initial states.

Given NFSM $S = \langle S, I, O, h_S, S' \rangle$, $S' \subseteq S$, $|S'| = m > 1$, in order to derive an AHS for S we construct an adaptive truncated successor tree (ATST). The difference between a TST described in Section 3 and an ATST is that each edge of the TST is labeled by an input while an edge of the ATST is labeled by an input/output pair. The root of an ATST is labeled with the set S' of initial states while other tree nodes are also labeled by subsets of states of the NFSM S. Given an intermediate node labeled by a subset P of states and an input/output pair i/o such that there is a transition labeled by i/o at least for one state of the set P, there is an edge labeled with i/o to the node labeled with the following subset P':

$$s \in P' \Leftrightarrow \exists s_i \in P \, ((s_i, i, o, s) \in h_S).$$

We note that since NFSM S is observable, for every state in P, there exists at most one state s with the above property. This is why ATST nodes are labeled by subsets of states differently from the TST where nodes are labeled by sets of state pairs. In addition, since S is observable, in the ATST, it holds that $|P'| \leq |P|$. Given a node labeled by a set P, the node is *terminal* if

Set P contains a subset of states that labels another node of the path from the root to the node labeled by the set P;

$|P| \leq 1$.

The number of different subsets with j items of the set with n states equals C_n^j. Since given a state s and an input/output pair i/o, there can be only one i/o–successor of state s, the following statement holds.

Proposition 6. If observable NFSM S has an AHS then there is a path in the ATST to a node labeled with a singleton. Moreover, in this case there exists an AHS with length at most $\displaystyle\sum_{j=2}^{m} C_n^j$.

Proof. Since the number of subsets which have at least two states of the set with n states does not exceed $\displaystyle\sum_{j=2}^{m} C_n^j$ the following statement holds.

$\square$

Corollary. Given observable NFSM $S = \langle S, I, O, h_S, S \rangle$, $| S | = n$, if NFSM S has an AHS then there exists an AHS with length at most $2^n - n - 1$.

5 Conclusion

In this paper, we have revisited the problem of deriving homing experiments for nondeterministic FSMs and have proposed necessary and sufficient conditions for checking whether a complete observable nondeterministic FSM possesses a preset or an adaptive homing sequence. Moreover, we have shown that length of a preset homing sequence if of order 2^{n^2} and there exist FSMs which have a preset homing sequence of this order. We have also discussed how an adaptive homing experiment can be derived. The upper bound on the height of an adaptive homing experiment for a NFSM with m initial states, $m \leq n$, states is of order $\displaystyle\sum_{j=2}^{m} C_n^j$ and this upper bound is of order 2^n when m tends to n.

In our future work, we are going to adapt the obtained results to non-observable possibly partial nondeterministic machines. It is also interesting to check whether the upper bounds on length of homing sequences are tight.

References

1. Moore, E.F.: Gedanken-experiments on sequential machines. In: Automata Studies (Annals of Mathematical Studies), vol. 1, pp. 129–153. Princeton University Press, Princeton (1956)
2. Kohavi, Z.: Switching and Finite Automata Theory. McGraw- Hill, New York (1978)
3. Mathur, A.: Foundations of Software Testing. Addison-Wesley (2008)
4. Lee, D., Yannakakis, M.: Testing finite-state machines: state identification and verification. IEEE Trans. on Computers 43(3), 306–320 (1994)
5. Ravikumar, B.: Parallel algorithms for finite automata problems. LNCS, vol. 1388, p. 373. Springer, Berlin (1998)
6. Sandberg, S.: 1 Homing and synchronizing sequences. In: Broy, M., Jonsson, B., Katoen, J.-P., Leucker, M., Pretschner, A. (eds.) Model-Based Testing of Reactive Systems. LNCS, vol. 3472, pp. 5–33. Springer, Heidelberg (2005)

7. Ginsburg, S.: On the length of the smallest uniform experiment which distinguishes the terminal states of a machine. Journal of the ACM 5(3), 266–280 (1958)
8. Hibbard, T.N.: Lest upper bounds on minimal terminal state experiments of two classes of sequential machines. Journal of the ACM 8(4), 601–612 (1961)
9. Gill, A.: State-identification experiments in finite automata. Information and Control, 132–154 (1961)
10. Milner, R.: Communication and Concurrency. Prentice-Hall (1989)
11. Rystsov, I.: Rank of finite automata. Cybernetics and Systems Analysis 28(3), 323–328 (1992)
12. Burkhard, H.V.: Zum Langenproblem homogener Experimente an determinierten und nicht-deterministischen. EIK 12, 301–306 (1976)
13. Imresh, B., Steinby, M.: Directable non-deterministic automata. Acta Informatica 14, 105–115 (1999)
14. Imresh, B., Imresh, C., Ito, M.: On directable non-deterministic trapped automata. Acta Informatica 16, 37–45 (2003)
15. Starke, P.: Abstract Automata, pp. 3–419. American Elsevier, Amsterdam (1972)
16. Alur, R., Courcoubetis, C., Yannakakis, M.: Distinguishing tests for nondeterministic and probabilistic machines. In: Proc. the 27th ACM Symposium on Theory of Computing, pp. 363–372 (1995)
17. Spitsyna, N., El-Fakih, K., Yevtushenko, N.: Studying the separability relation between finite state machines. Software Testing, Verification and Reliability 17(4), 227–241 (2007)

Towards More Expressive 2D Deterministic Automata

Violetta Lonati[1] and Matteo Pradella[2]

[1] Dipartimento di Scienze dell'Informazione, Università degli Studi di Milano
Via Comelico 39/41, 20135 Milano, Italy
`lonati@dsi.unimi.it`
[2] Dipartimento di Elettronica e Informazione, Politecnico di Milano
Via Ponzio 34/5, 20133 Milano, Italy
`matteo.pradella@polimi.it`

Abstract. REC defines an important class of picture languages that is considered a 2D analogous of regular languages. In this paper we recall some of the most expressive operational approaches to define deterministic subclasses of REC. We summarize their main characteristics and properties and try to understand if it is possible to combine their main features to define a larger deterministic subclass. We conclude by proposing a convenient generalization based on automata and study some of its formal properties.

1 Introduction

Generalizing string languages and related approaches to two dimensions has always been tempting, as pictures are an important part of our life, as well as text. One of the most successful classes of picture languages introduced in the literature is surely REC, the class of *tiling recognizable* languages [1], that aims at generalizing to 2D the class of regular string languages. REC is a robust class that has various characterizations: e.g. in terms of *online tessellation acceptors* [2], *tiling systems* [3], or *Wang systems* [4].

Unfortunately, many good properties of string languages are modified or are lost in the transition towards the two dimensions. One such property is related to *determinism*: all the proposed 2D analogous of regular languages lose expressivity if constrained to deterministic models.

Essentially two approaches are proposed in the literature for defining a deterministic model of finite state automaton within REC. The first one is presented in the seminal work [5], which clearly predates REC and actually define a subclass. In this approach, the input picture is seen as a read-only tape that can be visited freely, and finite states are exploited to propagate information (see also [6] for an account of the main properties of the model).

Another, orthogonal approach is based on fixing a *scanning strategy* to visit the input picture, and allowing to add marking information to its pixels, so that it is possible to propagate information locally. The first of such models is the one of the deterministic online tessellation acceptors (or DOTAs), a kind of cellular

B. Bouchou-Markhoff et al. (Eds.): CIAA 2011, LNCS 6807, pp. 225–237, 2011.

automata [2]. This approach was then generalized and extended by the following subclasses and models, presented in chronological order: class Diag-DREC [7,8], the closure by rotation of the class defined by DOTAs; *tiling automata* [9]; snake-deterministic tiling systems and class Snake-DREC [10], in which scanning strategies follow a boustrophedonic order; then the more recent μ-*directed Wang automata* and class Scan-DREC [11,12,13].

The literature has already considered and studied the relation among subclasses within the same basic approach, but our knowledge of the actual general situation is still quite partial. The aim of this paper is to consider and analyze the inherent characteristics of these two main families of approaches, in order to get a clearer idea of the picture (no pun intended), and obtain a larger deterministic class, yet still in REC.

The paper has the following structure: it first presents the preliminaries and basic notions. In Section 3 it then considers how to add expressivity and the problem of remaining within REC. In it a new deterministic model of automaton is presented. To conclude, Section 4 studies some properties of the model.

2   Preliminaries

The following definitions are taken and adapted from [1]. Let Σ be a finite alphabet. A two-dimensional array of elements of Σ is a *picture* over Σ. A picture having n rows and m columns has *size* (n, m). $\# \notin \Sigma$ is used when needed as a *boundary symbol*; $\hat{p}$ refers to the bordered version of picture p. For instance

$$p = \begin{array}{|c|c|c|} \hline p(1,1) & \cdots & p(1,m) \\ \hline \vdots & \ddots & \vdots \\ \hline p(n,1) & \cdots & p(n,m) \\ \hline \end{array} \qquad \hat{p} = \begin{array}{|c|c|c|c|c|} \hline \# & \# & \cdots & \# & \# \\ \hline \# & p(1,1) & \cdots & p(1,m) & \# \\ \hline \vdots & \vdots & \ddots & \vdots & \vdots \\ \hline \# & p(n,1) & \cdots & p(n,m) & \# \\ \hline \# & \# & \cdots & \# & \# \\ \hline \end{array}.$$

A *pixel* is an element $p(i, j)$ of p. We call (i, j) the *position* in p of the pixel. We say that $(i-1, j)$, $(i+1, j)$, $(i, j-1)$, and $(i, j+1)$ are *adjacent* to position (i, j).

The set of all pictures over Σ is Σ^{++}. A picture language is a subset of Σ^{++}. If D denotes some kind of picture-accepting device, then $\mathcal{L}(D)$ denotes the class of picture languages recognized by such devices.

We will sometimes consider the 90° clockwise *rotation*, the *horizontal mirror*, and the *vertical mirror* of a picture p. E.g. if $p = \begin{array}{|c|c|} \hline a & b \\ \hline c & d \\ \hline \end{array}$, then $\begin{array}{|c|c|} \hline c & a \\ \hline d & b \\ \hline \end{array}$, $\begin{array}{|c|c|} \hline c & d \\ \hline a & b \\ \hline \end{array}$, and $\begin{array}{|c|c|} \hline b & a \\ \hline d & c \\ \hline \end{array}$ are its rotation, horizontal mirror and vertical mirror, respectively. Naturally, the same operations can be applied to languages, and classes of languages, too.

2.1   Tiling Recognizable Picture Languages

An important class of two-dimensional languages is REC, i.e., the class of *tiling-recognizable languages*, originally defined in terms of tiling systems [3]. Another equivalent definition [14] is given by using online tessellation acceptors, OTAs, first introduced in [2]. Here we define REC by using the equivalent notation introduced in [4], which is based on a variant of Wang tiles.

Labeled Wang tiles. Let Σ be a finite alphabet and K be a set of colors, containing the special color # representing borders. A *labeled Wang tile* (or *tile* for short) is a unitary square with colored edges and a *label* in Σ. Formally, a tile is an element $A = (a, t, l, r, b) \in \Sigma \times K^4$, where t, b, r, l represent the colors at top, bottom, right and left edges, respectively. For better readability, we represent labeled Wang tiles as $A = l\,\boxed{a}\,r$ with t on top and b on bottom. For any direction $d \in Dirs = \{\uparrow, \rightarrow, \leftarrow, \downarrow\}$, A_d is the color of the edge of A towards direction d. We also use $-d$ for referring to the direction opposite to d. The set of tiles with labels in Σ and colors in K is Σ_{4K}. We also consider *partial* tiles, where some colors may be undefined: the set of partial tiles is denoted by Σ_K. The *domain* of a tile A is the set Δ_A of directions where A is defined. Given two partial tiles A, B bearing the same label, we say that B extends A if $B_d = A_d$ for every $d \in \Delta_A$. When we need to emphasize the fact that a tile is not partial, we will call it *complete*.

Wang pictures. Labeled Wang tiles in Σ_{4K} can be used to build pictures over Σ, by using colors to check compatibility: two tiles may be adjacent only if the color of the touching edges is the same. A picture $P \in \Sigma_{4K}^{++}$ is called a *Wang picture* if all borders are colored with # and $P(i, j)_\downarrow = P(i+1, j)_\uparrow$ for every $1 \le i < n$, and $P(i, j)_\rightarrow = P(i, j+1)_\leftarrow$ for every $1 \le j < m$, where (n, m) is the size of P. The *label* of a *Wang picture* P over Σ_{4K} is the picture having for pixels the labels of pixels of P. Next (on the left), the reader may find the example of a Wang picture of size $(2, 2)$ with its label (in the middle). For better readability, we represent Wang pictures by writing each common color only once, as in the figure on the right.

Sometimes we need to consider *partial* Wang pictures, whose pixels are partial tiles with compatible edges (some colors may be undefined). Any (partial) Wang picture is called a *(partial) Wang tiling* of its label.

Wang systems. A *Wang system* is a triple $\omega = \langle \Sigma, K, \Theta \rangle$, where Σ is a finite alphabet, K is a set of colors, Θ is a subset of Σ_{4K}. The language generated by ω is the language $L(\omega) \subseteq \Sigma^{++}$ of the labels of all Wang pictures built with tiles in Θ. Notice that a picture $p \in L(\omega)$ may have more than one Wang tiling in ω. REC is the class of picture languages generated by Wang systems.

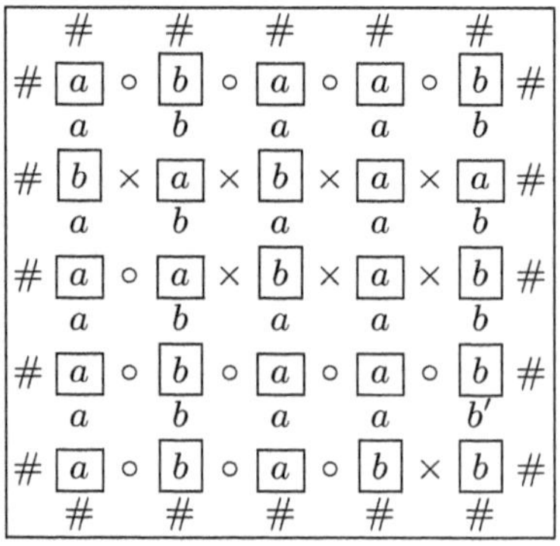

Fig. 1. A picture recognized by the Wang system of Example 1 and the corresponding tiling

Example 1. Let $L_{\exists r=1r}$ be the language of all pictures that have a row which equals the first row. L is recognized by the Wang system producing tilings as in Figure 1. In it, symbols from the first row are propagated downwards, and each row is examined from left to right to check its compatibility with the first row: if a wrong symbol is found, color $\times$ is propagated rightwards till the end of the row. If a row is found to be equal to the first one, a primed version of its rightmost symbol is propagated downwards. The picture is recognized only if the bottom-right corner is colored by a primed symbol (or the last row is checked as compatible and its rightmost symbol matches).

2.2   2D Automata Models

In the literature, several models of 2D automata have been proposed.

4-way automata. Historically, the first generalization of finite state automata to two dimensions is given by 4-way automata [5]. Soon after this paper, several other similar models have been proposed: a survey can be found in [6]. As the standard model of finite-state automata on strings, a 4-way automaton can be seen as a finite control having a head that visits the positions of a picture and can move in four directions. At each step, it reads the input symbol under the head, then it enters a new state and moves to an adjacent position: the direction to move towards and the new state to enter are determined by a transition function, according to the read symbol and the current state. The input picture is accepted if, starting from position $(1,1)$ in state q_0, the automaton eventually halts in state q_{yes}.

Definition 1. *A 4-way nondeterministic automaton (4NA) is a tuple $\langle \Sigma, Q, q_0, q_{yes}, q_{no}, \delta \rangle$ where: Σ is a finite input alphabet; Q is a finite set of states, containing in particular the initial state q_0, the accepting state q_{yes}, and the rejecting state q_{no}; $\delta : \Sigma \times (Q \setminus \{q_{yes}, q_{no}\}) \to 2^{Q \times Dirs}$ is the transition function.*

Example 2. In [15] it is proved that the language L of square pictures with the first row of the form $w\bar{w}$, where $\bar{w}$ is the reverse of w, is recognizable by a 4-way automaton. One can see that the same holds if L is generalized to pictures of size (n, m), with $n \geq m \geq 4$. We will call the latter language L_{half}.

Online tessellation acceptors. OTAs are defined in [2] as a restricted type of 2D cellular automaton in which cells do not make transitions at every time-step: rather a "transition wave" sweeps diagonally across the array. Each cell changes its state depending on the two neighbors to the top and to the left. A *run* of a OTA on a picture p of size (n, m) assigns a state (from a finite set) to each position (i, j) of p. Such state depends on the states already associated with positions $(i - 1, j)$ and $(i, j - 1)$ and on symbol $p(i, j)$. At time $t = 0$ an initial state q_0 is associated with all the positions of the first row and of the first column of $\hat{p}$. The computation starts at time $t = 1$ by reading $p(1, 1)$; at time $t = 2$, states are simultaneously assigned to positions $(1, 2)$ and $(2, 1)$, and so on, to the next diagonals. Picture p is recognized if there exists a run such that the state assigned to position (n, m) is final.

Wang automata directed by polite scanning strategies. Recently [11,12], we introduced μ-directed Wang automata (μ-NWA), a model of automata based on Wang tiles and leaded by a prefixed scanning strategy μ. A Wang automaton can be seen as having a head that visits the input picture, coloring at each step the edges of the position it is visiting. For each accepting computation, the automaton produces a complete Wang picture whose label is equal to the input picture. The coloring operations the automaton performs are determined by a finite control, whereas the movements of the head are lead by the scanning strategy μ; we requires that μ is *polite*, i.e., it has to satisfy some further properties. Fix any starting corner c_s and any starting direction $d_s \in Dirs$, and consider a *next-position function*, i.e., a partial function $\eta : 2^{Dirs} \times Dirs \to Dirs$ such that $\eta(D, d) = \bot$ if $-d \notin D$. The *scanning strategy* $\mu = \langle \eta, c_s, d_s \rangle$ determines how to visit any input picture. More precisely, let d be the current direction, representing the direction from the last considered position, and D represent the set of edges on the picture border together with the edges common with other already visited positions; then $\eta(D, d)$ is the direction towards the position to visit next.

In [13] we proved that any polite scanning strategy has to follow, except for some bootstrap steps, one of four kinds of movements, or their rotations and symmetrical, intuitively exemplified by the following pictures, where the number in each pixel denotes its scanning order.

1	6	7	12
2	5	8	11
3	4	9	10

(a) snake ($\mathcal{S}$)

1	10	11	12
2	9	8	7
3	4	5	6

(b) L-like ($\mathcal{L}$)

1	12	9	8
2	11	10	7
3	4	5	6

(c) U-like ($\mathcal{U}$)

1	10	9	8
2	11	12	7
3	4	5	6

(d) spiral ($\mathcal{C}$)

In the rest of the paper we will sometimes refer to rotations of the snake-like strategy (i.e. $\mathcal{S}$). The one depicted here is called the *left-to-right* version (denoted

by $\mathcal{S}^{l2r}$), while its rotation is called *top-to-bottom* ($\mathcal{S}^{t2b}$). Their respective 180°
rotations are called $\mathcal{S}^{r2l}$ and $\mathcal{S}^{b2t}$.

Definition 2. *A μ-directed nondeterministic Wang automaton (μ-NWA) is a
tuple $\langle \Sigma, K, \delta, \mu, F \rangle$ where: Σ is a finite input alphabet; K is a finite set of
colors; $\delta : \Sigma_K \times Dirs \rightarrow 2^{\Sigma_{4K}}$ is a partial function such that each tile in $\delta(A, d)$
extends A; $\mu = \langle \eta, c_s, d_s \rangle$ is a polite scanning strategy such that $\delta(A, d) \neq \emptyset$
implies $\eta(\Delta_A, d) \neq \bot$; $F \subset \Sigma_{4K}$ is the set of final tiles.*

Intuitively, the behavior of a μ-directed Wang automaton over an input picture
p is the following. At the beginning, the head of the automaton points at the
position in the starting corner c_s and the current direction is set to d_s. When
the current direction is d, the head is at position x, the pixel and the colors of
edges of $p(x)$ are summarized by A, then let $d' = \eta(\Delta_A, d)$ and $A' \in \delta(A, d)$.
The automaton can execute this move: color the edges of x according to A', set
the current direction to d', and move to the position adjacent to x following
direction d'. If no move is possible, the automaton halts. The input picture p is
accepted if there exists a computation such that the edges of the final position
are colored according to some tile in F.

The choice of the scanning strategy μ is not relevant from the point of view
of the recognizing power of μ-directed Wang automata: for every polite scanning
strategy μ, the class of picture languages recognized by μ-NWA equals REC [12].

Deterministic models. The deterministic versions DOTA, 4DA, μ-DWA of OTA,
4NA, and μ-NWA, respectively, are defined in the usual way, by making the transition functions deterministic. $\mathcal{L}$(DOTA) is characterized in terms of diagonal-
deterministic tiling systems [7]: here we use Diag-DREC to denote its closure
under rotation. For each polite scanning strategy μ, a subclass $\mathcal{L}(\mu$-DWA) of
REC is obtained; Scan-DREC is the union of all such deterministic classes [12].

All these models are strictly less powerful than their nondeterministic coun-
terparts, i.e., the corresponding classes of languages are properly included in
REC. Their inclusion relations are summarized as follows. First, DOTAs are
incomparable with both 4DA and 4NA [2]. Second, in [12], $\mathcal{L}(\mathcal{S}^{t2b}$-DWA) is
proved to coincide with t2b-UREC, a class introduced in [7] and, more gen-
erally, one can see that $\mathcal{L}(\mathcal{S}^d$-DWA) coincides with d-UREC for any direction
$d \in \{t2b, b2t, l2r, r2l\}$. Since Diag-DREC is properly contained in the union of
all classes d-UREC [7], we have that Scan-DREC properly extends Diag-DREC
and hence also $\mathcal{L}$(DOTA).

The relation between Wang automata and 4-way automata is not clear yet.
We do not know any example of language in 4DA that does not belong to
Scan-DREC; on the other hand, there exists a language $L \in$ Scan-DREC that
cannot be recognized by a 4NA (and a fortiori not by a 4DA), as shown in the
following example.

Example 3. Consider again $L_{\exists r=1r}$ as defined in Example 1. Such language cannot be recognized by a 4-way automaton [14]. But it is both in t2b-UREC and
in l2r-UREC [7], hence the corresponding $\mathcal{S}^{t2b}$-DWA and $\mathcal{S}^{l2r}$-DWA can be built:

they basically produce the tiling of Figure 1, except for a delay of one row in case of S^{t2b}-DWA, and one column in case of S^{l2r}-DWA.

It is interesting to note that, for every direction d, a necessary condition is known for a language to belong to d-UREC, and hence to $\mathcal{L}(S^d$-DWA), since those classes coincide [12]. Such condition is based on Matz's technique [16], that suggests to consider a picture as a string over the alphabet of columns (or rows), and Hankel matrices.

The Hankel matrix of a *string* language $S \subseteq \Omega^*$ is an infinite boolean matrix M_S indexed by words $\alpha, \beta \in \Omega^*$. M_S is defined by setting $M_S(\alpha, \beta) = 1$ if and only if $\alpha\beta \in S$. Let L be a picture language and, for every m, let $L(m)$ be the language of pictures in L having m rows. Then $L(m)$ can be seen as a string language over the alphabet $\Omega = \Sigma^m$ of columns of size m. In [7] it is proved that $L \in$ t2b-UREC implies that there exists an integer k such that, for every m, the number of distinct rows of the Hankel matrix $M_{L(m)}$ is lower than k^m. Similar properties can be given for any direction d.

Example 4. The 4NA cited in Example 2 and recognizing $L = L_{\text{half}}$ is deterministic. On the other hand, here we prove that L cannot be recognized by any S^{l2r}-DWA. For sake of simplicity, we assume that all rows except the first one are filled with symbol 0. Let us study the Hankel matrix $M_{L(m)}$ for a fixed m. One can verify that $\alpha \neq \alpha'$ implies that the rows of $M_{L(m)}$ indexed by α and α' differ. In other words, the number of distinct rows in $M_{L(m)}$ is not bounded w.r.t. m. Hence, the necessary condition stated above does not hold and this means that $L_{\text{half}} \notin \mathcal{L}(S^{l2r}$-DWA). Similarly, one has $L_{\text{half}} \notin \mathcal{L}(S^{r2l}$-DWA).

3   Adding Expressivity

In general, two main approaches are proposed in the literature in the attempt of defining a deterministic model of automaton. The first one considers the input picture as a read-only tape that can be visited freely, and uses finite states to propagate information [5,6]. The second one fixes a scanning strategy to visit the input picture, but allows the possibility to mark its positions, and use this markers to propagate information locally: each position is marked with a state by DOTAs [2]; it is rewritten with a symbol from a new alphabet in Diag-DREC [7,8]; its edges are colored by Wang automata [12,13].

In this section we try to combine these two apparently orthogonal approaches, in order to improve their expressive power: the idea is to use both states and colors assigned to positions. Clearly we want to stay inside REC, hence this combination must be done carefully.

The first idea is to imagine an automaton with a head that is able to move through the input picture according to its content, and depending on a finite control, changing its state at each step. Such head should move from a position to an adjacent one, and color at each step some edges of the position it is visiting.

In all models presented in the literature, the coloring operation in a given position is done once and for ever; similarly we do not admit the possibility to change a color previously assigned to a position. This leads to the following tentative definition. A 2D *free* deterministic automaton is defined by a tuple $\langle \Sigma, K, Q, q_0, q_{yes}, q_{no}, \delta \rangle$ where: Σ is a finite input alphabet; K is a finite set of colors; Q is a finite set of states, containing in particular the initial state q_0, the accepting state q_{yes}, and the rejecting state q_{no}; finally $\delta : \Sigma_K \times (Q \setminus \{q_{yes}, q_{no}\}) \rightarrow \Sigma_K \times Q \times Dirs$ is a partial function such that $\delta(A, q) \ni (A', q', d)$ implies that A' extends A.

The behavior of a 2D free deterministic automaton $\langle \Sigma, K, Q, q_0, q_{yes}, q_{no}, \delta \rangle$ over an input picture $p \in \Sigma^{++}$ would be described informally as follows. At the beginning, the head of the automaton points at the top-left corner and the current state is set to q_0. When the current state is q, the head is placed at position x, the pixel and the colors of edges of p at position x are summarized by $P(x)$, then let $(A', q', d) \in \delta(P(x), q)$. Hence the automaton may execute this move: if A is partial, color edges at position x according to A', then enter state q', move to the position adjacent to x towards direction d, and extend P to the Wang picture P' with $P'(x) = A'$.

Notice that the head can visit any cell any number of times, but colors cannot be changed (A' is a Wang tile that extends A). If no move is possible, the automaton halts. The input picture p is accepted if there is a computation such that the automaton eventually enters state q_{yes}.

Example 5. Language $L_{\exists r=1r}$ can be recognized by an automaton that visits the input picture following a sort of "comb-like" movement. Next you find an accepted input picture and the partial Wang picture obtained by the computation; the symbols never read by the automaton are omitted.

a	b	a	a	b
b	a	b	a	a
a	a	b	a	a
a	b	a	a	b
a	b	a	b	a

The automaton uses only two colors: the *reject* color $\times$ when it finds out that a row is different from the first one, and another symbol $\circ$ to mark the edges of the position it is visiting for the first time. The set of states is $Q = \{q_0, q_{yes}, q_{no}\} \cup \Sigma \cup \bar{\Sigma}$, where symbols in $\bar{\Sigma}$ are a barred version of symbols in Σ, and they are used to distinguish the part of the computation when the head moves leftwards. The transition function is summarized in Figure 2.

Considering an input picture of size (n, m), the automaton repeats the following sequence of moves m times. For every $j < m$, it visits the top row from left

Left table (coloring steps) — columns: tile, q_0, τ

tile (domain)	q_0	τ
# / # ⊡σ / ∘	∘, σ, ↓	
# / ∘ ⊡σ / ∘	∘, σ, ←	
# / ∘ ⊡σ # / ∘	#, σ, ←	
∘ / # ⊡σ		−, τ, ↓ (with ∘ below)
∘ / ∘ ⊡σ or ∘ / ∘ ⊡σ		−, τ̄, ←
∘ / # ⊡σ / #		−, q_0, ↑
∘ / ∘ ⊡σ / # or ∘ / ∘ ⊡σ / #		−, q_0, ←
∘ / ∘ ⊡σ # / # or ∘ / ∘ ⊡σ # / #, with σ ≠ τ		#, τ̄, ←
∘ / ∘ ⊡σ # / # or ∘ / ∘ ⊡σ # / #, with σ ≠ τ		q_{no}
∘ / ∘ ⊡τ # / # or ∘ / ∘ ⊡τ # / # or ∘ / ∘ ⊡τ # / # or ∘ / ∘ ⊡τ # / #		q_{yes}

Right table (revisiting steps) — columns: tile, q_0, τ, $\bar\tau$

tile (domain)	q_0	τ	$\bar\tau$
# / # ⊡σ ∘ / ∘	$q_0, \to$	$\tau, \downarrow$	
∘ / ∘ ⊡σ ∘ / ∘	$q_0, \to$	$\tau, \leftarrow$	
∘ / # ⊡σ ∘ / ∘	$q_0, \uparrow$	$\tau, \to$	$\tau, \downarrow$
∘ / # ⊡σ × / ∘	$q_0, \uparrow$	$\tau, \downarrow$	
∘ / ∘ ⊡σ ∘ / ∘		$\tau, \to$	$\bar\tau, \leftarrow$
∘ / ∘ ⊡σ × / ∘			$\bar\tau, \leftarrow$
∘ / # ⊡σ ∘ / #	$q_0, \uparrow$	$\tau, \to$	
∘ / # ⊡σ × / #	$q_0, \uparrow$		
∘ / ∘ ⊡σ ∘ / #	$q_0, \leftarrow$	$\tau, \to$	
∘ / ∘ ⊡σ × / #	$q_0, \leftarrow$		

Fig. 2. The transition function of the automaton for language $L_{\exists r = 1r}$: coloring steps (left), revisiting steps (right). Rows are indexed by partial Wang tiles, columns are indexed by states. For revisiting steps, the complete Wang tile in the codomain is omitted since equals the tile in the domain. Notation − stands for × if $\tau \neq \sigma$, for ∘ otherwise.

to right until it reaches the first unvisited position, i.e. $(1, j)$. The symbol found there is saved in the state. Then, the automaton goes back to $(1, 1)$ and moves downward, to find all rows that are compatible with the first j symbols of the first row.

This task is performed as follows. Each row, starting from the second, is scanned from left to right until one of the following two cases occurs. (1) A reject color is found, so the row was already marked as unsuitable, and the automaton has to go back to the first column and then move to the next row. (2) There is an unvisited position that, by construction, is position (i, j), so either it contains the symbol saved in the state and it has to be marked just as visited, or it contains a wrong symbol and it is hence marked with the reject color. As in the previous case, the automaton has to go back to the first column and then move to the next row. Once the bottom row has been examined, the automaton enters state q_0 and moves leftward to the first column, then goes back up to the first row and then repeats the cycle by considering position $(1, j + 1)$.

When the last position in the first row is scanned, to accept the input picture the automaton has only to check if there is at least one non-rejected row ending with the right symbol.

The previous example shows a critical feature of the model: whenever the symbol at position $(1, j)$ is considered, the automaton enters a sort of loop (it goes across each row i until it reaches position (i, j) or finds the reject color, then it comes back to the first column), whose outcome is different according to a piece of information which is not locally propagated (i.e., whether the row has already been rejected or not). Clearly this sort of cyclic computation cannot be removed and this prevents to apply a construction similar to the one used in [2] to prove that $\mathcal{L}(4\mathrm{NA})$ is included in REC. Actually, it turns out that this model is really too permissive, as next example illustrates.

Example 6. Consider the language $L_{a^n b^n}$ of pictures with one row of the form $a^n b^n$. Since $L_{a^n b^n}$, seen as a string language, is not regular, clearly $L \notin \mathrm{REC}$. However $L_{a^n b^n}$ is recognizable by the following free automaton: the set of states is $Q = \{q_0, q_1, q_{yes}, q_{no}\}$ and the set of colors is $K = \{ok\}$; it starts from the top-left border, if the current symbol is a, then it marks its right edge by color *yes*, enters state q_1, and move rightwards without changing states nor coloring, until it reaches a position having the right edge already marked (or bordered); if the current symbol is b, then it marks its left edge by color ok, enters state q_0, and moves leftwards without changing states nor coloring, until it reaches a position having the left edge already marked (or bordered). Such sequence of moves is repeated until all positions are marked, and in this case the input picture is accepted. Whenever one of the previous conditions is not satisfied, the input picture is rejected.

Hence, the definition of free automaton needs to be somehow constrained. For instance one should require that the first time a position is visited, a coloring step must be performed. This is obtained simply by replacing the codomain of δ by $\Sigma_{4K} \times Q \times Dirs$. The new condition would prevent the behavior of the automaton in the previous example. However this is not sufficient yet, as next example illustrates.

Example 7. Consider the language $L_{2a^n b^n}$ of pictures with two rows, the first one having the form $a^n b^n$. Again, it is easy to see that $L_{2a^n b^n} \notin \mathrm{REC}$ (otherwise it would be easy to build a Wang system for $L_{a^n b^n}$, too). However $L_{2a^n b^n}$ is recognizable by a free automaton that further respects the above condition. The behavior of such automaton is similar to the one described in Example 6, except that the first row is used only to mark the position under consideration, and the second row allows to go back and forth from left to right.

These problems suggest that the combination of coloring and revisiting steps should be simplified: first the automaton executes a sequence of coloring steps, then it performs a sequence of revisiting steps, using the information enclosed in the colors placed before. We will call the two phases *tiling* and *roaming*, respectively. In other words, the automaton simulates first the behavior of a μ-DWA (in the tiling phase), for some prefixed scanning strategy μ, and then the behavior of a 4DA (in the roaming phase).
This leads to this new definition:

Definition 3. *A 4-way deterministic μ-directed Wang automaton (μ-4DWA) is a tuple $\langle \Sigma, K, \gamma, \mu, Q, q_0, q_{yes}, q_{no}, \delta \rangle$ where:*

- *Σ is a finite input alphabet;*
- *K is a finite set of colors;*
- *$\gamma : \Sigma_K \times Dirs \to \Sigma_{4K}$, the tiling transition function, is a partial function such that each tile in $\gamma(A, d)$ extends A;*
- *$\mu = \langle \eta, c_s, d_s \rangle$ is a polite scanning strategy such that $\gamma(A, d) \neq \emptyset$ implies $\eta(\Delta_A, d) \neq \bot$;*
- *Q is a finite set of states, containing in particular the initial roaming state q_0, the accepting state q_{yes}, and the rejecting state q_{no};*
- *$\delta : \Sigma_{4K} \times Q \to Q \times Dirs$ is the roaming transition function.*

The formal semantics is not presented but it is a straightforward combination of μ-DWA and 4DA: when the μ-DWA component ends its picture scanning, the 4DA component starts working from the current position.

Example 8. For instance, a $\mathcal{S}^{\mathrm{t2b}}$-4DWA for the language $L_{\exists r=1r} \cap L_{\mathrm{half}}$ can be defined as follows: first visit the input picture row by row from top to bottom, simulating the $\mathcal{S}^{\mathrm{t2b}}$-DWA defined in Example 3, to check if the input picture is in $L_{\exists r=1r}$; then simulate the 4DA cited in Example 2, to check if the input picture is in L_{half}.

4   Properties of 4-Way Deterministic Wang Automata

Theorem 1. *For every polite μ, $\mathcal{L}(4DA) \cup \mathcal{L}(\mu\text{-DWA}) \subseteq \mathcal{L}(\mu\text{-4DWA}) \subseteq REC.$*

Proof. It is easy to verify that both 4DA and μ-DWA (for μ polite) are special kinds of μ-4DWA. On the one hand, 4DA are obtained by reducing the set of colors used in the tiling part to the empty set, i.e., when Σ_K is simply Σ. On the other hand, μ-DWA are obtained by omitting the roaming part of the transition function.

Let L be accepted by some μ-4DWA $\mathcal{A}$. Clearly, $\mathcal{A}$ can be seen as the combination sof a μ-4DWA $\mathcal{A}_1$ over alphabet Σ and a 4DA $\mathcal{A}_2$ over alphabet Σ_K. Now, let $L_1, L_2 \subseteq \Sigma_K^{++}$ be defined as follows. L_1 the set of (partial) Wang tilings produced by all accepting computations of $\mathcal{A}_1$; L_2 is the language accepted by $\mathcal{A}_2$. Then, by definition $L = \lambda(L_1 \cap L_2)$, where $\lambda : \Sigma_K \to \Sigma$ is the projection that maps each tile onto its label. Since REC is closed under intersection and alphabetic projection, we have that $L \in$ REC.   □

Theorem 2. *For every polite μ, $\mathcal{L}(\mu\text{-4DWA})$ is a boolean algebra.*

Proof. By definition, since both $\mathcal{L}(\text{4DA})$ and $\mathcal{L}(\mu\text{-DWA})$ are boolean algebras [12].   □

Theorem 3. *There exists a language in $\mathcal{L}(\mathcal{S}\text{-4DWA})$ which is not in $\mathcal{L}(\text{4DA})$ nor in Scan-DREC.*

Proof. Consider the language $L_{\text{half}} \cap L_{\exists r=2r}$ of pictures having the first row of the form $w\bar{w}$, and some row that equals the second row. Let L be the intersection of such language with its horizontal mirror.

Language L can be recognized by a $\mathcal{S}^{l2r}$-4DWA that simulates first a variant of the $\mathcal{S}^{l2r}$-DWA of Example 3 (such variant examines at the same time both the second and the second-last rows instead of the first one only), then the 4DA of Example 2, and then again a rotated version of the same 4DA.

On the other hand, it is known [14] that 4NA cannot recognize $L_{\exists r=2r}$, hence a fortiori L cannot be recognized by any 4DA. Now we show that L is not in Scan-DREC. Since a $\mathcal{S}^{t2b}$-DWA cannot recognize $L_{\exists r=2r}{}^h$, as proved in [12], then L cannot be recognized by neither a $\mathcal{S}^{t2b}$-DWA nor a $\mathcal{S}^{b2t}$-DWA. Moreover, since no $\mathcal{C}$-DWA can recognize $L_{\exists r=2r}$ (see [12]), then clearly L cannot be recognized by any $\mathcal{C}$-DWA. By a similar reasoning one can prove that L is neither in $\mathcal{L}(\mathcal{U}\text{-DWA})$, nor in $\mathcal{L}(\mathcal{J}\text{-DWA})$, or in any of their rotations. Finally, reasoning as in Example 4 we can verify that L cannot be recognized by neither a $\mathcal{S}^{r2l}$-DWA nor a $\mathcal{S}^{l2r}$-DWA.

Hence, the result follows from the fact that all polite scanning strategies are basically limited to $\mathcal{C}, \mathcal{U}, \mathcal{J}, \mathcal{S}$ and their rotations and mirrors [13].   □

Notice that the language used in the previous proof to separate classes Scan-DREC and $\mathcal{L}(\text{4DA})$ from $\mathcal{L}(\mathcal{S}\text{-4DWA})$ actually separates also $\mathcal{L}(\text{4NA})$ from $\mathcal{L}(\mathcal{S}\text{-4DWA})$. It is not clear if there exists any language recognizable nondeterministically by a 4NA but not by μ-4DWA.

Concludingly, these last results clearly show that the proposed approach of combining the free roaming of 4-way automata and coloring based on predefined scanning strategies is effective. Indeed, this yields to a concept of determinism which extends those presented in [5,2,7,8,10,11,12,13].

References

1. Giammarresi, D., Restivo, A.: Two-dimensional languages. In: Salomaa, A., Rozenberg, G. (eds.) Handbook of Formal Languages, vol. 3, pp. 215–267. Springer, Berlin (1997)
2. Inoue, K., Nakamura, A.: Some properties of two-dimensional on-line tessellation acceptors. Information Sciences 13, 95–121 (1977)

3. Giammarresi, D., Restivo, A.: Recognizable picture languages. International Journal Pattern Recognition and Artificial Intelligence, Special Issue on Parallel Image Processing 6, 241–256 (1992)
4. de Prophetis, L., Varricchio, S.: Recognizability of rectangular pictures by Wang systems. Journal of Automata, Languages and Combinatorics 2, 269–288 (1997)
5. Blum, M., Hewitt, C.: Automata on a 2-dimensional tape. In: Conference Record of 1967 Eighth Annual Symposium on Switching and Automata Theory, pp. 155–160. IEEE, Los Alamitos (1967)
6. Inoue, K., Takanami, I.: A survey of two-dimensional automata theory. Information Sciences 55, 99–121 (1991)
7. Anselmo, M., Giammarresi, D., Madonia, M.: From determinism to nondeterminism in recognizable two-dimensional languages. In: Harju, T., Karhumäki, J., Lepistö, A. (eds.) DLT 2007. LNCS, vol. 4588, pp. 36–47. Springer, Heidelberg (2007)
8. Anselmo, M., Giammarresi, D., Madonia, M.: Deterministic and unambiguous families within recognizable two-dimensional languages. Fundamenta Informaticae 98, 143–166 (2010)
9. Anselmo, M., Giammarresi, D., Madonia, M.: Tiling automaton: A computational model for recognizable two-dimensional languages. In: Holub, J., Žďárek, J. (eds.) CIAA 2007. LNCS, vol. 4783, pp. 290–302. Springer, Heidelberg (2007)
10. Lonati, V., Pradella, M.: Snake-deterministic tiling systems. In: Královič, R., Niwiński, D. (eds.) MFCS 2009. LNCS, vol. 5734, pp. 549–560. Springer, Heidelberg (2009)
11. Lonati, V., Pradella, M.: Picture recognizability with automata based on Wang tiles. In: van Leeuwen, J., et al. (eds.) SOFSEM 2010. LNCS, vol. 5901, pp. 576–587. Springer, Heidelberg (2010)
12. Lonati, V., Pradella, M.: Deterministic recognizability of picture languages with Wang automata. Discrete Mathematics and Theoretical Computer Science 4, 73–94 (2010)
13. Lonati, V., Pradella, M.: Strategies to scan pictures with automata based on Wang tiles. R.A.I.R.O. Theoretical Informatics and Applications 44 (2010)
14. Inoue, K., Takanami, I.: A characterization of recognizable picture languages. In: Nakamura, A., et al. (eds.) ICPIA 1992. LNCS, vol. 654, pp. 133–143. Springer, Heidelberg (1992)
15. Ito, A., Inoue, K., Takanami, I.: A note on three-way two-dimensional alternating Turing machines. Inf. Sci. 45, 1–22 (1988)
16. Matz, O.: On piecewise testable, starfree, and recognizable picture languages. In: Nivat, M. (ed.) FOSSACS 1998. LNCS, vol. 1378, pp. 203–210. Springer, Heidelberg (1998)

Complexity of Problems Concerning Reset Words for Cyclic and Eulerian Automata

Pavel Martyugin

Ural State University,
620083 Ekaterinburg, Russia
martugin@mail.ru

Abstract. A word is called a reset word for a deterministic finite automaton if it maps all states of this automaton to one state. We consider two classes of automata: cyclic automata and Eulerian automata. For these classes we study the computational complexity of the following problems: does there exist a reset word of given length for a given automaton? what is the minimal length of the reset words for a given automaton?

Keywords: Synchronization, Automata, Reset Words, Computational Complexity

1 Introduction

A *deterministic finite automaton* (DFA) $\mathscr{A}$ is a triple $\langle Q, \Sigma, \delta \rangle$, where Q is a finite set of *states*, Σ is a finite *alphabet*, and $\delta : Q \times \Sigma \to Q$ is a totally defined *transition function*. The function δ extends in a unique way to an action $Q \times \Sigma^* \to Q$ of the free monoid Σ^* over Σ; this extension is also denoted by δ. We denote $\delta(q, w)$ by $q.w$. For $S \subseteq Q$, $w \in \Sigma^*$, we also define $S.w = \{q.w \mid q \in S\}$.

A DFA $\mathscr{A}$ is called *synchronizing* if there exists a word $w \in \Sigma^*$ whose action synchronizes $\mathscr{A}$, that is, leaves the automaton in one particular state no matter at which state in Q it started: $\delta(q, w) = \delta(q', w)$ for all $q, q' \in Q$. Any word w with this property is said to be a *reset word* for the automaton.

In [3], Černý produced for each integer n a synchronizing automaton with n states, 2 input letters and the shortest reset word has length $(n - 1)^2$. He conjectured that these automata represent the worst possible case, that is, every synchronizing automaton with n states can be reset by a word of length $(n - 1)^2$. The conjecture is arguably the most longstanding open problem in the combinatorial theory of finite automata. Upper bounds within the confines of the Černý conjecture have been obtained for the maximum length of the shortest reset words for synchronizing automata in some special classes, see, e.g., [5, 1, 6, 4, 11]. Two of these classes are considered in the present paper. In general case there is only a cubic upper bound $(n^3 - n)/6$, see [9].

It is natural to consider computational complexity of various problems arising from the study of automata synchronization. The most important questions are:

B. Bouchou-Markhoff et al. (Eds.): CIAA 2011, LNCS 6807, pp. 238–249, 2011.

is a given automaton synchronizing or not, and what is the minimal length of the reset words for a given automaton?

It follous from [3] that there exists an algorithm that checks whether a given DFA $\mathscr{A} = \langle Q, \Sigma, \delta \rangle$ is synchronizing. This algorithm works within $O(|\Sigma| \cdot |Q|^2)$ time bound. In [5], Eppstein presented another algorithm which works within $O(|\Sigma| \cdot |Q|^2) + O(|Q|^3)$ time bound and finds some reset word (which need not to be the shortest reset word for $\mathscr{A}$). In [5,10] it was shown that the following problem SYN is NP-complete: given a DFA $\mathscr{A}$ and a positive integer L, is there a word of length at most L synchronizing the automaton $\mathscr{A}$. This problem remains NP-complete even if restricted to automata with 2-letter alphabet. Moreover, Berlinkov in [2] proved that no polynomial time algorithm approximates the length of the shortest synchronizing word within constant factor for a given DFA.

In [8], Olschewski and Ummels considered a problem MIN-SYN: given a DFA $\mathscr{A}$ and a positive integer L, is the minimum length of reset words for the automaton $\mathscr{A}$ equal to L? They proved that this problem is DP-complete, where DP is a class of all languages of the form $L = L_1 \setminus L_2$ with $L_1, L_2 \in$NP. The canonical DP-complete problem is SAT-UNSAT: given two Boolean formulae ϕ and ψ (in CNF), the problem is to decide whether ϕ is satisfiable and ψ is unsatisfiable. The problem MIN-SYN remains DP-complete even for 2-letter automata.

Since the problems SYN and MIN-SYN turn out to be computationally difficult in general, it is reasonable to consider their restrictions to some natural classes of automata. For any class C of automata, we define the restricted versions SYN(C) and MIN-SYN(C) of SYN and respectively MIN-SYN.

Instance: A DFA $\mathscr{A} \in C$ and an integer $L > 0$.
Question of SYN(C): Is there a reset word of length L for the automaton $\mathscr{A}$?
Question of MIN-SYN(C): Is the minimum length of reset words for the automaton $\mathscr{A}$ equal to L?

These problems have been considered in the literature for cyclically monotonic (see [5]), monotonic, commutative, aperiodic, $\mathscr{D}$-trivial automata, for automata with simple idempotents and for automata with a zero state (see [7]). In some cases they become solvable in polynomial time, in some other cases they remain computationally hard. In the present paper we consider these problems for two further natural classes of automata: the class $CYCLE$ of cyclic automata and the class $EULER$ of Eulerian automata. Let us define these classes and comment on their synchronization properties.

Let $\mathscr{A} = \langle Q, \Sigma, \delta \rangle$ be a DFA and $|Q| = n$. The letter $b \in \Sigma$ is said to be *cyclic* if it acts on the set Q as a cyclic permutation of order n. This means that for any $q \in Q$ and $i \in \{1, \ldots, n-1\}$, we have $\delta(q, b^n) = q \neq \delta(q, b^i)$. A DFA with cyclic letter is called *cyclic*. Dubuc [4] has proved that every n-state synchronizing cyclic DFA has a reset word of length $(n-1)^2$, thus, the Černý conjecture holds true for cyclic automata. Furthermore, this upper bound of the length of the shortest reset words is tight, because automata from the Černý series [3] are cyclic.

A DFA $\mathscr{A} = \langle Q, \Sigma, \delta \rangle$ is said to be *Eulerian* if its underlying digraph is Eulerian. It is well-known that the underlying digraph of $\mathscr{A}$ is Eulerian if and only if for every state $q \in Q$ there are exactly $|\Sigma|$ pairs $(p, a) \in Q \times \Sigma$ such that $p.a = q$. Kari [6] has proved that for any n-state synchronizing Eulerian DFA there exists a reset word of length $(n-2)(n-1)+1$. It means that the Černý conjecture is true for Eulerian automata.

For a class C of automata and a positive integer k, we denote by C_k the class of all automata in C with k input letters. Here we prove that each of the problems SYN($CYCLE$), SYN($CYCLE_k$) with $k \geq 2$, SYN($EULER$), SYN($EULER_k$) with $k \geq 3$ is NP-complete, and each of the problems MIN-SYN($CYCLE$), MIN-SYN($CYCLE_k$) with $k \geq 2$, MIN-SYN($EULER$), MIN-SYN($EULER_k$) with $k \geq 3$ is both NP-hard and co-NP-hard. The question about the complexity of the problems SYN($EULER_2$) and MIN-SYN($EULER_2$) remains open.

For the sequel, we need some notation. For a set Q, let $|Q|$ denote the cardinality of Q and let 2^Q stands for the set of all subsets of Q. For a word $w \in \Sigma^*$, we denote by $|w|$ the length of w and by $w[i]$, where $1 \leq i \leq |w|$, the i-th letter in w from the left. If $1 \leq i \leq j \leq |w|$, we denote by $w[i, j]$ the word $w[i] \cdots w[j]$.

2   Cyclic Automata

Theorem 1. *The problem SYN($CYCLE_2$) is NP-complete.*

Proof. It is easy to see that the general problem SYN belongs to NP, because any synchronizing automaton can be synchronizing by a word of polynomial length (of length at most $(n^3 - n)/6$, see [9]). Now we reduce the problem SAT to SYN($CYCLE_2$). Take an instance of SAT consisting of the clauses $c_1(x_1, \ldots, x_n), \ldots, c_p(x_1, \ldots, x_n)$ over the Boolean variables $x_1, \ldots, x_n \in \{0, 1\}$. We may (and will) assume that no clause contains both x_m and $\neg x_m$ for any $m \in \{1, \ldots, n\}$. We are going to construct a 2-letter automaton $\mathscr{A}_{cycle} = \langle Q, \Sigma, \delta \rangle$ and a number L such that there exists a reset word of length L for $\mathscr{A}_{cycle}$ if and only if $c_1 \wedge c_2 \wedge \cdots \wedge c_p$ is satisfiable.

Let $G = \{(i, m) \mid c_i \text{ contains } \neg x_m\}$. We put

$$\Sigma = \{a, b\}, \quad Q = (\bigcup_{i=1}^{p} Q_i) \cup (\bigcup_{i=1}^{p} D_i) \cup (\bigcup_{(i,m) \in G} S_i^m), \text{ where}$$

$$Q_i = \{q(i, 0), \ldots, q(i, n+2)\}, \quad D_i = \{d(i, 1), \ldots, d(i, n+4)\},$$
$$S_i^m = \{s(i, m, 1), \ldots, s(i, m, n+4)\}.$$

Now we define the action of the letters a and b. For all $i \in \{1, \ldots, p\}$ and $m \in \{1, \ldots, n\}$, we put

$$q(i, 0).a = q(i, 1); \quad q(i, 0).b = \begin{cases} q(i-1, 0) & \text{if } i > 1, \\ q(1, 1) & \text{if } i = 1; \end{cases}$$

$$q(i,m).b = \begin{cases} s(i,m,1) & \text{if } \neg x_m \text{ occurs in } c_i, \\ q(i,m+1) & \text{otherwise}; \end{cases}$$

$$q(i,m).a = \begin{cases} d(1,1) & \text{if } x_m \text{ occurs in } c_i, \\ q(i,m+1) & \text{otherwise}; \end{cases}$$

$$q(i,n+1).a = q(i,n+1).b = q(i,n+2);$$
$$q(i,n+2).a = d(1,1); \quad q(i,n+2).b = d(i,1).$$

For all $i \in \{1,\ldots,p\}$, $m \in \{1,\ldots,n\}$ such that $(i,m) \in G$, and for all $j \in \{1,\ldots,n+4\}$, we put

$$s(i,m,j).a = d(1,1); \quad s(i,m,j).b = \begin{cases} s(i,m,j+1) & \text{if } j < n+4, \\ q(i,m+1) & \text{if } j = n+4. \end{cases}$$

For all $i \in \{1,\ldots,p\}$ and $j \in \{1,\ldots,n+4\}$, we put

$$d(i,j).a = d(1,1); \quad d(i,j).b = \begin{cases} d(i,j+1) & \text{if } j < n+4, \\ q(i+1,1) & \text{if } j = n+4, i < p, \\ q(p,0) & \text{if } j = n+4, i = p. \end{cases}$$

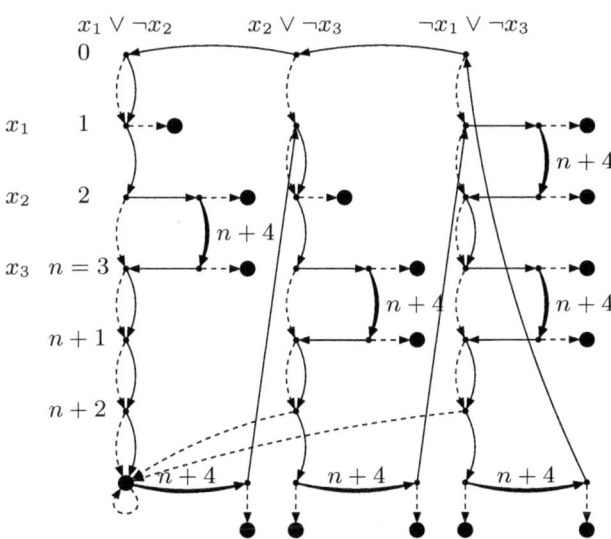

Fig. 1. The automaton $\mathscr{A}_{cycle}$ for the clauses $x_1 \vee \neg x_2$, $x_2 \vee \neg x_3$, $\neg x_1 \vee \neg x_3$

We put $L = n+2$. An example of the automaton $\mathscr{A}_{cycle}$ is presented in Fig. 1. The action of the letter b is shown with solid lines. The action of the letter a is shown with dashed lines. All large black circles represent the same state $d(1,1)$

(this way we try to improve the readability of the picture). Every bold arrow labelled by $n+4$ represents one of the sets S_i^m or D_i. Every set D_i, $i \in \{1, \ldots, p\}$, and every set S_i^m, $(i, m) \in G$, contains $n+4$ states. Fig. 1 contains three columns of states. The i-th column contains the states $q(0, i), \ldots, q(n+2, i)$ for fixed i. Every horizontal row contains the states $q(m, 1)$, $q(m, 2)$, $q(m, 3)$ for fixed m. There are some right arrows labelled by b from columns to the sets S_i^m. The set D_i is drawn under the corresponding set Q_i.

The size of the automaton $\mathscr{A}_{cycle}$ is a polynomial function of the size of the clauses $c_1, \ldots, c_p$. It easy to prove that the automaton $\mathscr{A}_{cycle}$ is cyclic, namely, the letter b acts on the set Q as a cyclic permutation of order $|Q|$.

We notice that the word a^{n+3} is a reset word for $\mathscr{A}_{cycle}$. We will prove that there is a reset word of length less than $n+3$ if and only if the formula $c_1 \wedge c_2 \wedge \ldots \wedge c_p$ is satisfiable.

Lemma 1. *If $q \in Q$, $w \in \Sigma^*$, $w[n+2] = a$ and there is an integer s such that $s \in \{1, \ldots, n+1\}$ and $q.w[1, s] = d(1, 1)$, then $q.w[1, n+2] = d(1, 1)$.*

Proof. We have

$$q.w[1, n+1] = q.w[1, s]w[s+1, n+1] = d(1, 1).w[s+1, n+1] \in D_1.$$

Therefore, $q.w[1, n+2] = d(1, 1)$.

Lemma 2. *If $w \in \Sigma^*$, $|w| = n+2$ and $w[1] = w[n+2] = a$, then*

$$\delta(Q \backslash \{q(1, 0), \ldots, q(p, 0)\}, w) = \{d(1, 1)\}.$$

Proof. The letter a maps all sets D_i and S_i^m to the state $d(1, 1)$. Therefore, if $q \in (\bigcup_{(i,j) \in G} S_i^j) \cup (\bigcup_{i=1}^p D_i)$, then $q.w[1] = d(1, 1)$, and we obtain from Lemma 1 that $q.w[1, n+2] = d(1, 1)$.

Let $q \in Q_i$ for some $i \in \{1, \ldots, p\}$. If for some $s \in \{1, \ldots, n+1\}$ the state $q.w[1, s]$ belongs to one of the sets S_i^m or D_i, then either $q.w[n+1]$ belongs to the same set S_i^m or D_i, or there is some $s' \in \{s+1, \ldots, n+1\}$ such that $q.w[1, s'] = d(1, 1)$ and we can apply Lemma 1. In any case, if for some $s \in \{1, \ldots, n+1\}$, one has $q.w[1, s] \in (\bigcup_{(i,j) \in G} S_i^j) \cup (\bigcup_{i=1}^p D_i)$, then $q.w[1, n+2] = d(1, 1)$, because $w[n+2] = a$.

Let $q \in Q_i \setminus \{q(0, i)\}$ and suppose that, for all $s \in \{1, \ldots, n+1\}$, one has

$$q.w[1, s] \notin (\bigcup_{(i,j) \in G} S_i^j) \cup (\bigcup_{i=1}^p D_i).$$

It means that $q.w[1, s] \in Q_i$ for $s \in \{1, \ldots, n+1\}$. Hence, $q = q(1, i)$. Therefore, $q.w[1, n+2] = d(1, 1)$. Thus, $\delta(Q \backslash \{q(1, 0), \ldots, q(p, 0)\}, w) = \{d(1, 1)\}$.

It follows from Lemma 2 that a word w with $|w| = n+2$ and $w[1] = w[n+2] = a$ is a reset word for $\mathscr{A}_{cycle}$ if and only if $q(1, 0).w = \ldots = q(p, 0).w = d(1, 1)$.

Lemma 3. *If $c_1 \wedge c_2 \wedge \ldots \wedge c_p$ is satisfiable, then there exists a reset word w of length $n + 2$ for the automaton $\mathscr{A}_{cycle}$.*

Proof. Let $w = a\alpha_1 \ldots \alpha_n a$, where for $i \in \{1, \ldots, n\}$, $\alpha_i = \begin{cases} a & \text{if } x_i = 1, \\ b & \text{if } x_i = 0. \end{cases}$ We are going to prove that $\{q(1,0), \ldots, q(p,0)\}.w = \{d(1,1)\}$. Let $i \in \{1, \ldots, p\}$. We have $q(i,0).a = q(i,1)$. We also have $c_i(x_1, \ldots, x_n) = 1$ (1 means true). Hence, there is $m \in \{1, \ldots, n\}$ such that $x_m = 1$ (in this case $w[m+1] = \alpha_m = a$) and the variable x_m occurs in c_i; or $x_m = 0$ (in this case $w[m+1] = \alpha_m = b$) and $\neg x_m$ occurs in c_i. Let m be the least number with such property. Then we have $q(i,1).\alpha_1 \ldots \alpha_{m-1} = q(i,m)$ and $q(i,m).\alpha_m \in \{s(i,m,1), d(1,1)\}$. If $q(i,m).\alpha_m = s(i,m,1)$ and $\alpha_{m+1} = \ldots = \alpha_n = b$, then $q(i,1).\alpha_1 \ldots \alpha_n \in S_i^m$ and $q(i,0).w = d(1,1)$. Otherwise, there is a number $m' \in \{m, \ldots, n+1\}$ such that $q(i,1).\alpha_1 \ldots \alpha_{m'} = d(1,1)$. In this case we obtain from Lemma 1 that $q(i,0).w = d(1,1)$. Therefore, by Lemma 2 we obtain $Q.w = \{d(1,1)\}$.

Lemma 4. *If there is a reset word $w \in \{a, b\}^*$ of length $n+2$ for the automaton $\mathscr{A}_{cycle}$, then $c_1 \wedge c_2 \wedge \ldots \wedge c_p$ is satisfiable.*

Proof. For any letter $w[1] \in \{a, b\}$, we have $\{q(1,1), \ldots, q(1,p)\} \subseteq Q.w[1]$. We consider the word $w[2, n+1]$. For $m \in \{1, \ldots, n\}$, we put

$$x_m = \begin{cases} 0 & \text{if } w[m+1] = b, \\ 1 & \text{if } w[m+1] = a. \end{cases}$$

Arguing by contradiction, suppose that $c_i(x_1, \ldots, x_n) = 0$ for some clause c_i. If for some $s \in \{2, \ldots, n+1\}$, we have $q(i,1).w[2,s] \in S_i^m$, then $c_i(x_1, \ldots, x_n) = 1$. Therefore, $q(i,1).w[2, n+1] = q(i, n+1)$. In both cases $w[n+2] = a$ and $w[n+2] = b$ we obtain $q(i,0).w = q(i, n+1).w[n+2] = q(i, n+2)$. Therefore, the word w resets the automaton $\mathscr{A}_{cycle}$ to the state $q(i, n+2)$. On the other hand, the state $q(i, n+2)$ cannot be reached from the state $q(j,1)$, $i \neq j$ by using the word of length $n + 2$. We obtain a contradiction. Therefore, for any $i \in \{1, \ldots, p\}$, $c_i(x_1, \ldots, x_n) = 1$. The lemma and the theorem is proved.

Corollary 1. *1. The problems $SYN(CYCLE)$ and $SYN(CYCLE_k)$ for $k \geq 2$ are NP-complete.*

 2. The problems $MIN\text{-}SYN(CYCLE)$ and $MIN\text{-}SYN(CYCLE_k)$ for $k \geq 2$ are NP-hard and co-NP-hard.

Proof. 1. The problem $SYN(CYCLE_2)$ is a special case of $SYN(CYCLE)$. Hence, the latter problem is NP-complete. To reduce the problem $SYN(CYCLE_2)$ to $SYN(CYCLE_k)$ for any $k \geq 2$, we add $k - 2$ letters that act as identical transformations to the construction in the proof above.

2. The NP-hardness of the problem $MIN\text{-}SYN(CYCLE_2)$ for $k \geq 2$ follows from the same reduction as in the proof of Theorem 1. To prove the co-NP-hardness, we use the same automaton $\mathscr{A}_{cycle}$ constructed from given clauses $c_1, \ldots, c_p$ but put $L = n + 3$. Then the shortest reset word for the automaton

$\mathscr{A}_{cycle}$ has length L if and only if there is no values for the variables $x_1, \ldots, x_n$ such that $c_1(x_1, \ldots, x_n) = \ldots = c_p(x_1, \ldots, x_n) = 1$. Therefore, the problem is co-NP-hard. The result extends to the problems MIN-SYN($CYCLE$) and MIN-SYN($CYCLE_k$) for $k \geq 2$ in an obvious way.

3   Eulerian Automata

Theorem 2. *The problem SYN($EULER_3$) is NP-complete.*

Proof. The problem SYN($EULER_3$) belongs to NP because general problem SYN belongs to NP. We use a reduction from SAT again. Take an instance of SAT consisting of the clauses $c_1(x_1, \ldots, x_n), \ldots, c_p(x_1, \ldots, x_n)$ over the Boolean variables $x_1, \ldots, x_n$. We assume that no clause contains both x_m and $\neg x_m$ for any $m \in \{1, \ldots, n\}$. We are going to construct a 3-letter automaton $\mathscr{A}_{euler} = \langle Q, \Sigma, \delta \rangle$ and an integer $L > 0$ such that there exists a reset word of length L for $\mathscr{A}_{euler}$ if and only if $c_1 \wedge c_2 \wedge \ldots \wedge c_p$ is satisfiable.

Let $\mathscr{A}_{euler} = \langle Q, \Sigma, \delta \rangle$, where

$$\Sigma = \{a, b, c\}, \ Q = Z \cup (\bigcup_{i=1}^{p} Q_i) \cup (\bigcup_{i=1}^{p} R_i) \cup (\bigcup_{i=1}^{p} S_i),$$

$$Z = \{z(m) \mid m \in \{2, \ldots, n+p+5\}\} \text{ and, for } i \in \{1, \ldots, p\},$$

$$Q_i = \{q(i,m) \mid m \in \{1, \ldots, n+3\}\}, \ R_i = \{r(i,m) \mid m \in \{2, \ldots, n+3\}\},$$

$$S_i = \{s(i,m) \mid m \in \{1, \ldots, p-i+1\}\}.$$

Now we define the action of the letters a and b. Let $i \in \{1, \ldots, p\}$. For all $m \in \{1, \ldots, n\}$, we put

$$q(i,m).a = \begin{cases} r(i, m+2) & \text{if } x_m \text{ occurs in } c_i, \\ q(i, m+1) & \text{otherwise;} \end{cases}$$

$$q(i,m).b = \begin{cases} r(i, m+2) & \text{if } \neg x_m \text{ occurs in } c_i, \\ q(i, m+1) & \text{otherwise.} \end{cases}$$

$$q(i,n+1).a = q(i,n+1).b = q(i,n+2); \ q(i,n+2).a = q(i,n+2).b = q(i,n+3);$$

$$q(i,n+3).a = r(i,n+3).a = s(i,1); \ q(i,n+3).b = q(i,n+3);$$

$$r(i,n+3).b = r(i,n+3).$$

For $m \in \{2, \ldots, n+2\}$, we put $r(i,m).a = r(i,m).b = r(i, m+1)$.
For $m \in \{1, \ldots, p-i+1\}$, we put

$$s(i,m).b = s(i,m); \ s(i,m).a = \begin{cases} s(i, m+1) & \text{if } m < p+i-1, \\ s(i-1, p-i+2) & \text{if } m = p+i-1, i>1, \\ z(n+p+4) & \text{if } m = p, i=1. \end{cases}$$

For $m \in \{2, \ldots, n+p+2\}$, we put $z(m).a = z(m).b = z(m+1)$. We also put

$$z(n+p+3).a = z(n+p+4); \; z(n+p+3).b = z(n+p+3);$$

$$z(n+p+4).a = z(n+p+5); \; z(n+p+4).b = z(n+p+4);$$

$$z(n+p+5).a = z(n+p+5).b = z(n+p+5).$$

The letters a and b encode satisfiability of the clauses. Now we define the action of the letter c such that the automaton $\mathscr{A}$ becomes Eulerian. For $q \in Q$, we put $q.c = q$ except the following cases. Let $i \in \{1, \ldots, p\}$, $m \in \{1, \ldots, n\}$. If either x_i or $\neg x_i$ occurs in c_m, then we put $r(i, m+2).c = q(i, m+1)$. Besides that, we put

$$q(i, n+3).c = s(i, 1).c = q(i, 1); \; r(i, n+3).c = r(i, 2) \text{ for } i \neq p;$$

$$s(i, p-i+1).c = r(i+1, 2); \; z(n+p+3).c = z(n+p+4).c = z(2);$$

$$z(n+p+5).c = r(1, 2).$$

An example of the automaton $\mathscr{A}_{euler}$ is presented in Fig. 2. The action of the letters a, b, c is shown by solid, dashed and dotted lines respectively. The states in Fig. 2 are organized in several columns containing respectively the sets Z, $Q_1 \cup S_1$, R_1, $Q_2 \cup S_2$, R_2, and so on. We put $L = n+p+3$.

In general, the states of the automaton $\mathscr{A}_{euler}$ can be partitioned in $n+p+5$ "rows" $T_1, \ldots, T_{n+p+5}$. We put $T_1 = \{q(1,1), \ldots, q(p,1)\}$; for $m \in \{2, \ldots, n+3\}$ we put $T_m = \{z(m), q(1,m), r(1,m), \ldots, q(p,m), r(p,m)\}$ and for $m \in \{n+4, \ldots, n+p+4\}$ we put $T_m = \{z(m), s(m-n-3, 1), \ldots, s(m-n-3, p-m+n+4)\}$; we also put $T_{n+p+5} = \{z(n+p+5)\}$. Clearly, the size of the automaton $\mathscr{A}_{euler}$ is a polynomial function of the size of the clauses $c_1, \ldots, c_p$.

Lemma 5. *The DFA $\mathscr{A}_{euler}$ is Eulerian.*

Proof. It is easy to check that for any state $q \in Q$ there exist exactly 3 pairs $(r, \alpha) \in Q \times \Sigma$ such that $r.\alpha = q$.

Let $U \subseteq Q$, $\Theta \subseteq \Sigma$. We denote by $d_\Theta(U)$ the minimum length of words $w \in \Theta^*$ such that $|U.w| = 1$. Thus, the minimum length of reset words for $\mathscr{A}_{euler}$ is equal to $d_\Sigma(Q)$.

Lemma 6. $d_\Sigma(Q) = d_{\{a,b\}}(Q) \in \{n+p+3, n+p+4\}$.

Proof. It is immediate to check that the word a^{n+p+4} is a reset word for DFA $\mathscr{A}_{euler}$. Therefore $d_\Sigma(Q) \leq n+p+4$. We notice that $d_{\{a,b\}}(\{z(2), z(n+p+5)\}) = n+p+3$. Therefore $d_{\{a,b\}}(Q) \geq n+p+3$.

Now let w be a shortest reset word for the $\mathscr{A}_{euler}$ and suppose that the letter c occurs in w. We aim to prove that $|w| \geq n+p+4$. It is not difficult to verify that $d_\Sigma(\{z(2), r(1,2)\}) = n+p+2$ and there is no word $u \in \Sigma^* \setminus \{a,b\}^*$ of length $n+p+2$ such that $z(2).u = r(1,2).u$. Let i be the position of the leftmost occurrence of the letter c in w. If $i \geq n+p+4$, then $|w| \geq n+p+4$. Let $i \leq n+p+3$, then we have $z(n+p+4), z(n+p+5) \in Q.w[1, i-1]$. Hence $z(2), r(1,2) \in Q.w[1, i]$.

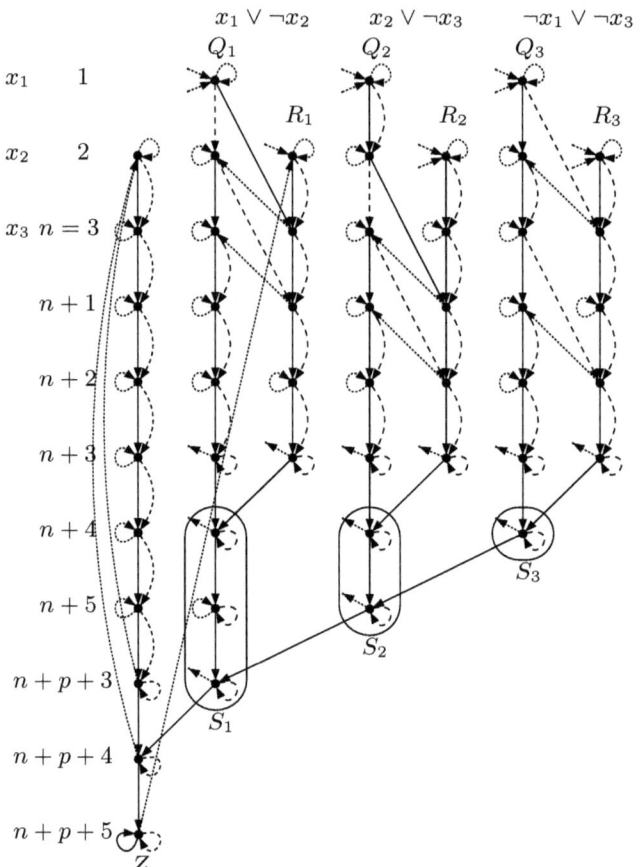

Fig. 2. The automaton $\mathscr{A}_{euler}$ for the clauses $x_1 \vee \neg x_2$, $x_2 \vee \neg x_3$, $\neg x_1 \vee \neg x_3$

Therefore $|w| \geq i + d_\Sigma(\{z(2), r(1,2)\}) = i + n + p + 2$. If $i \geq 2$, then $|w| \geq n + p + 4$. If $i = 1$, then $w[1] = c$ and $\{z(2), z(3), r(1,2)\} \subseteq Q.w[1]$. It is not difficult to prove that $d_\Sigma(\{z(2), z(3), r(1,2)\}) \geq n + p + 3$. Hence, $|w| \geq n + p + 4$ in any case. Therefore $d_\Sigma(Q) = d_{\{a,b\}}(Q)$.

In particular, the lemma claims that there exists a reset word of the minimum length in which the letter c does not occur. We notice that $T_{n+p+5}.a = T_{n+p+5}.b = T_{n+p+5}$, and $T_m.a, T_m.b \subseteq (T_m \cup T_{m+1})$ for $m \in \{n+1, \ldots, n+p+4\}$ while $T_m.a, T_m.b \subseteq T_{m+1} \cup T_{m+2}$ for $m \in \{1, \ldots, n\}$. The following lemma is an immediate corollary of this property.

Lemma 7. 1. *Every reset word from $\{a,b\}^*$ resets the DFA $\mathscr{A}_{euler}$ to the state $z(n+p+5)$.*

2. $d_{\{a,b\}}(T_{n+2} \cup \ldots \cup T_{n+p+5}) = p + 3$.

3. *If a word $w \in \{a,b\}^*$ of length $n + p + 3$ is a reset word for $\mathscr{A}_{euler}$, then $Q.w[1,n] \subseteq T_{n+2} \cup \ldots \cup T_{n+p+5}$.*

4. *If a word $w \in \{a, b\}^*$ of length $n + p + 3$ is a reset word for $\mathscr{A}_{euler}$, then $T_1.w[1, n] \subseteq T_{n+2}$.*
5. *If there is a word w of length n such that $Q.w \subseteq T_{n+2} \cup \ldots \cup T_{n+p+5}$, then the word wa^{p+3} is a reset word of length $n + p + 3$ for $\mathscr{A}_{euler}$.*
6. *If there is a word w of length n such that $T_1.w \subseteq T_{n+2}$, then there is a reset word of length $n + p + 3$ for $\mathscr{A}_{euler}$.*
7. $d_{\{a,b\}}(Q) = d_{\{a,b\}}(T_1)$.

Lemma 8. *If $c_1 \wedge c_2 \wedge \ldots \wedge c_p$ is satisfiable, then there exists a reset word w of length $n + p + 3$ for the automaton $\mathscr{A}_{euler}$.*

Proof. Let $w = \alpha_1 \ldots \alpha_n a^{p+3}$, where for $i \in \{1, \ldots, n\}$, $\alpha_i = \begin{cases} a & \text{if } x_i = 1, \\ b & \text{if } x_i = 0. \end{cases}$ We are going to prove that $T_1.w = \{z(n + p + 5)\}$. Let $i \in \{1, \ldots, p\}$. We have $c_i(x_1, \ldots, x_n) = 1$. Hence, there is $m \in \{1, \ldots, n\}$ such that $x_m = 1$ (in this case $w[m] = a$) and x_m occurs in c_i or $x_m = 0$ (in this case $w[m] = b$) and $\neg x_m$ occurs in c_i. Let m be the least number with this property. We obtain from the definition of the action of a and b that

$$q(i, 1).w[1, m-1] = q(i, m) \quad \text{and} \quad q(i, m).w[m] = r(i, m+2).$$

For any word $w[m+1, n] \in \{a, b\}^*$ we have $r(i, m+2).w[m+1, n] = r(i, n+2) \in T_{n+2}$. Hence, from Lemma 7, w is a reset word for the automaton $\mathscr{A}_{euler}$.

Lemma 9. *If there is a reset word $w \in \{a, b\}^*$ of length $p + n + 3$ for the automaton $\mathscr{A}_{euler}$, then $c_1 \wedge c_2 \wedge \ldots \wedge c_p$ is satisfiable.*

Proof. The word w resets DFA $\mathscr{A}_{euler}$. Therefore, we have $T_1.w[1, n] \in T_{n+2}$. For $m \in \{1, \ldots, n\}$ we put

$$x_m = \begin{cases} 0 & \text{if } w[m] = b, \\ 1 & \text{if } w[m] = a. \end{cases}$$

Arguing by contradiction, suppose that $c_i(x_1, \ldots, x_n) = 0$ for some clause c_i. If, for some $m \in \{1, \ldots, n\}$, we would have $q(i, 1).w[1, m] = r(1, m+2)$, then $c_i(x_1, \ldots, x_n) = 1$. Therefore, $q(i, 1).w[1, n] = q(i, n+1) \in T_{n+1}$. By Lemma 7, the word w is not a reset word for $\mathscr{A}_{euler}$. We obtain a contradiction. Therefore, for any $i \in \{1, \ldots, p\}$, we have $c_i(x_1, \ldots, x_n) = 1$.

Corollary 2. 1. *The problems SYN(EULER) and SYN($EULER_k$) for $k \geq 3$ are NP-complete.*
2. *The problems MIN-SYN(EULER) and MIN-SYN($EULER_k$) for $k \geq 3$ are NP-hard and co-NP-hard.*

Proof. The proof is the same as the proof of Corollary 1.

4   Conclusion and Conjectures

We proved that problems SYN($EULER$) and SYN($CYCLE$) are NP-complete. This means that these problems have the same complexity as the general problem SYN stated for a class of all DFA. At the same time we proved that the problems MIN-SYN($EULER$) and MIN-SYN($CYCLE$) are NP-complete and co-NP-complete. But it is only the lower bound and it does not seem to be a tight. The general problem MIN-SYN is DP-complete for a class of all DFA (see. [8]). It is natural to conjecture the following.

Conjecture 1. For any integer $k \geq 2$ the problems MIN-SYN($EULER$), MIN-SYN($CYCLE$), MIN-SYN($EULER_k$), MIN-SYN($CYCLE_k$) are DP-complete.

The NP-completeness of the problem SYN($EULER_2$) now is also unproved. It may happen that if problem SYN($EULER_2$) can be solved in a polynomial time, then the problem MIN-SYN($EULER_2$) can be also solved in polynomial time.

It follows from [2] that no polynomial time algorithm approximates the length of the shortest synchronizing word within constant factor for a given DFA. There is no such algorithm even for automata over the binary alphabet. So we can conjecture the following.

Conjecture 2. No polynomial time algorithm approximates the length of the shortest synchronizing word within constant factor for a given cyclical or Eulerian DFA. There is no such algorithm even for DFA over the binary alphabet.

Acknowledgement. The author is grateful to Dr. D. Ananichev and Prof. M. Volkov for valuable help. He also acknowledges support from the Federal Education Agency of Russia, project 2.1.1/3537, and from the Russian Foundation for Basic Research, grant 09-01-12142.

References

1. Ananichev, D.S., Volkov, M.V.: Synchronizing monotonic automata Theoret. Theoret. Comput. Sci. 327, 225–239 (2004)
2. Berlinkov, M.V.: Approximating the Minimum Length of Synchronizing Words Is Hard. In: Ablayev, F., Mayr, E.W. (eds.) CSR 2010. LNCS, vol. 6072, pp. 37–47. Springer, Heidelberg (2010)
3. Černý, J.: Poznámka k homogénnym eksperimentom s konecnými avtomatami, Mat.-Fyz. Čas. Slovensk. Akad. Vied. 14, 208–216 (1964) (in Slovak)
4. Dubuc, L.: Surles automates circulaires et la conjecture de Černý, RAIRO Inform. Theor. Appl. 32, 21–34 (1998) (in French)
5. Eppstein, D.: Reset sequences for monotonic automata. SIAM J. Comput. 19, 500–510 (1990)
6. Kari, J.: Synchronizing finite automata on Eulerian digraphs. Theoret. Comput. Sci. 295, 223–232 (2003)
7. Martyugin, P.: Complexity of problems concerning reset words for some partial cases of automata. Acta Cybernetica 19, 517–536 (2009)

8. Olschewski, J., Ummels, M.: The Complexity of Finding Reset Words in Finite Automata. In: Hliněný, P., Kučera, A. (eds.) MFCS 2010. LNCS, vol. 6281, pp. 568–579. Springer, Heidelberg (2010)
9. Pin, J.-E.: On two combinatorial problems arising from automata theory. Ann. Discrete Math. 17, 535–548 (1983)
10. Salomaa, A.: Composition sequences for functions over a finite domain. Theor. Comput. Sci. 292, 263–281 (2003)
11. Volkov, M.V.: Synchronizing automata preserving a chain of partial orders. Theoret. Comput. Sci. 410, 3513–3519 (2009)

Distributed Event Clock Automata
Extended Abstract

James Ortiz[1] and Axel Legay[2,3] and Pierre-Yves Schobbens[1]

[1] Computer Science Faculty, University of Namur
[2] INRIA/IRISA, Rennes
[3] Institut Montefiore, University of Liège
james.ortizvega@fundp.ac.be
alegay@irisa.fr
pierre-yves.schobbens@fundp.ac.be

Abstract. In distributed real-time systems, we cannot assume that clocks are perfectly synchronized. To model them, we use independent clocks and define their timed semantics. The universal timed language, and the timed language inclusion of icTA are undecidable. Thus, we propose Recursive Distributed Event Clock Automata (DECA). DECA are closed under all boolean operations and their timed language inclusion problem is decidable (more precisely PSPACE-complete), allowing stepwise refinement. We also propose Distributed Event Clock Temporal Logic (DECTL), a real-time logic with independent time evolutions. This logic can be model-checked by translating a DECTL formula into a DECA automaton.

1 Introduction

Real-Time Distributed Systems (RTDS) take an increasingly important role in our society, including in aircrafts and spacecrafts, satellite telecommunication networks or positioning systems. Distributed Systems consist of computer systems at different locations, that communicate through a network to achieve their function. Real-Time Systems have to obey strict requirements about the time of their actions. To ensure these, they rely on clocks. When systems are widely distributed, we cannot assume that their clocks are perfectly synchronized.

One of the most successful techniques for modeling real-time systems are Timed Automata (TA) [2]. A timed automaton is a finite automaton augmented with real-valued clocks. Constraints on these clocks are used to restrict the behaviors of the automaton. The model of TA assumes perfect clocks: all clocks have infinite precision and are perfectly synchronized.

This causes TA to have an undecidable language inclusion problem [2]. The situation contrasts strongly with the one of automata without real time, where the problems of complementation, language inclusion, emptiness, union and intersection are decidable, as well as the satisfiability and validity of propositional linear temporal logic (LTL). These properties are the basis of the success of model-checking. When all these problems are decidable, we call the formalism

B. Bouchou-Markhoff et al. (Eds.): CIAA 2011, LNCS 6807, pp. 250–263, 2011.

(automata or logic) fully decidable. These negative results spurred a quest for expressive but still fully decidable formalisms. To restore decidability, [4] proposed to restrict the behavior of clocks. The key idea is that the problematic clocks of TA are reset by non-deterministic transitions. In contrast, an event clock (EC) x_p is reset when a given atomic proposition p occurs. The event clock values are deterministic and thus Event Clock Automata ECA are determinizable, making language inclusion decidable and thus enabling refinement based development. Event clocks can also be introduced in temporal logic [20]. An event clock constraint is naturally translated into a proposition $\vartriangleleft_I p$, that means "the last time that a p occurred was d time units ago, where d lies in Γ".

However, the expressiveness of ECA is rather weak. Furthermore, this logic violates the substitution principle: Any proposition should be replaceable by a formula. Therefore [12] introduced the notion of "recursive" event. In a recursive event model, the reset of a clock is decided by a lower-level automaton (or formula). This automaton cannot read the clock that it is resetting. Clock resets are thus still deterministic, but the concept of "event" is now much more expressive. $\vartriangleright_I$ and $\vartriangleleft_I$ are modalities that can contain any subformulas, and can be nested. The temporal logic of recursive event clocks (variously called SCL [20] or Event-ClockTL [12]) has the same expressiveness as Metric Interval Temporal Logic MITL [3] (a decidable fragment of Metric Temporal Logic MTL where punctual constraints $\mathcal{U}_{\{k\}}$ are forbidden) in the interval semantics. First-and second-order monadic logics with matching expressiveness have been provided [12], yielding a natural, robust, fully decidable level of real-time expressiveness.

In this paper, we remove the assumption of perfect clock synchronization. Here, inspired by [6,14,1,10], we study the worst case: the clocks can advance totally independently if they are in different processes. [18,8] studied the opposite case, where the difference between clocks (drift) is infinitesimally small.

While [1] only studied untimed languages of their timed automata, namely the universal and existential languages, our first contribution is to define and study the corresponding timed languages (Section 4).

Our second contribution is to extend the Recursive Event Clock Automata (RECA) with distributed (a.k.a independent) clocks, yielding the Distributed Recursive Event Clock Automata (DECA). We will show that DECA are determinizable, thus closed under complementation, and thus that their language inclusion problem is decidable (more exactly, PSPACE-complete). We also show the decidability and regularity of their existential and universal timed languages (Section 5).

Our third contribution is to define a temporal logic with multiple observers, each with its own time evolution. This gives us the (Recursive) Distributed Event Clock Temporal Logic (DECTL), which is also PSPACE-complete (Section 6).

Structure of the paper. The rest of the paper is organized as follows. Sections 2 and 3 recall preliminary notions. Section 4 extends the semantics to timed languages. Section 5 defines DECA and studies their properties. Section 6 examines real-time temporal logics: it recalls EventClockTL [20], then introduces and studies DECTL. Due to space constraints, we only sketch proofs.

2   Preliminaries

We briefly recall the various models of time that are used in the literature [5]. We present our results in the interval semantics, that is the richest and most natural (but also most difficult) model. We also recall clocks and their constraints.

2.1   Models of Time

Models of time can be linear, considering a single future, or branching, considering several alternative futures. We only consider linear time in this paper. Our goal here is to model real-time systems, and thus we use the real numbers as our model of time. This avoids a premature commitment to a specific discretization of time. In this paper, we use the *interval semantics*, where the state of the model is known at any point in time, as opposed to *point semantics*, where it is known only at transitions.

Let $\mathbb{P}$ be a finite set of *(propositional) atoms*. A *letter* is an element of a finite set Σ. In this paper, we choose to define a letter as a propositional valuation over $\mathbb{P}$, so we pose $\Sigma = 2^{\mathbb{P}}$. Let $\mathbb{N}$ be the set of natural numbers, $\mathbb{R}$ denote the set of real numbers, $\mathbb{R}_{\geq 0}$ the set of non-negative real numbers. We denote by $\mathcal{I}(\mathbb{R}_{\geq 0})$ the set of real intervals whose bounds are in $\mathbb{R}_{\geq 0}$. An interval $I \in \mathcal{I}(\mathbb{R}_{\geq 0})$ is a convex subset of $\mathbb{R}_{\geq 0}$. An interval I is *contiguous* to I' when they are ordered: $I < I'$, and $I \cup I'$ is convex. An (alternating) interval sequence (AIS) is a monotone sequence $I = I_0 I_1 \cdots$ of non-empty intervals of $\mathcal{I}(\mathbb{R}_{\geq 0})$ where : (i) singular and open intervals alternate; (ii) $I_0 = \{0\}$; (iii) I_j is contiguous to I_{j+1}; (iv) if infinite, the sequence of intervals is *progressive*, i .e., for every $t \in \mathbb{R}_{\geq 0}$, there exists $j \in \mathbb{N}$ such that $t \in I_j$. An interval state sequence (ISS) on Σ is a pair $\theta = (\sigma, I)$ where $\sigma = \sigma_0 \sigma_1 \cdots$ is a (possibly infinite) word $\sigma \in \Sigma^{\leq \omega}$, and $I = I_0 I_1 \cdots$ is an AIS of the same length. This is the analog of a *timed word* [2]. An ISS can equivalently be seen as a sequence of pairs in $\Sigma \times \mathcal{I}(\mathbb{R}_{\geq 0})$. It can also be seen as a *signal*, i.e. a function from $\mathbb{R}_{\geq 0} \to \Sigma$: given $t \in \mathbb{R}_{\geq 0}$, let $i \in \mathbb{N}$ be the interval such that $t \in I_i$: We define $\theta(t)$ as σ_i. A signal derived from an ISS will always have finite variability. Below, our automata will consider two ISS θ_1, θ_2 that define the same signal as equivalent, noted $\theta_1 \equiv \theta_2$, even if the intervals might be split differently.

2.2   Clocks

A clock is a variable that increases with time. Thus, the value of a clock is the time elapsed since its last reset. When we use continuous time, there is not always a "last" reset, e.g. when the reset holds in an open interval. For this case, we will use non-standard clock values of the form v^+, intuitively meaning that the clock was reset v units before. The set of non-standard real numbers, noted $\mathbb{R}_{\geq 0}^+$, is the set of $\{v, v^+ \mid v \in \mathbb{R}_{\geq 0}\}$, ordered by $<_{ns}$ as following: $v_1 <_{ns} v_2^+$ iff $v_1 \leq v_2$. The addition is commutative, and $v_1^+ + v_2 = (v_1 + v_2)^+$. $\mathbb{R}_{\perp}^+$ is $\mathbb{R}_{\geq 0}^+$ plus a special value $\perp$ for uninitialized clocks. $\perp$ is not comparable to other values, and is absorbing for addition.

Let X be a finite set of clock names. A clock valuation over X is a mapping $\nu : X \to \mathbb{R}_{\perp}^{+}$. The set of constraints over X, denoted $\Phi(X)$, is defined by the grammar $\phi ::= true \mid x \sim c \mid \phi_1 \wedge \phi_2$ where $x \in X$, $c \in \mathbb{N}$, and $\sim \in \{<, \leq, =, >, \geq\}$. We write $\nu \models \phi$ when the valuation ν satisfies the constraint ϕ. When x has the value $\perp$, we evaluate $x \sim c$ to false.

3 Automata Background

Based on time and clocks, several variants of timed automata have been proposed after the seminal Timed Automata (TA) [2]. Below, we review briefly icTA [1] and RECA [12], that are the basis of our DECA.

We use an interval semantics throughout the paper, i.e. predicates (or letters) are functions of time with finite variability. In particular, we do not allow to be in two locations, or to make two transitions, at the same time. Transitions are taken in a single instant; therefore we have to stay in a location during an open interval. Thus, we have to label both locations and transitions (together, we call them *locansitions*) to ensure that predicates are always defined. Time thus strictly increases along an ISS, as in [2]. We allow unobservable transitions [7], that were absent from [2]: here, they are expressed as a transition with the same label as the previous and next location.

3.1 Timed Automata

A Timed Automaton (TA) [2] is a finite state automaton augmented with clocks: real variables that can be reset to 0, and otherwise increase at a uniform rate. Time is thus global, and clocks are perfectly precise and synchronized.

Definition 1. *A Timed Automaton is a tuple* $\mathcal{A} = (\Sigma, X, S, s_0, \to_{ta}, Inv, \lambda, F)$, *with:*

(i) Σ, *a finite alphabet. In this paper, we take* $\Sigma = 2^{\mathbb{P}}$.
(ii) X, *a finite set of positive real variables called clocks.*
(iii) S, *a finite set of locations.*
(iv) $S_0 \subseteq S$, *the initial locations.*
(v) $\to_{ta} \subseteq S \times \Phi(X) \times 2^X \times S$, *a finite set of transitions, each with a guard and a reset.*
(vi) $Inv : S \to \Phi(X)$ *gives the invariant.*
(vii) $\lambda : (S \cup \to_{ta}) \to \Sigma$, *a labelling of locations and transitions.*
(viii) F, *an acceptance condition. For instance, for finite acceptance, we have* $F \subseteq S$, *a set of final locations. We also use Büchi acceptance (where* $F \subseteq S$) *or parity conditions (where* $F : S \to \mathbb{N}$).

TA are neither determinizable nor complementable. Their emptiness problem can be solved using the region construction, but their inclusion problem is undecidable [2].

3.2   Timed Automata with Independent Clocks

Distributed Timed Automata (DTA) [14,1] consist of a number of local timed automata, called processes. Each processes owns clocks. The clocks of a same process evolve synchronously, but independently of the clocks of the other processes. The idea is that the clocks of the same process are all computed from a same hardware clock. A clock can be read by any process, but can only be reset by its owner process.

The homonymous Distributed Timed Automata of [10] work differently: they model processes whose execution is interleaved by a scheduler. Thus, only one process increases its (perfect) clocks at a time. They are a subclass of stopwatch automata.

In [1], DTA are not much studied. Instead, their product is first computed, giving rise to the class of Timed Automata with independent clocks (icTA).

Definition 2. *An icTA is a pair* $(\mathcal{A}, \pi)$*, where* $\mathcal{A}$ *is a TA and* $\pi : X \to Proc$ *maps each clock to a process.*

Definition 3. *A Rate is a tuple* $\tau = (\tau_q)_{q \in Proc}$ *of local time functions. Each local time function* τ_q *maps the reference time to the time of process* q*, i.e,* $\tau_q :$ $\mathbb{R}_{\geq 0} \to \mathbb{R}_{\geq 0}$*. The functions* τ_q *must be continuous, strictly increasing, divergent, and satisfy* $\tau_q(0) = 0$*.*

Note that the reference time is arbitrary, and thus not meaningful.

Definition 4. *Given a clock valuation* $\nu : X \to \mathbb{R}_{\geq 0}$*, a rate* τ*, and two reference times* $t_1 > t_2$*, the valuation* $\nu + (t_1 - t_2)$ *maps* x *to* $\nu(x) + \tau_{\pi(x)}(t_1) - \tau_{\pi(x)}(t_2)$*.*

Definition 5. *A run of an icTA* $\mathcal{A}$ *for* τ *is an ISS alternating states and transitions* $(\beta; I)$ *where* $\beta = \zeta_0, q_1, \zeta_1, q_2, \ldots,$ $I = \{0\},]0, t_1[, \{t_1\},]t_1, t_2[, \ldots,$ *the states* $q_i \in Q = \{(s_i, \nu_i) \in S \times \mathbb{R}_{\geq 0}^X \mid \nu_i \models Inv(s_i)\}$*. It must satisfy:*

1. *the starting state is* $q_0 = (s_0, \nu_0)$*, where* ν_0 *assigns 0 to all the clocks, and* $s_0 \in S_0$*.*
2. *When spending time* $]t_{i-1}, t_i[$ *in* $q_i = (s_i, \nu_i)$*, the invariant must stay continuously true:* $\forall t \in]t_{i-1}, t_i[: \nu_i + (t - t_{i-1})) \models Inv(s_i)$*.*
3. *When following a transition* $\zeta_i = (s_i, \phi, Y, s_{i+1}) \in \to_{icTA}$*, the clock constraint* ϕ *must be satisfied:* $\nu_i + (t_i - t_{i-1}) \models \phi$*. The clocks in* Y *are then reset:* $\nu_{i+1} = (\nu_i + (t_i - t_{i-1}))[Y \to 0]$*. This transition is instantaneous.*
4. *The acceptance condition is verified, e.g. for a finite automaton,* $s_n \in F$*.*

Definition 6. *Given a run* $\rho = (\zeta_0, (s_1, \nu_1), \zeta_1, (s_2, \nu_2), \ldots, I)$ *we define its ISS, noted* $\lambda(\rho)$*, as* $(\lambda(\zeta_0), \lambda(s_1), \lambda(\zeta_1), \lambda(s_2), \ldots, I)$*.*

Definition 7. *The language* $\mathcal{L}(\mathcal{B}, \tau)$ *is the set of ISS of accepting runs of* $\mathcal{B}$ *for* τ*, closed under* $\equiv$*.*

3.3   Recursive Event Clocks Automata

Recursive Event Clock Automata (RECA) [19,12] extend ECA [5]. "Recursive" refers to the fact that the resets of an event clock $x_\mathcal{B}$ are controlled by a lower-level automaton $\mathcal{B}$: When $\mathcal{B}$ visits a monitored locansitions (location or transition), it resets $x_\mathcal{B}$. Symmetrically, *prediction clocks* of the form $y_\mathcal{B}$ measure the time until $\mathcal{B}$ can next visit one of its monitored locansitions. Distributed Real-Time Automata [9] are a special case of RECA where only the time since the last change of labelling can be measured.

Definition 8. *A RECA $\mathcal{A}$ of level $l \in \mathbb{N}$ is a tuple composed of:*

   (i) Σ is a finite alphabet.
   (ii) S is a finite set of locations.
   (iii) $S_0 \subseteq S$ are the initial locations.
   (iv) $\rightarrow_{reca} \subseteq S \times S$ are the transitions.
   (v) C is a finite set of clocks, of the form $x_\mathcal{B}$ or $y_\mathcal{B}$, with $\mathcal{B}$ a lower-level RECA.
   (vi) $\lambda : (S \cup \rightarrow_{reca}) \rightarrow \Sigma$ is a labelling function.
   (vii) $\mathsf{Inv} : (S \cup \rightarrow_{reca}) \rightarrow \Phi(C)$ gives the guard or invariant.
   (viii) $M \subseteq (S \cup \rightarrow_{reca})$ is the set of monitored locansitions: when the automaton
         visits them, it resets its two associated clocks $x_\mathcal{A}, y_\mathcal{A}$.
   (ix) F is an acceptance condition.

RECA can be determinized and thus complemented: They are *fully decidable* [19,12]. They are quite expressive, since they can express the logic MITL [3], but less expressive than TA (otherwise we would lose full decidability).

Below, we assume the uniform naming conventions defined in this section.

4   Timed Languages

Surprisingly, Akshay et al. [1] only consider untimed languages for their timed automata. We are interested in timed languages, but we have a different time scale for each process; thus each process p will determine a timed language observed by p. These languages only differ by their timings. Let τ_p be the rate of process p. τ_p extends naturally to intervals, to interval sequences, to ISS, and to timed languages: Given an ISS $\theta = (\sigma, I)$ expressed in the reference time, $\tau_p(\theta)$ is $(\sigma, \tau_p(I))$. The timed language for τ observed by p is $\tau_p(\mathcal{L}(\mathcal{B}, \tau))$. When there is only one process, i.e. $Proc = \{q\}$, the timed language observed by q is the usual timed language $\mathcal{L}(\mathcal{A})$ of its TA. When τ is a vector of identity functions, we also obtain the usual timed language whatever the observer process chosen.

When we want to avoid some forbidden timed behaviours (ISS) , we consult the existential timed semantics: we consider time evolutions as non-deterministic, since this gives the worst-case assumption. If we want a given timed behaviour to be possible whatever the evolution of local times, we look at the universal semantics.

Definition 9. *For an automaton $\mathcal{B}$ and one of its processes p, we define:*

- *the existential timed language observed by* $p : \mathcal{L}_\exists(\mathcal{B}, p) = \bigcup_{\tau \in Rates} \tau_p(\mathcal{L}(\mathcal{B}, \tau))$
- *the universal timed language observed by* $p : \mathcal{L}_\forall(\mathcal{B}, p) = \bigcap_{\tau \in Rates} \tau_p(\mathcal{L}(\mathcal{B}, \tau))$

This variety of languages leads to three generalisations of the classical problems of emptiness, inclusion, intersection and union. First, the τ-wise definitions:

Definition 10. *Given icTA* $\mathcal{A}, \mathcal{B}, \mathcal{C}$,

1. $\mathcal{C}$ *is an intersection of* $\mathcal{A}, \mathcal{B}$ *iff* $\forall \tau \in Rates, \mathcal{L}(\mathcal{C}, \tau) = \mathcal{L}(\mathcal{A}, \tau) \cap \mathcal{L}(\mathcal{B}, \tau)$
2. $\mathcal{C}$ *is a union of* $\mathcal{A}, \mathcal{B}$ *iff* $\forall \tau \in Rates, \mathcal{L}(\mathcal{C}, \tau) = \mathcal{L}(\mathcal{A}, \tau) \cup \mathcal{L}(\mathcal{B}, \tau)$
3. $\mathcal{C}$ *is a complement automaton of* $\mathcal{A}$ *iff* $\forall \tau \in Rates, \mathcal{L}(\mathcal{C}, \tau) = \mathcal{L}(\mathcal{A}, \tau)^c$, *where* c *is the complement operator.*
4. $\mathcal{A}$ *is language-included in* $\mathcal{B}$ *iff* $\forall \tau \in Rates, \mathcal{L}(\mathcal{A}, \tau) \subseteq \mathcal{L}(\mathcal{B}, \tau)$
5. *The emptiness problem for* $\mathcal{A}$ *is* $\forall \tau \in Rates, \mathcal{L}(\mathcal{A}, \tau) = \emptyset$

The p-existential and p-universal variants use respectively the existential and universal timed languages observed by p.

4.1 Timed Languages of icTA

The existential timed languages of icTA are *timed regular* [2]:

Theorem 1. *For any icTA* $\mathcal{B}$, $\mathcal{L}_\exists(\mathcal{B}, p)$ *is the language of a TA.*

Proof. (sketch) This TA can be computed by a variant of the region construction, of which the construction of [1] is a special case. Let $q \in Proc \setminus \{p\}$ be a process whose clocks we want to eliminate, i.e. we have an icTA $\mathcal{B}$ on $Proc$ and we would like to construct an icTA on $Proc \setminus \{q\}$ whose existential language is preserved for any observer but q. We construct the region automaton, but on the clocks of q only. If a locansition had invariant $\bigwedge_{p \in Proc} \phi_p$, its associated regions have invariant $\bigwedge_{p \in Proc\{q\}} \phi_p$. The constraints on clocks of q are not lost, they become part of the region constraint. This gives a region icTA without the clocks of q, and where the locations are now a pair of an original location and a region constraint on clocks of q, which has the required languages. If we want to eliminate several processes, we eliminate them one by one: eliminating several processes together would give a result that does not reflect the independence of their clocks.

However, the emptiness of their universal timed languages is undecidable, and thus cannot be the language of a TA.

Theorem 2. *icTA are closed under* τ*-wise and* p*-existential intersection and union, and under* p*-universal intersection.*

However, icTA are not determinizable, nor closed under timed complement, and their inclusion problem is undecidable (whether τ-wise, p-existential, or p-universal), essentially because TA [2] are a special case of icTA.

5   Distributed Event Clock Automata

To restore full decidability, we use event clocks [5]. For expressiveness, we use RECA [12] with independent clocks [1]. The distributed event clock (DEC) $x_{\mathcal{A}}^q$ (or $y_{\mathcal{A}}^q$) records the time since the last (resp. next) time that the automaton $\mathcal{A}$ could visit a monitored locansition, measured in the local time of process q.

Definition 11. *A distributed recursive event clock automaton (DECA) is a pair $(\mathcal{A}, \pi)$ where $\mathcal{A}$ is a RECA and $\pi : C \to Proc$ maps each clock to a process.*

For better readability, we write the owner process in the clock name: $\pi(x_{\mathcal{A}}^q) = q$.

Definition 12. *A run ρ of a DECA $\mathcal{A}$ for a rate τ is an ISS alternating transitions and locations $(\zeta_0, s_1, \zeta_1, s_2, \ldots, I)$, such that:*

(i) The run starts from an initial location: $\zeta_0 \in S_0 \times S$.
(ii) The run follows discrete transitions: $\zeta_i = (s_i, s_{i+1}) \in \to_{reca}$
(iii) The clock constraints (invariant or guard) are satisfied by the valuation of the clocks (defined below): $\forall t \in \mathbb{R}_{\geq 0}, \nu(\lambda(\rho), t, \tau) \models Inv(\rho(t))$.
(iv) It satisfies the acceptance condition.

Definition 13. *The ISS of a run $\rho = (s, I)$, noted $\lambda(\rho)$, is the pair $(\lambda(s), I)$.*

Definition 14. *$\mathcal{A}$ accepts an ISS θ at t with τ, if there is a run ρ for θ that visits a monitored location at t. This is noted $(t, \theta) \in \mathcal{L}^+(\mathcal{A}, \tau)$, its anchored language.*

This acceptance time will be used to reset the associated clocks $x_{\mathcal{A}}^q$ below.

Definition 15. *The DEC valuation depends on the ISS θ, on the reference time of evaluation t, and on the rate τ. It assigns a (non-standard) positive real, or undefined, to each clock variable.*

$$\nu(\theta, t, \tau, x_{\mathcal{B}}^q) = \begin{cases} \tau_q(t) - \tau_q(r) & if\ r = \max\{s < t | (s, \theta) \in \mathcal{L}^+(\mathcal{B}, \tau)\}\ exists \\ (\tau_q(t) - \tau_q(r))^+ & else,\ if\ r = \sup\{s < t | (s, \theta) \in \mathcal{L}^+(\mathcal{B}, \tau)\}\ exists \\ \bot & else \end{cases}$$

$$\nu(\theta, t, \tau, y_{\mathcal{B}}^q) = \begin{cases} \tau_q(l) - \tau_q(t) & if\ l = \min\{s > t | (s, \theta) \in \mathcal{L}^+(\mathcal{B}, \tau)\}\ exists \\ (\tau_q(l) - \tau_q(t))^+ & else,\ if\ l = \inf\{s > t | (s, \theta) \in \mathcal{L}^+(\mathcal{B}, \tau)\}\ exists \\ \bot & else \end{cases}$$

Definition 16. *The timed language of a DECA $\mathcal{A}$, noted $\mathcal{L}(\mathcal{A}, \tau)$, are the ISS of its runs for τ, closed under $\equiv$.*

Example 1. The example of Fig.1 from [1] is in fact both a DECA and an icTA $\mathcal{A}$ over $Proc = \{p, q\}$, and the set of propositions $\mathbb{P} = \{a, b, c\}$. Locations have no invariant and an ϵ labelling. Both clocks are reset by the initial monitored transition of $\mathcal{B}$. After this, they may diverge. The existential timed languages, here, are read from the automaton:

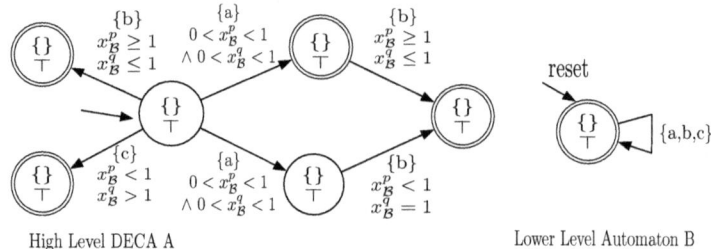

High Level DECA A   Lower Level Automaton B

Fig. 1. Example of DECA from [1]

$$\mathcal{L}_\exists(\mathcal{A}, p) = \mathsf{ITL}^1(\{(a, t_1^p) \mid 0 < t_1^p < 1\} \cup \{(b, t_1^p) \mid t_1^p \geq 1\} \cup \{(c, t_1^p) \mid 0 < t_1^p < 1\}$$
$$\cup \{(a, t_1^p), (b, t_2^p) \mid 0 < t_1^p < 1 \wedge t_1^p < t_2^p\})$$
$$\mathcal{L}_\exists(\mathcal{A}, q) = \mathsf{ITL}(\{(a, t_1^q) \mid 0 < t_1^q < 1\} \cup \{(b, t_1^q) \mid 0 < t_1^q \leq 1\} \cup \{(c, t_1^q) \mid t_1^q > 1\}$$
$$\cup \{(a, t_1^q), (b, t_2^q) \mid 0 < t_1^q < 1 \wedge t_1^q < t_2^q \leq 1\})$$

Here, all universal timed languages are empty: $\mathcal{L}_\forall(\mathcal{A}, p) = \emptyset = \mathcal{L}_\forall(\mathcal{A}, q)$. For instance, we cannot have $(a, t_a) \in \mathcal{L}_\forall(\mathcal{A}, p)$, because there are some τ where the time of q increases steeply, and gets over 1 before the time of p could reach t_a. However, the universal untimed language $\mathcal{L}_\forall(\mathcal{A})$ is $\{a, ab\}$.

5.1   Timed Languages of DECA

DECA inherit the main property of RECA: they are determinizable. The theorems below are valid for the finite version, but also for the infinite ones, e.g. for Büchi automata, which are determinized to a parity automaton [17].

Definition 17. *A DECA $\mathcal{A}$ is deterministic iff all the following conditions hold:*

(i) $\mathcal{A}$ has exactly one initial location $\{s_0\} = S_0$;

(ii) It has no ϵ-transitions: There are no two successive locations $s_1 \rightarrow s_2$, with the same labellings: $\lambda(s_1) = \lambda(s_1, s_2) = \lambda(s_2)$.

(iii) Any two distinct successor locations $s_2 \neq s_3, s_1 \rightarrow s_2, s_1 \rightarrow s_3$ with same labellings: $\lambda(s_2) = \lambda(s_3)$ and $\lambda(s_1, s_2) = \lambda(s_1, s_3)$, have mutually exclusive clock constraints: $\nu \not\models \mathsf{Inv}(s_1, s_2) \wedge \mathsf{Inv}(s_1, s_3)$.

Theorem 3. *Given a deterministic DECA, a rate τ, an ISS θ, there is at most one accepting run on τ for θ, i.e. $\lambda(\rho) \equiv \theta$.*

As for RECA, prediction clocks render a deterministic DECA dependent on the future, and thus unsuitable for realizability [11].

We don't have space to present our determinization construction [19], but its complications rather stems from continuous time than from DEC.

---

[1] ITL will add the missing intervals between time points.

Theorem 4. *Determinization preserves the τ-wise, existential and universal languages.*

Theorem 5. *DECA are closed under union, intersection and complementation, whether τ-wise, p-existential and p-universal.*

Theorem 6. *For all DECA $\mathcal{B}, p \in Proc, \mathcal{L}_\exists(\mathcal{B}, p)$ is the language of a RECA.*

Proof. (sketch) We eliminate each process $q \neq p$ in turn from the DECA while preserving the existential language of the remaining processes. We first complete and determinize automata appearing in the clocks of this process q. We then make their product with the main automaton. We then perform the region construction [3] on the clocks of q. Remember that the clocks are constrained to be 0 in the respective monitored locansition, i.e. when at least one original monitored locansition appears in this construction, and that prediction clocks run backwards so that it is the complement of their fractional part that participates in the region construction [3]. The region construction for prediction clocks is non-deterministic and is not a bisimulation quotient, unlike the one of TA, but preserves the language [19]. Note that the elimination of the clocks of one process only, allows independent evolution of the other clocks.

Theorem 7. *The τ-wise and p-existential emptiness, universality and language inclusion problem for DECA are PSPACE-complete.*

Finite automata have the same complexity, thus the added expressiveness is "for free".

Theorem 8. *For all DECA $\mathcal{B}, q \in Proc, \mathcal{L}_\forall(\mathcal{B}, q)$ is the language of a RECA.*

Proof. (sketch) We complete and determinize the main automaton $\mathcal{A}$. Then we apply the region construction for independent clocks [1]. The automaton becomes non-deterministic, because each region has several successors, depending on τ. Transitions are also considered as regions. A region constraint is expressed as a conjunction $\bigwedge_{p \in Proc} \phi_p$. We choose as invariant of each region locansition ϕ_q. The other constraints are part of the identity of the region, but are not kept as an invariant. Then we determinize it again but we mark as final the locations where all members are final (which, in turn, means that one of their members is an original final location), to represent that the ISS must be accepted under *all* evolutions of time τ.

In contrast, the universal language of DTA and icTA is undecidable [1].

6   Recursive Distributed Event Clocks Temporal Logic

The aim of this section is to construct a fully decidable distributed logic to specify real-time requirements when time scales can be independent.

6.1 Syntax

Our Distributed Event Clock Temporal Logic (DECTL) extend the Event Clock
Temporal Logic (EventClockTL) [20,12] with distributed (a.k.a. independent)
real-time modalities. As in Section 3.2, we assume a set of processes *Proc*. The
clocks of each process will evolve according to its local time given by a Rate τ.

DECTL is based on LTL, and adds two local real-time modalities. The record-
ing modality $\vartriangleleft_I^q \phi$ means that ϕ was true for the last time at reference time t_1
and that the distance, as measured by the time scale of q, is within the interval
I: $\tau_q(t_0) - \tau_q(t_1) \in I$. Symmetrically, the predicting modality $\vartriangleright_I^q \phi$ says that
the next ϕ will occur within I according to the local time of q. With only one
process, we find back EventClockTL [20].

Definition 18. *The formulas of DECTL are defined by the grammar:*

$$\phi ::= true \mid p \mid \vartriangleright_I^q \phi \mid \vartriangleleft_I^q \phi \mid \phi_1 \wedge \phi_2 \mid \neg\phi \mid \phi_1 \,\mathcal{U}\, \phi_2 \mid \phi_1 \,\mathcal{S}\, \phi_2$$

where $p \in \mathbb{P}$ is an atom, $I \in \mathcal{I}(\mathbb{N})$ is an interval and $q \in Proc$ is a process.

6.2 Semantics

Definition 19. *The satisfaction of a DECTL formula ϕ is noted $(t, \theta) \models_\tau \phi$.
We omit τ and θ below, since they are fixed.*

$$
\begin{aligned}
&t \models p && \text{iff } p \in \theta(t)\\
&t \models \neg\phi && \text{iff } t \not\models \phi\\
&t \models \phi_1 \wedge \phi_2 && \text{iff } t \models \phi_1 \text{ and } t \models \phi_2\\
&t \models \phi_1 \mathcal{U} \phi_2 && \text{iff } \exists t' > t, t' \models \phi_2 \text{ and } \forall t'' \in (t, t'), t'' \models \phi_1\\
&t \models \phi_1 \mathcal{S} \phi_2 && \text{iff } \exists t' < t, t' \models \phi_2 \text{ and } \forall t'' \in (t', t), t'' \models \phi_1\\
&t \models \vartriangleleft_I^q \phi && \text{iff } \exists t' < t, \tau_q(t) - \tau_q(t') \in I \wedge t' \models \phi\\
&&& \text{and } \forall t'' < t, \tau_q(t) - \tau_q(t'') < I, t'' \not\models \phi\\
&t \models \vartriangleright_I^q \phi && \text{iff } \exists t' > t, \tau_q(t') - \tau_q(t) \in I \wedge t' \models \phi\\
&&& \text{and } \forall t'' > t, \tau_q(t'') - \tau_q(t) < I, t'' \not\models \phi
\end{aligned}
$$

Example 2. The formula $\neg(\mathcal{F}b \wedge \neg \vartriangleright_{\leq 1}^q b)$, where $\mathcal{F}b = true\,\mathcal{U}b$ says that the
first b, if any, must occur within 1 second, as measured by q. It holds on the
automation of Fig.1. However, the formula measured by p, $\neg(\mathcal{F}b \wedge \neg\vartriangleright_{\leq 1}^p b)$, does
not hold.

6.3 Decidability

Theorem 9. *For any DECTL formula ϕ, there is a DECA automaton $\mathcal{A}_\phi$ with
the same anchored language: $(t, \theta) \in \mathcal{L}^+(\mathcal{A}, \tau)$ iff $(t, \theta) \models_\tau \phi$.*

Proof. (sketch) The translation to a Generalised Büchi tableau is done level by
level, where the level of a formula is the nesting depth of real-time modalities
[19]. A formula $\vartriangleright_I^q \phi$ is translated as constraint $x_{\mathcal{A}_\phi}^q \in I$. The monitored locations of $\mathcal{A}_\phi$ are those containing ϕ. The initial locations are those containing

$\neg true\ \mathcal{S}true$. The transitions are the sets of closure formulae that entail instantaneity. Each location has the Hintikka property: the conjunction of its formulae is satisfied exactly by the ISS of the runs visiting it, at the time they visit it.

The construction is exponential in the size of the non-real time part of the formula, but linear in the real-time part. The test of emptiness is done by the region construction presented in Section 5, that is exponential in the real-time part but linear for the rest.

Theorem 10. *Satisfiability and validity of DECTL are PSPACE-complete.*

The axiomatisation of this logic happens to be given in [21]. There, *shift and order axioms* express the pairwise synchronisation of real-time modalities. We restrict them to modalities of the same process.

6.4   Extensions

(1) We can extend the known expressive equivalence of EventClockTL and MITL+ Past [12] to construct a distributed version of MITL (DMITL) with independent modalities $\mathcal{U}_I^p, \mathcal{S}_I^p$.

(2) DECTL is expressively equivalent to DQTL, a new first-order monadic logic with a metric quantifiers $\exists t \in^p]t_0, t_0 + k[\ . \ \phi, \ \exists t \in^p]t_0 - k, t_0[\ . \ \phi$, where ϕ has only the free variable t (see [13] for QTL).

(3) The more expressive logic DMECTL allows to observe not only the last ϕ, but also the last but n ϕ [15]. This logic is still translatable in DECA.

(4) This logic is expressively equivalent to DQ2MLO, a new first-order monadic logic with a metric quantifier $\exists t \in^p]t_0, t_0 + k[\ . \ \phi, \ \exists t \in^p]t_0 - k, t_0[\ . \ \phi$, where ϕ has only the free variables t_0, t (see [13] for Q2MLO).

(5) We can add DECA automata operators [22].

(6) We can add second-order quantification on predicates that are not subjected to a real-time constraint.

(7) We can also introduce these independent modalities $\mathcal{U}_I^p, \mathcal{S}_I^p$ in a linear μ-calculus.

The last three extensions are expressively equivalent.

7   Conclusions

We have proposed the basis of a framework for analyzing distributed real-time systems through of the introduction of independent (or distributed) event clocks, inspired by icTA [1]. In contrast to [1], we have given a real-time semantics, and thus we can specify real-time properties. We have defined DECA and showed that they are fully decidable, and that their language inclusion problem is PSPACE-complete, as for classical automata. This give us an algorithm to verify real-time properties. Since the number of regions is reduced wrt. ECA, we can even expect faster verification. They are also a good basis for partial-order techniques [6]. The universal (timed) languages of DECA are decidable and (timed) regular, unlike the universal languages of icTA [1]. We proposed the logic DECTL to specify real-time properties with distributed observers. The problems of satisfiability, validity and model-checking of DECTL are PSPACE-complete, as for LTL - we cannot hope better.

Acknowledgements

This work was funded by the Interuniversity Attraction Poles Programme (IAP) of the Belgian State, Belgian Science Policy (MoVES project), by the Belgian Science Foundation (FNRS) under FRFC project CFV, by the Hubert Curien Grants (Tournesol), and by the European Science Foundation (ESF) under EUROCORES project LogiCCC/GASICS.

References

1. Akshay, S., Bollig, B., Gastin, P., Mukund, M., Narayan Kumar, K.: Distributed timed automata with independently evolving clocks. In: van Breugel, F., Chechik, M. (eds.) CONCUR 2008. LNCS, vol. 5201, pp. 82–97. Springer, Heidelberg (2008)
2. Alur, R., Dill, D.L.: A theory of timed automata. Theor. Comput. Sci. 126(2), 183–235 (1994)
3. Alur, R., Feder, T., Henzinger, T.A.: The benefits of relaxing punctuality. ACM 43(1), 116–146 (1996)
4. Alur, R., Fix, L., Henzinger, T.A.: A determinizable class of timed automata. In: Dill, D.L. (ed.) CAV 1994. LNCS, vol. 818, pp. 1–13. Springer, Heidelberg (1994)
5. Alur, R., Henzinger, T.A.: Logics and models of real time: A survey. In: Huizing, C., de Bakker, J.W., Rozenberg, G., de Roever, W.-P. (eds.) REX 1991. LNCS, vol. 600, pp. 74–106. Springer, Heidelberg (1992)
6. Bengtsson, J.E., Jonsson, B., Lilius, J., Yi, W.: Partial order reductions for timed systems. In: Sangiorgi, D., de Simone, R. (eds.) CONCUR 1998. LNCS, vol. 1466, pp. 485–500. Springer, Heidelberg (1998)
7. Bérard, B., Petit, A., Diekert, V., Gastin, P.: Characterization of the expressive power of silent transitions in timed automata. Fundam. Inform. 36(2-3), 145–182 (1998)
8. De Wulf, M., Doyen, L., Markey, N., Raskin, J.-F.: Robustness and implementability of timed automata. In: Lakhnech, Y., Yovine, S. (eds.) FORMATS 2004 and FTRTFT 2004. LNCS, vol. 3253, pp. 118–133. Springer, Heidelberg (2004)
9. Dima, C.: Distributed real-time automata. In: Martin-Vide, C., Mitrana, V. (eds.) Essays in honor of Gheorghe Păun, pp. 131–140. Taylor & Francis, Abington (2003)
10. Dima, C., Lanotte, R.: Distributed time-asynchronous automata. In: Jones, C.B., Liu, Z., Woodcock, J. (eds.) ICTAC 2007. LNCS, vol. 4711, pp. 185–200. Springer, Heidelberg (2007)
11. Doyen, L., Geeraerts, G., Raskin, J.-F., Reichert, J.: Realizability of real-time logics. In: Ouaknine, J., Vaandrager, F.W. (eds.) FORMATS 2009. LNCS, vol. 5813, pp. 133–148. Springer, Heidelberg (2009)
12. Henzinger, T.A., Raskin, J.-F., Schobbens, P.-Y.: The regular real-time languages. In: Larsen, K.G., Skyum, S., Winskel, G. (eds.) ICALP 1998. LNCS, vol. 1443, pp. 580–591. Springer, Heidelberg (1998)

13. Hirshfeld, Y., Rabinovich, A.: An expressive temporal logic for real time. In: Královič, R., Urzyczyn, P. (eds.) MFCS 2006. LNCS, vol. 4162, pp. 492–504. Springer, Heidelberg (2006)
14. Krishnan, P.: Distributed timed automata. Electr. Notes Theor. Comput. Sci. 28 (1999)
15. Jerson Ortiz, J., Legay, A., Schobbens, P.-Y.: Memory event clocks. In: Chatterjee, K., Henzinger, T.A. (eds.) FORMATS 2010. LNCS, vol. 6246, pp. 198–212. Springer, Heidelberg (2010)
16. Ortiz, J., Legay, A., Schobbens, P.-Y.: Distributed event clock automata. Tech. rep., FUNDP University, Belgium (2011), http://www.info.fundp.ac.be/~jor/DecaReport
17. Piterman, N.: From nondeterministic Büchi and Streett automata to deterministic parity automata. In: Proceedings of the 21st Annual IEEE Symposium on Logic in Computer Science, pp. 255–264. IEEE Computer Society, Washington, DC, USA (2006), http://portal.acm.org/citation.cfm?id=1157735.1158062
18. Puri, A.: Dynamical properties of timed automata. In: Ravn, A.P., Rischel, H. (eds.) FTRTFT 1998. LNCS, vol. 1486, pp. 210–227. Springer, Heidelberg (1998)
19. Raskin, J.-F.: Logics, Automata and Classical Theories for Deciding Real Time. PhD thesis, FUNDP University, Belgium (1999), http://www.ulb.ac.be/di/ssd/jfr/
20. Raskin, J.-F., Schobbens, P.-Y.: State clock logic: A decidable real-time logic. In: Maler, O. (ed.) HART 1997. LNCS, vol. 1201, pp. 33–47. Springer, Heidelberg (1997)
21. Schobbens, P.-Y., Raskin, J.-F., Henzinger, T.A.: Axioms for real-time logics. Theoretical Computer Science 274, 151–182 (2002)
22. Wolper, P.: Temporal logic can be more expressive. Information and Control 56(1-2), 72–99 (1983)

Fly-Automata, Their Properties and Applications

Bruno Courcelle and Irène A. Durand

LaBRI, CNRS, Université de Bordeaux, Talence, France
{idurand, courcell}@labri.fr

Abstract. We address the concrete problem of implementing huge bottom-up term automata. Such automata arise from the verification of Monadic Second Order propositions on graphs of bounded tree-width or clique-width. This applies to graphs of bounded tree-width because bounded tree-width implies bounded clique-width. An automaton which has so many transitions that they cannot be stored in a transition table is represented be a fly-automaton in which the transition function is represented by a finite set of meta-rules.

Fly-automata have been implemented inside the Autowrite[1] software and experiments have been run in the domain of graph model checking[2].

1 Introduction

The following theorem connects the problem of verifying graph properties with term (tree) automata.

Theorem 1. *Monadic second-order model checking is* fixed-parameter tractable *for tree-width [Courcelle (1990)] and clique-width [Courcelle, Makowski, Rotics (2001)].*

Tree-width and *clique-width* are graph complexity measures based on graph decompositions. A *decomposition* produces a term representation of the graph. For a graph property expressed in monadic second order logic (MSO), the *algorithm* verifying the property takes the form of a term automaton which recognizes the terms denoting graphs satisfying the property.

In [2], we have given two methods for finding such an automaton given a graph property. The first one is totally general; it computes the automaton directly from the MSO formula; it starts with ad-hoc automata corresponding to atomic formulas and combines them with boolean operations, relabellings and inverse relabellings; however this method it is not practically usable because the intermediate automata that are computed along the construction can be very big even if the final one is not. The second method is very specific: it is a direct construction of the automaton; one must describe the states and the transitions of the automaton. Although the direct construction avoids the bigger intermediate automata, we are still faced with the hugeness of the automata. For instance, one can construct an automaton recognizing graphs which

---

[1] http://dept-info.labri.fr/~idurand/autowrite/
[2] http://dept-info.labri.fr/~idurand/autograph/

B. Bouchou-Markhoff et al. (Eds.): CIAA 2011, LNCS 6807, pp. 264–272, 2011.
© Springer-Verlag Berlin Heidelberg 2011

are acyclic has 3^{3^k} states where k is the clique-width of the graph. Even for $k = 2$, which yields the very restricted class of co-graphs, it is unlikely that we could store the transition table of such an automaton.

The solution to this last problem is to use *fly-automata*. In a fly-automaton, the transition function is represented, not by a table (that would use too much space), but by a finite set of meta-rules. Little space is then required to represent the transition function. In addition, fly-automata are more general than finite bottom-up term automata; they can be infinite in two ways: they can work on an infinite (countable) signature. they can have an infinite (countable) number of states. They are more powerful: a fly-automaton can recognize $\{t \in \mathcal{T}(\mathcal{F}) \mid t = f(t_1, t_2)$ and $|t_1| = |t_2|\}$ where $\mathcal{F}$ is a finite signature.

The purpose of this article is to present in detail the concept of fly-automaton and some experiments done with these automata for the verification of properties of graphs of bounded clique-width.

2 Preliminaries: Terms

We recall some basic definitions concerning terms. The formal definitions can be found in the on-line book [1]. We call *signature* $\mathcal{F}$ a set of symbols equiped with a function arity : $\mathcal{F} \to \mathbb{N}$. We denote by $\mathcal{F}_n$ the subset of symbols of $\mathcal{F}$ with arity n. So $\mathcal{F} = \bigcup_n \mathcal{F}_n$. $\mathcal{T}(\mathcal{F})$ denotes the set of (ground) *terms* built upon the signature $\mathcal{F}$. Given a term t, $\mathcal{P}\mathrm{os}(t)$ denotes the set of positions of the term. The position of the root of a term is denoted by ϵ. A term t can also be viewed as a map from its set of positions $\mathcal{P}\mathrm{os}(t)$ to $\mathcal{F}$.

Example 1. Let $\mathcal{F}$ be a signature containing the symbols $\{a, b, add_{a\_b}, rel_{a\_b}, rel_{b\_a}, \oplus\}$ with

arity$(a) =$ arity$(b) = 0$ arity$(\oplus) = 2$
arity$(add_{a\_b}) =$ arity$(rel_{a\_b}) =$ arity$(rel_{b\_a}) = 1$

We will see in Section 3 that this signature is suitable to build terms representing graphs of clique-width at most 2.

t_1, t_2, t_3 and t_4 are terms built upon the signature $\mathcal{F}$ of Example 1.

$t_1 = \oplus(a, b)$
$t_2 = add_{a\_b}(\oplus(a, \oplus(a, b)))$
$t_3 = add_{a\_b}(\oplus(add_{a\_b}(\oplus(a, b)), add_{a\_b}(\oplus(a, b))))$
$t_4 = add_{a\_b}(\oplus(a, rel_{a\_b}(add_{a\_b}(\oplus(a, b)))))$

We will see in Table 1 their associated graphs.

3   Application Domain

All this work will be illustrated through the problem of verifying properties of graphs of bounded clique-width. We present here the connection between graphs and terms and the connection between graph properties and term automata.

3.1   Graphs as Logical Structures

We consider finite, simple, loop-free, undirected graphs (extensions are easy)[3]. Every graph can be identified with the relational structure $\langle V_G, edg_G \rangle$ where V_G is the set of vertices and edg_G the binary symmetric relation that describes edges: $edg_G \subseteq V_G \times V_G$ and $(x, y) \in edg_G$ if and only if there exists an edge between x and y. Properties of a graph G can be expressed by sentences of relevant logical languages. For instance, G is *complete* can be expressed by $\forall x, \forall y, edg_G(x, y)$ or G is *stable* by $\forall x, \forall y, \neg edg_G(x, y)$. Monadic Second order Logic is suitable for expressing many graph properties like k-colorability, acyclicity, ...

3.2   Term Representation of Graphs of Bounded Clique-Width

Let $\mathcal{L}$ be a finite set of vertex labels and let us consider graphs G such that each vertex $v \in V_G$ has a label $label(v) \in \mathcal{L}$. The operations on graphs are $\oplus$[4], the union of disjoint graphs, the unary edge addition add_{a_b} that adds the missing edges between every vertex labeled a and every vertex labeled b, the unary relabeling rel_{a_b} that renames a to b (with $a \neq b$ in both cases). A constant term a denotes a graph with a single vertex labeled by a and no edge. Let $\mathcal{F}_\mathcal{L}$ be the set of these operations and constants. Every term $t \in \mathcal{T}(\mathcal{F}_\mathcal{L})$ defines a graph $G(t)$ whose vertices are the leaves of the term t. Note that, because of the relabeling operations, the labels of the vertices in the graph $G(t)$ may differ from the ones specified in the leaves of the term. A graph has *clique-width* (*cwd* for short) at most k if it is defined by some $t \in \mathcal{T}(\mathcal{F}_\mathcal{L})$ with $|\mathcal{L}| \leq k$.

Table 1. The graphs corresponding to the terms of Example 1

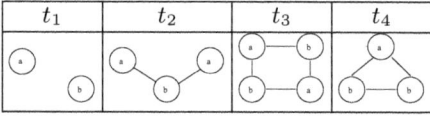

We will express graph properties using MSO formulas that formalize coloring and partitioning problems to take a few examples.

4   Term Automata

We recall some basic definitions concerning term automata. Again, much more information can be found in the on-line book [1].

---

[3] We consider such graphs for simplicity of the presentation but we can work as well with directed graphs, loops, labeled vertices and edges

[4] oplus will be used instead of $\oplus$ inside the software Autowrite

4.1   Finite Bottom-Up Term Automata

Definition 1. *A finite (bottom-up) term automaton*[5] *$\mathcal{A}$ is a quadruple $(\mathcal{F}, Q_{\mathcal{A}}, Q_{\mathcal{A}}^{Acc}, \Delta_{\mathcal{A}})$ consisting of a finite signature $\mathcal{F}$, a finite set $Q_{\mathcal{A}}$ of states, disjoint from $\mathcal{F}$, a subset $Q_{\mathcal{A}}^{Acc} \subseteq Q_{\mathcal{A}}$ of accepting states, and a set of transitions rules $\Delta_{\mathcal{A}}$. Every transition is of the form $f(q_1, \ldots, q_n) \to q$ with $f \in \mathcal{F}$, arity$(f) = n$ and $q_1, \ldots, q_n, q \in Q_{\mathcal{A}}$.*

Example 2. Figure 1 shows an example of such an automaton. It recognizes terms representing graphs of clique-width 2 which are stable (do not contain edges). State <a> (resp. <b>) means that we have found at least a vertex labeled a (resp. b). State <ab> means that we have at least a vertex labeled a and at least a vertex labeled b but no edge. State error means that we have found at least an edge so that the graph is not stable. Note that when we are in the state <ab>, an add_a_b operation creates at least an edge so we reach the <error> state.[6]

```
Automaton 2-STABLE
Signature: a b ren_a_b:1 ren_b_a:1 add_a_b:1 oplus:2*
States: <a> <b> <ab> <error>
Accepting States: <a> <b> <ab>
Transitions   a -> <a>              b -> <b>
  add_a_b(<a>) -> <a>           add_a_b(<b>) -> <b>
  ren_a_b(<a>) -> <b>           ren_b_a(<a>) -> <a>
  ren_a_b(<b>) -> <b>           ren_b_a(<b>) -> <a>
  ren_a_b(<ab>) -> <b>          ren_b_a(<ab>) -> <a>
  oplus*(<a>,<a>) -> <a>        oplus*(<b>,<b>) -> <b>
  oplus*(<a>,<b>) -> <ab>       oplus*(<b>,<ab>) -> <ab>
  oplus*(<a>,<ab>) -> <ab>      oplus*(<ab>,<ab>) -> <ab>
  add_a_b(<ab>) -> <error>      ren_a_b(<error>) -> <error>
  add_a_b(<error>) -> <error>   ren_b_a(<error>) -> <error>
  oplus*(<error>,q) -> <error>  for all states q
```

Fig. 1. An automaton recognizing terms representing stable graphs

Finite term automata recognize *regular* term languages[7]. The class of regular term languages is closed under the Boolean operations (union, intersection, complementation) on languages which have their counterpart on automata. For all details on terms, term languages and finite term automata, the reader should refer to [1]. Figure 2 shows in a graphical way the run of the automaton 2-STABLE on a term representing a graph of clique-width 2. Below we show a successful run of the automaton on a term representing a stable graph.

```
add_a_b(ren_a_b(oplus(a,b))) -> add_a_b(ren_a_b(oplus(<a>,b)))
-> add_a_b(ren_a_b(oplus(<a>,<b>)) -> add_a_b(ren_a_b(<ab>))
-> add_a_b(<b>) -> <b>
```

---

[5] Term automata are frequently called tree automata, but it is not a good idea to identify trees, which are particular graphs, with terms.

[6] Our software Autowrite takes into account the notion of commutative symbols. The star in oplus* means that this symbol is commutative. When we have a rule like oplus*(q1,q2) -> q the rule oplus*(q2,q1) -> q is implicitly defined.

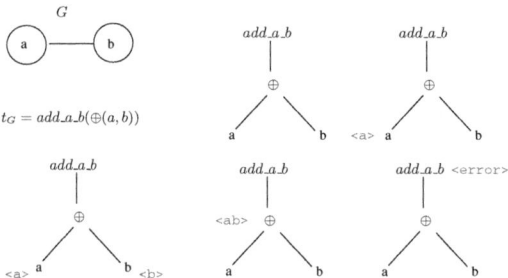

Fig. 2. Graphical representation of an (unsuccessful) run of the automaton on a term

To distinguish these automata from the infinite automata defined in the next section (4.2) and as we only deal with terms in this paper we will refer to the previously defined term automata as *table-automata*.

4.2 Infinite (Bottom-up) Term Automata

Definition 2. *From now on, a* term automaton $\mathcal{A}$ *will be given by* $(\mathcal{F}, \delta, \mathsf{acf})$ *or* $(\mathcal{F}, \delta)$ *where the signature* $\mathcal{F}$ *may be countably infinite,* δ *is the transition relation defined as a function*

$$\delta : \bigcup_n \mathcal{F}_n \times Q^n \to \mathcal{P}_f(Q)$$
$$f q_1 \ldots q_n \quad \mapsto \{q \in Q \mid f(q_1, \ldots, q_n) \to q\}$$

where the set Q *of states accessible using* δ *may be countably infinite,* $\mathcal{P}_f(Q)$ *is the set of* **finite** *subsets of* Q *and* acf *is the accepting state function* $\mathsf{acf} : Q \to Boolean$ *which indicates whether a state is accepting or not. Note that the set of states* Q *is given implicitly by* δ*. The notions of a run, of an accepting run, the sets* $\mathcal{L}(\mathcal{A}, q)$ *and* $\mathcal{L}(\mathcal{A})$ *are the same. Term automata may be complete and/or deterministic in an obvious way. We will shortly consider effectivity conditions insuring that membership of a term to* $\mathcal{L}(\mathcal{A})$ *is decidable.*

Sometimes, in the case where the number of states is infinite, these automata will have no accepting state function. It is the case for instance, for counting automata as shown in the following example.

Example 3. The automaton COUNTING presented p.269 is an example of an infinite automaton. Given a term, it counts the number of vertices of the associated graph of any clique-width. State <i> means that we have found i vertices. The set of states $Q = \{$<i> $\mid i \in \mathbb{N}\}$. There is no accepting state function. However, if we want an automaton recognizing terms corresponding to graphs having an *prime* number (or a multiple of some integer) of vertices, we may add an accepting state function $\mathsf{acf} :$ <i> $\mapsto T$ if i is prime, F otherwise. Note that as the automaton works on graphs of any clique-width, we need a countable set of labels, so we use numbers instead of letters in the finite examples.

4.3 Fly Term Automata

Definition 3. *A* fly-automaton *is an automaton* $(\mathcal{F}, \delta, \mathsf{acf})$ *such that* δ *and* acf *are computable functions.*

Theorem 2. *Let $\mathcal{A}$ be a fly-automaton. Membership to $\mathcal{L}(\mathcal{A})$ is decidable. The emptiness of $\mathcal{L}(\mathcal{A})$ is not decidable.*

```
Automaton COUNTING
Signature: 0 1 2 ...
 ren_0_1:1 ren_1_0:1 add_0_1:1 ren_0_2:1 ren_2_0:1 add_0_2:1 ...
 oplus:2*
States: <0> <1> <2> ...
Metarules:
 x -> <1> for all x
 add_x_y(<i>) -> <i> for all x,y s.t. x < y
 ren_x_y(<i>) -> <i> for all x,y
 oplus*(<i>,<j>) -> <i+j> for all i,j
```

Theorem 3. *Fly-automata are closed under Boolean operations, arity-preserving re-labellings and inverse-relabellings.*

Proof. Let $\mathcal{A} = (\mathcal{F}, \delta, \mathrm{acf})$ be a deterministic and complete fly-automaton. The complement of $\mathcal{A}$ is $(\mathcal{F}, \delta, \mathrm{acf}^c)$ where $\mathrm{acf}^c(q) = \neg\mathrm{acf}(q)$ for every $q \in Q_{\mathcal{A}}$.
Given two fly-automata $\mathcal{A}_1 = (\mathcal{F}, \delta_1, \mathrm{acf}_1)$ and $\mathcal{A}_2 = (\mathcal{F}, \delta_2, \mathrm{acf}_2)$, one can easily define a computable transition function δ corresponding to the product of the two automata whose states are in $Q_{\mathcal{A}} \times Q_{\mathcal{B}}$. The following accepting state functions are suitable (and computable) for union and intersection respectively.

$$\mathrm{acf}^u : Q_{\mathcal{A}} \times Q_{\mathcal{B}} \to Boolean \qquad \mathrm{acf}^i : Q_{\mathcal{A}} \times Q_{\mathcal{B}} \to Boolean$$
$$\{q_1, q_2\} \;\mapsto\; \mathrm{acf}_1(q_1) \vee \mathrm{acf}_2(q_2) \qquad \{q_1, q_2\} \;\mapsto\; \mathrm{acf}_1(q_1) \wedge \mathrm{acf}_2(q_2)$$

Then $\mathcal{A}_1 \cup \mathcal{A}_2 = (\mathcal{F}, \delta, \mathrm{acf}^u)$ and $\mathcal{A}_1 \cap \mathcal{A}_2 = (\mathcal{F}, \delta, \mathrm{acf}^i)$ are fly-automata. The proofs are similar for arity-preserving relabellings and inverse-relabellings.

In the same spirit, fly-automata may be determinized and completed. The determinized version of $\mathcal{A}$ is an automaton $d(\mathcal{A}) = (\mathcal{F}, \delta', \mathrm{acf}')$. If $Q_{\mathcal{A}}$ is the domain of δ (the set of states of $\mathcal{A}$), let $d(Q_{\mathcal{A}})$ denote the set of states of $d(\mathcal{A})$. Each subset $\{q_1, \ldots, q_p\}$ of $Q_{\mathcal{A}}$ yields a state $[q_1, \ldots, q_p]$ in $d(Q_{\mathcal{A}})$. δ' is defined by with

$$\delta' : \bigcup_n \mathcal{F}_n \times d(Q_{\mathcal{A}})^n \to d(Q_{\mathcal{A}})$$
$$f, S_1, \ldots, S_n \;\mapsto\; S$$

with $q \in S$ if and only if $\exists q_1, \ldots, q_b \in S_1 \times \ldots S_n$ such that $q \in \delta(f, q_1, \ldots, q_n)$.

When a fly-automaton $(\mathcal{F}, \delta, \mathrm{acf})$ is finite, it can be compiled into a table-automaton $(\mathcal{F}, Q, Q^{Acc}, \Delta)$, provided that the resulting table is not too big. The transition table Δ can be computed from δ starting from the constant transitions and then saturating the table with transitions involving new accessible states until no new state is computed. The set of (accessible) states Q is obtained during the construction of the transitions table. The set of accepting states Q^{Acc} is obtained by removing the non accepting states (according to the accepting state function acf) from the set of states. A table-automaton is a particular case of a fly-automaton. It can be seen as a compiled version of a fly-automaton whose transition function δ is described by the transitions table Δ and whose accepting state function acf corresponds to membership to Q^{Acc}. It follows that the automata operations defined for fly-automaton will work for table-automata. Table-automata are faster for recognizing a term but they use space for storing the transitions table and the access time may be important in case of a very large table. Fly-automata use a much smaller space (the space corresponding to the code of the transition function) but are slower for term recognition when the transition function is complex. A table-automaton should be used when the transition table can be computed in reasonable space and a fly-automaton otherwise.

5   Implementation of Fly-Automata

We will call *basic* fly-automata the ones that are built from scratch in order to distinguish them from the ones that are obtained by combinations of existing automata using the operations cited in Theorem 3, determinization and completion. We call the later *composed* fly-automata. Fly-automata have been implemented inside the software Autowrite [5] (entirely written in Common Lisp) which already had table-automata. States are not stored in the representation. For basic fly-automata, they are created on the fly by calls to the transition function. For composed automata, the states returned by the transition function are constructed from the ones returned from the transition functions of the combined automata. For operations like determinization, inverse-relabellings, sets of states are involved. The implementation of fly-automata use intensively the functional paradigm to represent and combine transition and accepting states functions. More details about the implementation can be found in [6]. The main operations that are implemented on fly-automata are: run of an automaton $\mathcal{A}$ on a term t, recognition of a term t by an automaton $\mathcal{A}$, decision of emptiness for $\mathcal{L}(\mathcal{A})$ (when $\mathcal{A}$ is finite), completion, determinization, complementation of an automaton $\mathcal{A}$, union, intersection of two (or more) automata, relabellings and inverse-relabellings of constants.

For table-automata, we have also implemented reduction (removal of inaccessible states), minimization but this is not discussed in this paper. Because a table-automaton can always be transformed into a fly-automaton and a finite fly-automaton back to a table-automaton we get the corresponding operations for table-automata for free once we have implemented them for fly-automata. However, for efficiency reasons, it might be interesting to implement some of these operations at the level of table-automaton. For instance, the complementation which consists in inverting non accepting and accepting states is easily performed directly on a table-automaton. Implementing operations directly at the level of table-automaton has the drawback that it depends on the representation chosen for the transitions table. Whenever, we would want to change this representation we would have to re-implement these operations. The only advantage is a gain in efficiency. Some operations on table-automata may give a blow-up in terms of the size of the transition table (determinization, intersection). In this case, the solution is to omit to compile the resulting operation back to a table-automaton. It is though possible to deal uniformly with table and fly-automata.

6   Experiments

Most of our experiments have been run in the domain of verifying graph properties. Many construction of basic automata can be found in [4,3] and have been implemented with Autowrite. In order to compare the running time of a fly-automaton and that of the corresponding table-automaton, we have chosen a property and a clique-width for which the automaton is compilable. This is the case for the *connectedness property*. We have a direct construction of an automaton verifying whether a graph is connected. The corresponding table automaton has $2^{2^{cwd}-1} + 2^{cwd} - 2$ states. It is compilable up to $cwd = 3$. For $cwd = 4$, which gives $|Q| = 32782$, we run out of memory. It is possible to show that the number of states of the minimal automaton is $|Q| > 2^{2^{\lfloor cwd/2 \rfloor}}$. So there is no hope of having a table-automaton for this property and $cwd > 3$.

We have direct constructions of the automata for properties like $\text{Edge}(X_1, X_2)$, k-Cardinality(), k-Coloring$(X_1, \ldots, X_k)$, Connectedness(), Acyclic() among others. With these properties and using relabellings and Boolean operations, we obtain automata for properties like k-Colorability(), k-Acyclic-Colorability(), k-Vertex-Cover() among others. The Vertex-Cover property can be expressed by a combination (intersection, homomorphisms) of already defined basic automata (stablility, k-cardinality).

Many problems that where unthinkable to solve with table-automata could be solved with fly-automata. For very difficult (NP-complete) problems we still reach time or space limitations.

7   Conclusion and Perspectives

In the near future, we plan to implement more graph properties and to run tests on real and random graphs. We cannot hope to check arbitrary Monadic Second Order formulas because, even on words, the problem is intractable if the formula is part of the input. However, many interesting graph properties seem to be reachable. We did not address the problem of finding terms representing a graph, that is, to find a clique-width decomposition of the graph. In some cases, the graph of interest comes with a "natural decomposition" from which the clique decomposition of bounded clique-width is easy to obtain but for the general case the known algorithms are not practically usable.

The concept of fly-automaton is general and could be applied to other domains where big automata are needed.

References

1. Comon, H., Dauchet, M., Gilleron, R., Jacquemard, F., Lugiez, D., Tison, S., Tommasi, M.: Tree Automata Techniques and Applications, draft (2002),
   http://tata.gforge.inria.fr
2. Courcelle, B., Durand, I.: Verifying monadic second order graph properties with tree automata. In: Proceedings of the 3rd European Lisp Symposium, pp. 7–21 (May 2010)
3. Courcelle, B., Durand, I.: Automata for the verification of monadic second-order graph properties (2011) (in preparation)

4. Courcelle, B., Engelfriet, J.: Graph structure and monadic second-order logic, a language theoretic approach. Cambridge University Press, Cambridge (2011),
   `http://www.labri.fr/perso/courcell/Book/CourGGBook.pdf`
5. Durand, I.: Autowrite: A tool for term rewrite systems and tree automata. Electronics Notes in
   Theorical Computer Science 124, 29–49 (2005)
6. Durand, I.: Implementing huge term automata. In: Proceedings of the 4th European Lisp Symposium, pp. 17–27 (March 2011)
7. Thatcher, J., Wright, J.: Generalized finite automata theory with an application to a decision
   problem of second-order logic. Mathematical Systems Theory 2, 57–81 (1968)

Tree Template Matching in Ranked Ordered Trees by Pushdown Automata[*]

Tomáš Flouri[1], Jan Janoušek[1], Bořivoj Melichar[1], Costas S. Iliopoulos[2,3], and Solon P. Pissis[2]

[1] Dept. of Theoretical Computer Science, Faculty of Information Technology, Czech Technical University in Prague
[2] Dept. of Informatics, King's College London, London, UK
[3] DEBII, Curtin University of Technology, Perth, Australia

Abstract. We consider the problem of tree template matching in ranked ordered trees, and propose a solution based on the bottom-up technique. Specifically, we transform the tree pattern matching problem to a string matching problem, by transforming the tree template and the subject tree to strings representing their postfix notation, and then use pushdown automata as the computational model. The method is analogous to the construction of string pattern matchers. The given tree template is preprocessed once, by constructing a nondeterministic pushdown automaton, which is then transformed to the equivalent deterministic one. Although we prove that the space required for preprocessing is exponential to the size of the tree template in the general case, the space required for a specific class of tree templates is linear. The time required for the searching phase is linear to the size of the subject tree in both cases.

1 Introduction

Tree pattern matching, the process of finding all occurrences of a given tree pattern in a subject tree, is an important operation in computer science on which a number of tasks are based on, e.g. mechanical theorem proving, term-rewriting, instruction selection and non-procedural programming languages [8]. In addition, tree pattern matching has direct applications in computational biology, e.g. glycan classification [10], exact and approximate pattern matching and discovery in RNA secondary structure [11].

We distinguish among two types of tree pattern matching: the *subtree* and the *tree template* matching. While subtrees consist of only specific fixed labeled nodes, tree templates have some of their leaves denoted as "don't care", representing arbitrary subtrees – such nodes match any subtree. In this paper, we focus on the problem of tree template matching in ranked ordered trees.

Since 1960, many methods have been described in the literature for solving the tree pattern matching problem [1], [2], [6], [7], [8], [14]. However, most of them

---

[*] This research has been partially supported by the Ministry of Education, Youth and Sports of Czech Republic under research program MSM 6840770014, and by the Czech Science Foundation as project No. 201/09/0807.

B. Bouchou-Markhoff et al. (Eds.): CIAA 2011, LNCS 6807, pp. 273–281, 2011.

lack clear references to a systematic approach of the standard theory of formal languages, grammars and automata. In general, there exist two such approaches using automata. Linearising trees and using string automata represents the first approach [6, 7, 14]. Usage of finite automata is not sufficient, as linear notations of trees are context-free languages. Therefore, the pushdown automaton (PDA) seems to be an appropriate model of computation. The second approach does not reside on tree linearisation, but represents a generalisation from string automata to tree automata [3]. [3] presents a systematic approach for solving the tree pattern matching problem, by utilising finite tree automata, which accept regular tree languages, as the computational model.

Recently, it has been proved that the deterministic PDA can accept a proper superclass of regular tree languages in a linear notation [9]. Based on this, [5] presents a new systematic approach for solving the subtree matching problem using deterministic PDA, with the preprocessing phase requiring time and space linear to the size of the tree pattern; the searching phase runs in time linear to the size of the subject tree.

In this paper, we continue from [5], using notions from [8], to propose and prove a new class of deterministic PDA for solving the tree template matching problem. This is directly analogous to the finite automata based string matching approaches [4]. Notice that methods which use tree pattern matching and are described by PDA, are known [2,7]. However, these methods work in an LR-parser-like fashion, where the parser is constructed for an ambiguous grammar, and some heuristics are used for the tree pattern matching to be deterministic. Our method does not use any grammar or such heuristics, but instead a deterministic PDA is constructed, similarly as in the case of the string pattern matchers. This agrees with a systematic approach for designing algorithms whose computational model is the deterministic PDA [12].

2   Preliminaries

2.1   Basic Definitions

We denote the set of nonnegative natural numbers by $\mathbb{N}$. An *alphabet* Σ is a finite, nonempty set of symbols. A *string* is a succession of zero or more symbols from an alphabet Σ. The string with zero symbols is denoted by ε. The set of all strings over Σ, including ε, is denoted by Σ^*, and $\Sigma^+ = \Sigma^* \setminus \{\varepsilon\}$. A string x of length m is represented by $x_1 x_2 \ldots x_m$, where $x_i \in \Sigma$ for $1 \le i \le m$. The length of a string x is denoted by $|x|$. A string w is a *factor* of x if $x = uwv$ for $u, v \in \Sigma^*$, and is represented as $w = x_i \ldots x_j$, $1 \le i \le j \le |x|$. A *ranked alphabet* is a couple $\mathcal{A} = (\Sigma, \varphi)$, where Σ is an alphabet and φ is a mapping $\varphi : \Sigma \mapsto \mathbb{N}$. The *arity* (rank) of a symbol $x \in \Sigma$ is $\varphi(x)$. The cardinality of a set X is denoted by $\sigma(X)$ and its *powerset* by $\mathcal{P}(X)$. The number of nodes of a tree t is denoted by $|t|$. The *postfix notation* $post(t)$ of a labeled, ordered, ranked tree t is obtained by applying *Step* recursively, beginning at the root of t:

Step: Let this application of *Step* be node v. If v is a leaf, list v and halt. If v is an internal node having descendants $v_1, v_2, \ldots, v_{\varphi(v)}$, apply *Step* to $v_1, v_2, \ldots, v_{\varphi(v)}$ in that order and then list v.

An (extended) *nondeterministic pushdown automaton* is a seven-tuple $M = (Q, \mathcal{A}, G, \delta, q_0, Z_0, F)$, where Q is a finite set of *states*, $\mathcal{A}$ is the *input alphabet*, G is the *pushdown store alphabet*, δ is a mapping from $Q \times (\mathcal{A} \cup \{\varepsilon\}) \times G^*$ into a set of finite subsets of $Q \times G^*$, $q_0 \in Q$ is the initial state, $Z_0 \in G$ is the initial content of the pushdown store, and $F \subseteq Q$ is the set of final (accepting) states. The triplet $(q, w, x) \in Q \times \mathcal{A}^* \times G^*$ denotes the configuration of a PDA. In this paper we write the top of the pushdown store x on its left hand side. The initial configuration of a PDA is a triplet (q_0, w, Z_0) for the input string $w \in \mathcal{A}^*$. The relation $\vdash_M \subset (Q \times \mathcal{A}^* \times \Gamma^*) \times (Q \times \mathcal{A}^* \times \Gamma^*)$ is a transition of a PDA M. It holds that $(q, aw, \alpha\beta) \vdash_M (p, w, \gamma\beta)$ if $(p, \gamma) \in \delta(q, a, \alpha)$. For simplicity, in the rest of the text, we use the notation $p\alpha \xmapsto[M]{a} q\beta$ when referring to the transition $\delta_1(p, a, \alpha) = (q, \beta)$ of a PDA M. A PDA is *deterministic*, if:

1. $|\delta(q, a, \gamma)| \leq 1$ for all $q \in Q$, $a \in \mathcal{A} \cup \{\varepsilon\}$, $\gamma \in G^*$.
2. If $\delta(q, a, \alpha) \neq \emptyset$, $\delta(q, a, \beta) \neq \emptyset$ and $\alpha \neq \beta$ then α is not a suffix of β and β is not a suffix of α.
3. If $\delta(q, a, \alpha) \neq \emptyset$, $\delta(q, \varepsilon, \beta) \neq \emptyset$, then α is not a suffix of β and β is not a suffix of α.

2.2 Properties of Trees in Postfix Notation

Lemma 1. *Given a tree t and its postfix notation $post(t)$, the postfix notations of all subtrees of t are factors of $post(t)$.*

However, not every factor of the postfix notation of a tree represents a subtree. This is obvious due to the fact that there can be $\mathcal{O}(n^2)$ distinct factors of a given postfix notation of some tree with n nodes, but the tree consists of only n subtrees – each node of the tree is the root of one subtree. Only the factors which themselves are trees in postfix notation represent subtrees. This property is formalised by the following definition and theorem.

Definition 1. *Let $x = x_1 x_2 \ldots x_m$, $m \geq 1$, be a string over a ranked alphabet $\mathcal{A} = (\Sigma, \varphi)$. Then, the* arity checksum $ac(x) = \varphi(x_1) + \varphi(x_2) + \ldots + \varphi(x_m) - m + 1 = \sum_{i=1}^m \varphi(x_i) - m + 1$.

Theorem 1. *Let $post(t)$ and x be a tree t in postfix notation and a factor of $post(t)$, respectively, over a ranked alphabet $\mathcal{A} = (\Sigma, \varphi)$. Then, x is the postfix notation of a subtree of t, if and only if $ac(x) = 0$, and $ac(y) \geq 1$ for each y, where $x = zy$, $y, z \in \Sigma^+$.*

2.3 Problem Definition

Definition 2 (Set of all trees). *Given a ranked alphabet $\mathcal{A} = (\Sigma, \varphi)$, $T(\mathcal{A})$ denotes the set of all trees over $\mathcal{A}$, and is defined as follows:*

$$T(\mathcal{A}) = \{ x : x \in \Sigma^+ \wedge ac(x) = 0 \wedge ac(y) \geq 1, x = zy, y, z \in \Sigma^+ \}$$

We also introduce a new nullary symbol S, not in Σ, serving as a placeholder for any tree t, where $post(t) \in T(\mathcal{A})$. We denote the set $\Sigma \cup \{S\}$ as Σ_S, and define $\mathcal{A}_S = (\Sigma_S, \varphi_S)$, where:

$$\varphi_S(a) = \begin{cases} \varphi(a) & : & a \in \Sigma \\ 0 & : & a = S \end{cases}$$

Definition 3 (Tree Pattern). *Given a ranked alphabet $\mathcal{A}_S = (\Sigma_S, \varphi_S)$ and the set of all trees $T(\mathcal{A}_S)$, a tree pattern is any tree in $T(\mathcal{A}_S)$.*

Definition 4 (Tree Template). *Tree templates are the elements of the set $T(\mathcal{A}_S) \setminus T(\mathcal{A})$, i.e. trees having at least one "don't care" node.*

Definition 5 (Tree template matching). *A tree template P over a ranked alphabet $\mathcal{A}_S = (\Sigma_S, \varphi_S)$ with k occurrences of the unary placeholder symbol S matches a subject tree T in $T(\mathcal{A})$ at node v, if there exist trees $t_1, t_2, \ldots, t_k$ in $T(\mathcal{A})$, such that the tree p', obtained by substituting t_i with the i-th occurrence of S in P, is equal to the subtree of T rooted at v. Two trees are equal if, for example, their postfix notations are equal strings.*

While not necessary in general, a new identifier can be encoded for each node of the subject tree, based on its attributes (such as label) and rank. These identifiers, along with the arity of the respective nodes, form the ranked alphabet. In this way, the case when the tree consists of nodes having the same label but different arity, can easily be handled.

3   Algorithm

In this section, we present an algorithm for tree template matching based on PDA. The algorithm preprocesses the tree template once, by computing the so-called match-sets, which are required for the construction of a PDA matching the given tree template. The constructed PDA then reads the postfix notation of the subject tree, and matches each read subtree with the corresponding subtrees of the tree template. Indication that a read subtree matches the tree template is provided by the final state of the PDA. The rest of this section is divided in three parts: first, we formally introduce the notion of match-sets; then, we show a method for computing match-sets; finally, the algorithm for preprocessing the tree template is presented.

3.1   Match-Sets

Definition 6 (Set of subtrees). *Given a tree t such that $post(t) = x_1 x_2 \ldots x_m$ over a ranked alphabet $\mathcal{A}_S = (\Sigma_S, \varphi_S)$, the set of subtrees of t is the set $Sub(t)$ consisting of the postfix notations of all subtrees of t, and is formally defined as:*

$$Sub(t) = \{\, x : post(t) = yxz, \; y, z \in \Sigma^*, \; x \in \Sigma^+, x \neq S \,\}$$

such that Theorem 1 holds for each $x \in Sub(t)$.

We are now in a position to formally define the notion of match-sets. Each tree $t \in T(\mathcal{A}_S)$ can be mapped to a set consisting of all subtrees of the given tree template P that match t. We call this particular set a match-set.

Definition 7 (Match-set). *Given a tree template P over $\mathcal{A}_S = (\Sigma_S, \varphi_S)$, a match-set is the mapping:*

$$\mu : T(\mathcal{A}) \mapsto R$$

where $R \subseteq \mathcal{P}(Sub(P) \cup \{S\})$, and is defined as:

1. *For each $v \in \Sigma$, where $\varphi(v) = 0$:*

$$\mu(v) = \begin{cases} \{v, S\} & : \quad v \in Sub(P) \\ \{S\} & : \quad v \notin Sub(P) \end{cases}$$

2. *For each $x = post(t_1)post(t_2)\ldots post(t_q)v$, $v \in \Sigma$, $\varphi(v) = q$, $x \in T(\mathcal{A})$,*

$$\mu(x) = \{S\} \cup \{y : y = post(t_1')\ldots post(t_q')v \wedge y \in Sub(P) \wedge post(t_i') \in \mu(post(t_i))\}$$

Using general terms, $T(\mathcal{A})$ is the domain of the match-set function, while R is the range of the mapping. For simplicity, throughout the paper we will refer to match-sets as the range R of the defined mapping.

Definition 8. *Let p and p' be subtrees of the tree template P over a ranked alphabet $\mathcal{A}_S = (\Sigma_S, \varphi_S)$, i.e. $p, p' \in Sub(P)$. Then p is* inconsistent *with p' ($p \mid p'$) if there is no tree $t \in T(\mathcal{A})$ such that $p, p' \in \mu(t)$. p and p' are* independent *($p \sim p'$) if there are trees $t_1, t_2, t_3 \in T(\mathcal{A})$, such that $p \in \mu(t_1)$, $p' \notin \mu(t_1)$, $p \notin \mu(t_2)$, $p' \in \mu(t_2)$, $p, p' \in \mu(t_3)$. p* subsumes *p' ($p > p'$) if, for all $t \in T(\mathcal{A})$, $p \in \mu(t) \Rightarrow p' \in \mu(t)$.*

Example 1. Let p_1, p_2, p_3, p_4 be trees, where $post(p_1) = a_0Sa_2$, $post(p_2) = b_0Sa_2$, $post(p_3) = Sb_0a_2$, $post(p_4) = SSa_2$. p_1 and p_2 are inconsistent ($p_1 \mid p_2$) as nodes a_0 and b_0 cannot be matched in the same position. Trees p_1 and p_3 are independent ($p_1 \sim p_3$), since there exist trees t_1, t_2, t_3, where $post(t_1) = a_0a_0a_2$, $post(t_2) = b_0b_0a_2$, $post(t_3) = a_0b_0a_2$, holding that $p_1 \in \mu(t_1)$ and $p_3 \notin \mu(t_1)$, $p_3 \in \mu(t_2)$ and $p_1 \notin \mu(t_2)$, and $p_1, p_3 \in \mu(t_3)$. Finally, $p_1 > p_4$, $p_2 > p_4$, $p_3 > p_4$.

Lemma 2 (Size of match-sets). *Given a tree template P, the upper theoretical bound of the number of possible match-sets is $\mathcal{O}(2^{|P|})$, and is reached only if there exist sets of pairwise independent subtrees in the tree template.*

Definition 9 (Combination of tree templates). *The combination of two pairwise independent tree templates P and P' (denoted by $P \circ P'$) with $post(P) = post(p_1)\ldots post(p_{\varphi(v)})v$ and $post(P') = post(p_1')\ldots post(p_{\varphi(v)}')v$, respectively, is the tree t where $post(t) = post(t_1)\ldots post(t_{\varphi(v)})v$ is defined as*

$$post(t_j) = \begin{cases} post(p_j) & : \quad post(p_j) > post(p_j') \vee post(p_j) = post(p_j') \\ post(p_j') & : \quad post(p_j') > post(p_j) \\ post(p_j) \circ post(p_j') & : \quad post(p_j) \sim post(p_j') \end{cases}$$

The combination of a set $\mu = (t_1, t_2, \ldots, t_{\sigma(\mu)})$ of pairwise independent tree templates is defined as $C(\mu) = t_1 \circ t_2 \circ \ldots \circ t_{\sigma(\mu)}$.

3.2 Computing Match-Sets

In this section, we present an approach for computing the match-sets of a given tree template P. The method takes as input the set $Sub(P)$ of all unique subtrees of P, and for each $t \in Sub(P)$ it computes two sets: I_t and S_t (which are subsets of $Sub(P)$). Set S_t consists of the trees which t subsumes (including t), while set I_t consists of the trees with which t is pairwise independent. For each distinct set I_t, at least $\sigma(I_t) + 1$ and at most $2^{\sigma(I_t)} - 1$ match-sets need to be constructed. Those match-sets correspond to elements of $\mathcal{P}(I_t)$ (with the exception of the empty set) unified with S_t, as proved in Lemma 2. Note that for any element $X \in \mathcal{P}(I_t)$, the corresponding match-set Y will be $Y = X \cup S_t \cup \{t\} \cup \{t' : t' \in I_t \wedge C(X \cup \{t\} \cup S_t) > t'\}$. The lower bound is for the case all trees in I_t are pairwise inconsistent between themselves, and the upper bound is for the case all trees in I_t are pairwise independent between themselves.

Algorithm 1. Computing the relation between two nonidentical trees

Input : $x = post(t_1)\ldots post(t_{\varphi(v)})v$ and $y = post(t'_1)\ldots post(t'_{\varphi(u)})u$

Output: The relation between two trees t and t' represented by their postfix notation x and y, respectively

1 **if** $v \neq u$ **then return** INCONSISTENT **else** $(p, q, r) \leftarrow$ TREEREL(x, y)
2 **if** $p =$ **true** **then return** INCONSISTENT
3 **else if** $q = r =$ **true** **then return** INDEPENDENT
4 **else if** $q =$ **true** **then return** T-SUBSUMES-T'
5 **else return** T'-SUBSUMES-T

Function TREEREL(x,y)

Input : $x = post(t_1)\ldots post(t_{\varphi(v)})v$ and $y = post(t'_1)\ldots post(t'_{\varphi(v)})v$

Output: Triplet (p, q, r)

1 Let $p \leftarrow$ **false**, $q \leftarrow$ **false**, $r \leftarrow$ **false**, $q' \leftarrow$ **false**, $r' \leftarrow$ **false**
2 **for** $i \leftarrow 1$ **to** k **do**
3   **if** $r(post(t_i)) \neq r(post(t'_i))$ **and** $r(post(t_i)) \neq S$ **and** $r(post(t'_i)) \neq S$ **then** $p \leftarrow$ **true**
4   **if** $r(post(t_i)) \neq r(post(t'_i))$ **and** $r(post(t_i))$ **then** $r \leftarrow$ **true**
5   **if** $r(post(t_i)) \neq r(post(t'_i))$ **and** $r(post(t'_i)) = S$ **then** $q \leftarrow$ **true**
6   **if** $r(post(t_i)) = r(post(t'_i))$ **and** $post(t_i) \neq post(t'_i)$ **then**
7       $(p, q', r') \leftarrow$ TREEREL$(post(t_i), post(t'_i))$
8   **if** $p =$ **true** **then return** $($**true, false, false**$)$
9    $q \leftarrow q$ **or** q'
10    $r \leftarrow r$ **or** r'
11 **return** (p, q, r)

Algorithm 1 computes the relation between two, nonidentical, subtrees of P (i.e. elements of $Sub(P)$). An auxiliary function $r(post(t))$ is used for returning the root of t (i.e. the last symbol of $post(t)$). The whole process of computing match-sets takes time $\mathcal{O}(|P|^2)$ in case there do not exist sets I_t with their elements being pairwise independent, or $\mathcal{O}(2^{|P|})$ in the worst case (Lemma 2).

3.3   The Algorithm

The method for tree pattern matching works in a similar fashion as the finite automata based algorithms for string pattern matching: Given a tree template P, Algorithm 3 constructs a nondeterministic PDA M that can match all occurrences of P in a given subject tree T, by final state. The constructed PDA belongs to the class of height-deterministic PDA and can be determinised [13]. Algorithm 4 presents a novel method that takes as input the nondeterministic PDA obtained from Algorithm 3, computes the match-sets, and constructs an equivalent deterministic PDA M_D, serving as the tree template matcher.

Algorithm 3. Construction of a nondeterministic tree template matching

Input : Tree template $post(P) = post(p_1)post(p_2)\ldots post(p_{\varphi(v)})v$ over $\mathcal{A}$
Output: Nondeterministic PDA $M = (\{q_I, q_F\}, \mathcal{A}, \Gamma, \delta, \{q_I\}, \varepsilon, \{q_F\})$

1 Let $\Gamma \leftarrow \{\{x\} : x \in Sub(P)\} \cup \{\{S\}\}$

2 For each $x \in \Sigma$, let $q_I T^{\varphi(x)} \xmapsto{x}{M} q_I T$, where $T = \{S\}$

3 Let $q_I X_{\varphi(x)}\ldots X_2 X_1 \xmapsto{x}{M} q_I X$, where $X_i = \{post(t_i)\}$ and $X = \{post(t)\}$,
   for each $post(t) = post(t_1)post(t_2)\ldots post(t_{\varphi(x)})x \in Sub(P) \setminus \{post(P)\}$

4 Let $q_I X_{\varphi(v)}\ldots X_2 X_1 \xmapsto{v}{M} q_F X$, where $X_i = \{post(p_i)\}$ and $X = \{post(P)\}$

Algorithm 4. Determinisation

Input : Nondeterministic PDA $M = (Q, \mathcal{A}, \Gamma, \delta, q_I, \varepsilon, F)$
Output: Deterministic PDA $M' = (Q', \mathcal{A}, \Gamma', \delta', \{q_I\}, \varepsilon, F')$

1 **for each** $t \in Sub(P)$ **do**

2   **for each** $X \in \mathcal{P}(I_t)$ *in ascending order of cardinality* **do**

3          $\Gamma' \leftarrow \Gamma' \cup \{X \cup \{t\} \cup S_t \cup \{t' : y = C(X \cup \{t\} \cup S_t) \wedge t' \in I_t \wedge y > t'\}\}$

4 Let $Q' \leftarrow \{\{q_I\}, \{q_I, q_F\}\}$ and $F' \leftarrow \{\{q_I, q_F\}\}$

5 For each $x \in \Sigma$, let $q'\gamma'_1\gamma'_2\ldots\gamma'_{\varphi(x)} \xmapsto{x}{M'} p'X'$ for all $\gamma'_i \in \Gamma'$, where
   $q', p' \in Q'$, $1 \leq i \leq \varphi(x)$, $p' = \bigcup_{j \leftarrow 1}^{l}\{p_j\}$ and $X' = \bigcup_{j \leftarrow 1}^{l}\{\theta_j\}$, such that
   there exist l transitions of form $q_j\gamma_1\gamma_2\ldots\gamma_{\varphi(x)} \xmapsto{x}{M} p_j\theta_j$, $\gamma_i \in \gamma'_i$, $q_j \in q'$

Lemma 3. *Given a nondeterministic PDA constructed using Algorithm 3 by preprocessing a given tree template P, Algorithm 4 constructs an equivalent deterministic PDA matching all occurrences of P in a subject tree T.*

Theorem 2. *For tree template P, the space needed for preprocessing is $\mathcal{O}(2^{|P|})$.*

Proof. In general, there can be $\mathcal{O}(2^{|P|})$ pushdown store symbols (see Lemma 2). The PDA can be implemented as a table and thus $\mathcal{O}(2^{|P| \times k} \times |\mathcal{A}|)$ space is required for preprocessing, where $k = \max\{\varphi(x) : \forall x \in \mathcal{A}\}$. □

Theorem 3. *The deterministic template matching PDA constructed using Algorithms 3 and 4 matches all occurrences of a tree template P in a subject tree T in time $\mathcal{O}(|T|)$.*

Proof. For each input symbol x of the subject tree, $\varphi(x) + 1$ operations are performed: $\varphi(x)$ **pop** operations from the pushdown store and one **push**. The sum of arities of all nodes of the input tree t is $n - 1$ (number of edges). Thus, $n - 1$ pop and n push operations are performed, a total of $2n - 1$ operations. □

4 Conclusion

In this paper, we have formally defined the tree template matching problem for ordered ranked trees, and presented a new class of PDA, which serve as tree template matchers and can be determinised. The main contribution of this paper is a systematic approach for constructing deterministic PDA, which match tree templates in time linear to the size of the subject tree. Although we prove that the space required for preprocessing is exponential to the size of the tree template in the general case, the space required for a specific class of tree templates – the tree templates that do not consist of pairwise independent subtrees – is linear. The time for the searching phase is linear to the size of the subject tree in both cases. The implementation of the proposed algorithm is available at the website `http://www.dcs.kcl.ac.uk/pg/pississo/`.

References

1. Aho, A.V., Ganapathi, M.: Efficient tree pattern matching (extended abstract): an aid to code generation. In: POPL, pp. 334–340. ACM, New York (1985)
2. Chase, D.R.: An improvement to bottom-up tree pattern matching. In: POPL 1987, pp. 168–177. ACM, New York (1987)
3. Cleophas, L.G.W.A.: Tree Algorithms: Two Taxonomies and a Toolkit. Ph.D. thesis, Eindhoven University of Technology (April 2008)
4. Crochemore, M., Rytter, W.: Jewels of Stringology. World Scientific, Singapore (1994)
5. Flouri, T., Janoušek, J., Melichar, B.: Subtree matching by pushdown automata. Computer Science and Information Systems 7, 331–358 (2010)
6. Fraser, C.W., Henry, R.R., Proebsting, T.A.: Burg: fast optimal instruction selection and tree parsing. SIGPLAN Notices 27(4), 68–76 (1992)
7. Glanville, R.S., Graham, S.L.: A new method for compiler code generation. In: POPL, pp. 231–240 (1978)
8. Hoffmann, C.M., O'Donnell, M.J.: Pattern matching in trees. J. ACM 29, 68–95 (1982)
9. Janoušek, J., Melichar, B.: On regular tree languages and deterministic pushdown automata. Acta Informatica 46, 533–547 (2009)

10. Kuboyama, T.: Matching and Learning in Trees. Phd thesis (2007)
11. Mauri, G., Pavesi, G.: Algorithms for pattern matching and discovery in rna secondary structure. Theor. Comput. Sci. 335(1), 29–51 (2005)
12. Melichar, B.: Arbology: Trees and pushdown automata. In: Dediu, A.-H., Fernau, H., Martín-Vide, C. (eds.) LATA 2010. LNCS, vol. 6031, pp. 32–49. Springer, Heidelberg (2010)
13. Nowotka, D., Srba, J.: Height-deterministic pushdown automata. In: Kučera, L., Kučera, A. (eds.) MFCS 2007. LNCS, vol. 4708, pp. 125–134. Springer, Heidelberg (2007)
14. Shankar, P., Gantait, A., Yuvaraj, A.R., Madhavan, M.: A new algorithm for linear regular tree pattern matching. Theor. Comput. Sci. 242(1-2), 125–142 (2000)

Information Extraction from Semi-structured Resources: A Two-Phase Finite State Transducers Approach

Vesna Pajić[1], Gordana Pavlović Lažetić[2], and Miloš Pajić[1]

[1] Faculty of Agriculture, University of Belgrade,
Nemanjina 6, 11080 Zemun, Belgrade, Republic of Serbia
[2] Faculty of Mathematics, University of Belgrade,
Studentski trg 17, 11000 Belgrade, Republic of Serbia

Abstract. The paper presents a new method for extracting information from semi-structured resources, based on finite state transducers. The method has two clearly distinguished phases. The first phase - pre-processing phase - strongly relies upon the analysis of the document structure and it is used for locating records of data in the text. The second phase is based on the finite state transducers created for extracting information. The transducers can be modified so that preferred efficiency is achieved and can be reused for extracting information from other pre-processed documents. We conclude that even untagged text can be treated as a semi-structured one, providing its structure can be successfully pre-processed. As a result, we extracted data from free form encyclopedia text and created a fully structured database with genotype and phenotype characteristics of the organisms.

Keywords: information extraction, finite state transducer, semi-structured resource, linguistic resource, bioinformatics, genome.

1 Introduction

Information Extraction (IE) is part of artificial intelligence which studies and develops techniques used to detect and extract relevant information from larger text documents and present it in a structured form. Depending on the manner and the form in which information is stored in some document, the documents being processed in IE tasks can be structured, semi-structured and unstructured.

In the up-to-date literature, web pages are the most commonly processed semi-structured resources ([1] and [2]). In this paper, we argue that there are textual resources whose structure is not defined by tags, as in HTML or XML text, but still could be considered as semi-structured. The structure of a document could be determined by its logical structure elements, such as headings and paragraphs. If these elements are in a relation with the content so that they can be used by a researcher to conclude something about the information they wish to extract, then we considered such documents as semi-structured ones.

B. Bouchou-Markhoff et al. (Eds.): CIAA 2011, LNCS 6807, pp. 282–289, 2011.

We present a two-phase method for information extraction, based on finite state transducers (FST). Finite state transducers are commonly used in Natural Language Processing for different tasks, and the idea of using FST for information extraction is not new ([3] and [4]), but it has been suppressed lately by methods based on probability and statistics ([5] and [6]). The method we present uses FST first for pre-processing the text, then for describing the context of information in specific text segments, and finally for extracting the information. The great advantage of the method is its reusability and precision. Transducers used for extracting the data, which are created for one resource, can be used again for any other resource of the same domain, i.e. for the same kind of information. Also, transducers are created by human experts so that their precision could be increased until it reaches the preferred level.

We used the proposed method for extracting data from encyclopedia "Systematic Bacteriology" [7] which is organized in such a way that can be treated as a semi-structured resource. As a result we created a fully structured database of microbes, which contains information about genomic and ecological characteristics, such as habitat or shape of bacterial organisms.

2   Finite State Transducers in NLP

Finite state transducers (FST) are finite state machines which define relations between two sets of strings in the way that they transform one string to another [8]. FST are being used in many fields of computational linguistics. Their use is justified from the standpoint of linguistics as well as from the standpoint of computer science ([8], [9] and [10]).

The basic property of FST is that they produce some output and this property determines the way transducers are being used in Natural Language Processing. Also, they can be visually presented by graphs, which make them convenient for human use. FSTs are being used in computational linguistics for morphological parsing, describing orthographic rules, describing inflectional rules etc. Detailed review of theoretical and practical use of finite state transducers in natural language processing is given in [3], [4], [9], [11], [12] and [13].

Finite State Transducers and their corresponding graphs can be very complex and difficult to maintain, which, in practice, leads to some problems. So, instead of one big graph, we use a collection of sub graphs. This method has a strong theoretical background in theory of Recursive Transition Networks (RTN). RTN are an extension of context free grammars ([14]). The arcs in RTN are labeled with corresponding grammars, while the states are labeled arbitrarily. There are several computer tools for linguistic research based on FST and RTN ([15], [16] and [17]).

3   Resources and Tools Used

3.1   Software System for Linguistic Tasks

In our research, we used Unitex [16] as a tool for creating and applying FST graphs, and also for pre-processing the text. Unitex is a collection of programs developed for analyzing natural language text using linguistics resources and tools, such as electronic dictionaries.

Electronic dictionaries contain simple and compound words, together with their lemmas and a set of grammatical codes. Unitex uses electronic dictionaries in DELA format, where each entry is a line of text terminated by a new line, which conforms to the following syntax:

```
apples,apple.N+conc:p
```

The first word (apples) is an inflected form of the entry and it is followed by the canonical form (lemma). The sequence of codes N+conc gives the grammatical and semantic information about the entry. Code N stands for noun and conc indicates that this noun designates a concrete object.

After applying these resources to the text, the user can refer to the dictionary entry from the Unitex by using lexical masks. For example, the query <be.V> will match all entries having *be* as canonical form and the grammatical code V. Thus, all occurrences of verb *to be* (*am*, *is*, *being* etc.) will be recognized by this query.

Using this kind of linguistic resources is the main advantage of Unitex system, because the researcher can define classes of words and phrases with very simple patterns, just by using the information from the dictionary.

3.2 Semi-structured Resource: Encyclopedia

In our research our main goal was to extract information about genomic and ecological characteristics of microbes from a free form text and put them into a relational database. As a resource, we used the electronic form of the encyclopedia "Systematic Bacteriology" [7]. The very structure of the encyclopedia makes it possible to use it for information extraction process, so we treated it as a semi-structured document. The analysis of this structure was one of the most important tasks in the research.

Class I. *Alphaproteobacteria*	1
Order I. *Rhodospirillales*	1
Family I. *Rhodospirillaceae*	1
Genus I. *Rhodospirillum*	1
Genus II. *Azospirillum*	7
Genus III. *Levispirillum*	27
Genus IV. *Magnetospirillum*	28
Genus V. *Phaeospirillum*	32
Genus VI. *Rhodocista*	33
Genus VII. *Rhodospira*	35

Fig. 1. An excerpt from the content of the encyclopedia

The content of the document is as follows. The chapters of the encyclopedia correspond to systematic categories of the bacteria. Each chapter with the family description is followed by the chapters of the genera in this family. The excerpt from the content is given in Figure 1.

Descriptions of the species, containing information we want to extract, are given inside the chapters about genera, located at the end of the chapters. Described structure of the document was used to discover data records, as will be explained in Section 4, where each record corresponds to one systematic category.

4   The Two Phase FST Method

The method we have developed for extracting the information from semi-structured resources, based on the finite state transducers, distinguishes two phases of IE process. Both phases were implemented through a software system using programming language Java.

4.1   The First Phase: Creating Records of Data

The first phase strongly relies on the structure of the document from which the extraction is to be done. Therefore, this phase differs for different documents and has to be adjusted to the structure of particular text. During the first phase, the main goal is to locate pieces of the text in which the information about one record is situated. Those pieces of text are being put in a relational database, for further analysis.

Species : Table										
ID	Genus	Genusl	SpeciesName	SpeciesDesc	Size	GC	GenBankNmbr	TypeStrain	Gram	Habitat
1	Genus	1	Rhodospirillum rubrum	(Esmarch 1887) Molisch 1907, 25AL (Spirillum rubrum Esmarch 1887, 230.) rub'rum. M.L. neut. Adi. rubrum red. Cells are vibrioid						
2	Genus	1	Rhodospirillum photometricum	Molisch 1907, 24AL pho.to.me'tri.cum. Gr. n. phos light; Gr. adj. metricus measuring; M.L. neut. adi. photometricum light measuring.						
3	Genus	2	Azospirillum lipoferum	(Beijerinck 1925) Tarrand, Krieg and Do¨bereiner 1979, 79AL (Effective publication:Tarrand. Krieg and Do¨bereiner						
4	Genus	2	Azospirillum amazonense	Magalha~es, Baldani, Souto, Kuykendall and Do¨bereiner 1984, 355VP (Effective publication: Magalha~es. Baldani. Souto.						
5	Genus	2	Azospirillum brasilense	Tarrand, Krieg and Do¨bereiner 1979, 79AL (Effective publication: Tarrand, Krieg and Do¨bereiner 1978. 979.) bra.si.len'.se. M.L.						

Fig. 2. Excerpt from the table "*Species*" after the first phase is finished

In our research, having the "Systematic Bacteriology" as a resource, we used the fact that each chapter of the text corresponds to one systematic category. The description of specific bacteria species, containing the data we wanted to extract, is located at the end of the chapters about genera. It is preceded by the line beginning with "List of species of the genus ...". There are a different number of species descriptions for the different genera, but each one begins with the number, followed by name of the species and description in a free text form. Based on that fact, we developed an algorithm for extracting species descriptions and putting them into the database. After the first phase had been finished, we had the database containing free form descriptions about bacteria species. A part of the data in the table Species is shown in Figure 2.

4.2   The Second Phase: Extracting Particular Attributes

In the second phase, the system takes unstructured text with data about a record from the database (the field "*SpeciesDesc*") and analyzes it with FST graphs. The piece of text that contains some information of interest is being recognized by a particular transducer. The output of this transducer is information which is put into the database.

After the structure of the encyclopedia had been analyzed and processed, we had a text with species description for each species inserted into the database. The individual attributes we wanted to extract, such as size, G+C content, GenBank accession number etc., were all contained in the species descriptions, but in a free, unstructured form.

For extracting this kind of data from the text, we used finite state transducers specifically designed to fulfill this task. For every attribute we wanted to extract (e.g., bacteria growth temperature, habitat, pH value, oxygen requirement etc.), we created a separate transducer using Unitex. Each transducer recognizes the context of an attribute (information) and produces the output which represents the value of the attribute, i.e., the information itself. This output is inserted into the database.

We were motivated to use transducers for our task of describing context and extracting information by the fact that in biological texts there is a limited number of possible phrases for describing some properties of an organism. For example, not many different ways exist to tell that some bacteria is Gram negative. Therefore, by constructing a transducer which recognizes part of the text about Gram stain and produces the output "*positive*" or "*negative*", depending on the information in the text, we can process not only descriptions from the encyclopedia, but also we can process any other text resource about bacteria.

As mentioned in Section 3.1., we used the Unitex software system for creating graphs that correspond to transducers, and also for pre-processing the text. Beside the pre-processing tasks which are required by Unitex's programs for locating patterns in the text, such as normalization and tokenization of the text, Unitex allows applying linguistics resources to the text. We used this possibility and applied English electronic dictionary to the species descriptions, so we could use lexical masks in transducers.

As an example of using lexical masks, the transducer for extracting the genome size is given in Figure 3. This transducer uses lexical masks such as <be.V> and <estimate.V>, which recognize any inflected form of the verbs *to be* or *to estimate*, to describe the context in which the information about genome size could occur. In patterns recognized by this transducer, the part of the text which corresponds to the part of the graph inside the brackets (marked with the label *Size*) will be produced as the output of this transducer, i.e. will be extracted from the text and inserted into the database, into the field *Size*.

The transducer (a) in the Figure 3 has two calls to sub graphs *SizeRange* and *SizeUnits*. These two sub graphs are also shown on Figure 3, part (b) and (c). Some of the expressions recognized by this transducer are:

"*genome sizes of four G. oxydans strains were estimated to be between 2240 and 3787 kb*"
"*genome size of R. australis is 1256–1276 kbp*"
"*Genome size: 2.73 X 109 Da*"
"*genome size is 1.713 Mbp*"
"*genome size was estimated to be approximately 4061 kb*"

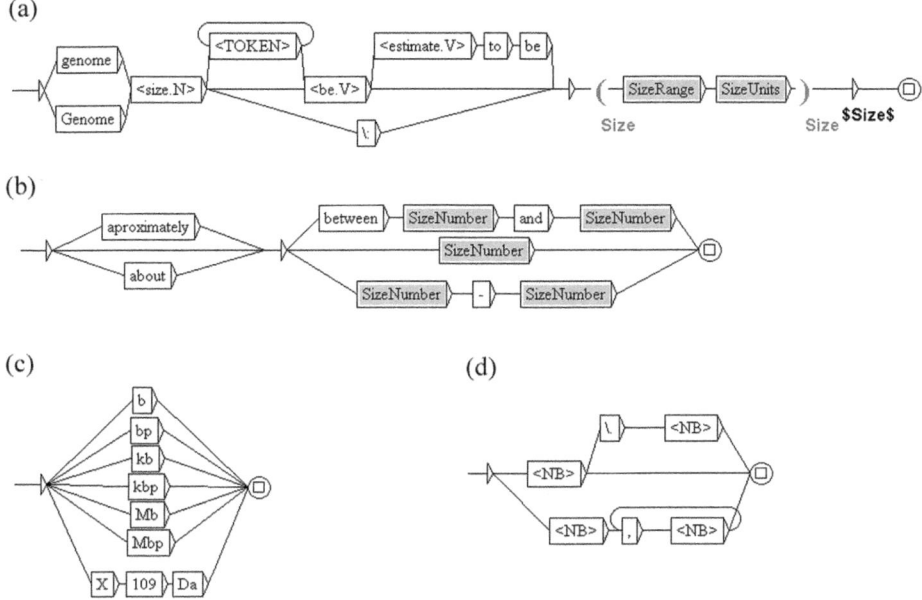

Fig. 3. (a) Transducer for extracting information about the genome size; it has calls to sub graphs *SizeRange* and *SizeUnits*; (b) *SizeRange* sub graph for describing possible ways of specifinig size value; (c) *SizeUnits* sub graph which recognizes the units for size of the genome; (d) *SizeNumber* sub graph for describing different formats of numbers

5   Results and Evaluation of the Method

The Part C of the Volume 2 of the encyclopedia "Systematic Bacteriology" that we experimented with, contains descriptions of 643 species of bacteria, grouped by the genus they belong to. The algorithm we used for the first phase of the proposed method was very efficient and it extracted all the 643 descriptions. The reason for achieving such a good efficiency is thorough analysis of the document structure. The initial algorithm was tuned and modified by the researchers until it has reached such an excellent level of efficiency.

After the second phase was finished, we had data about microbes extracted and inserted into the database. The table *Species*, previously shown in the Figure 2 in Section 4.2, at the end of the second phase looked as shown in the Figure 4. The extracted information was inserted in the corresponding fields of the database.

In order to evaluate efficiency of transducers, we manually analyzed species description and calculated precision and recall of the method. Precision was the highest possible, i.e. all of the extracted information was relevant. This is a consequence of the fact that transducers were designed by human experts to extract particular attributes, and therefore they recognize only sequences of text in which the information is stored.

ID	Genus	Genus	SpeciesName	SpeciesDesc	Size	GC	GenBankNmbr	TypeStrain	Gram	Habitat
1	Genus	1	Rhodospirillum rubrum	(Esmarch 1887) Molisch 1907, 25AL (Spirillum rubrum Esmarch 1887, 230.) rub'rum. M.L. neut. Adj. rubrum red. Cells are vibrioid shaped to spiral. 0.8-1.0 lm wide: one complete turn of		63.8-65.8	D30778, M32020	ATCC 11170, DSM 467, NCIB 8255		stagnant and anoxic freshwater habitats that
2	Genus	1	Rhodospirillum photometricum	Molisch 1907, 24AL pho.to.me'tri.cum. Gr. n. phos light; Gr. adj. metricus measuring; M.L. neut. adj. photometricum light measuring. TABLE BXII. .2. Carbon sources and electron		64.8-65.8	AJ222662	ATCC 49918, DSM 122, NTHC 132		stagnant and anoxic freshwater habitats that
3	Genus	2	Azospirillum lipoferum	(Beijerinck 1925) Tarrand, Krieg and Do'bereiner 1979, 79AL (Effective publication:Tarrand, Krieg and Do'bereiner 1978. 978 (Soirillum liooferum Beijerinck 1925.		69-70	M59061	BR11080, Sp 59b, ATCC 29707, DSM 1691	neg	
4	Genus	2	Azospirillum amazonense	Magalha~es, Baldani, Souto, Kuvkendall and Do'bereiner 1984, 355VP (Effective publication: Magalha~es, Baldani, Souto, Kuvkendall and Do'bereiner 1983. 417.)		67-68	Z29616, X79735	BR 11142, Am14, Y1, ATCC 35 119, DSM GenBank		Soil and tissues + + + mainly of nonlegumes

Fig. 4. Table *Species* with data after the second phase of IE process

Recall differs for different transducers, depending on the complexity of the information context. For example, the transducer for Gram stain property was very efficient; it properly extracted attributes from all descriptions which contained that kind of information. Some other transducers, especially those for extracting information that occur in a complex context, weren't that efficient. For example, the initial transducer for genome size extracted 14 out of 18 data about genome size. Some expressions weren't recognized by this transducer, such as:

"genome has a size of 1,231,204 bp"
"genome size is distinctly larger (1.49 X 109 Da)"
"genome is 1,257,710 bp in size"

Nevertheless, with slight modification of the transducer and extending it in order to recognize the former expressions as well, the recall can be increased. This is a key point and a major advantage of methods based on FST over methods based on probability. Efficiency of methods based on FST can be increased to the preferred level by modifying transducers. This fact, together with the fact that transducers can be reused for other resources of the same domain makes this method justified and suitable to use for IE tasks.

6 Conclusion

In our research we successfully applied the proposed method to a resource whose structure is not explicitly tagged (as in HTML or XML documents). Nevertheless, considering that there were certain regularities between the structure of the document and the data content, which were noticed by analyzing the structure, this document is treated and processed as a semi-structured one. We are convinced that this approach could be applied to other documents with similar characteristics.

The second phase of the method involves creating and applying transducers to the text from which the information is to be extracted. We showed that using this method is very efficient, especially when applied to texts from some specific science or domain, in which case the transducer has to describe specific and relatively simple context of information. The use of transducers is also justified by their reusability on other texts from the same domain.

The advantages of the proposed two-phase FST based method is its conceptual simplicity, efficiency, possibility to adjust precision of the transducers, reusability of the transducers and no need for large sets of training data.

We hope that this method will attract more attention from the research community in the future, and that spreading its use will lead to creation of transducers libraries, which can be reused by other researchers. We plan to make our collection of transducers, as well as databases of extracted information, available to others.

References

1. Carlson, A., Schafer, C.: Bootstrapping information extraction from semi-structured web pages. In: Daelemans, W., Goethals, B., Morik, K. (eds.) ECML PKDD 2008, Part I. LNCS (LNAI), vol. 5211, pp. 195–210. Springer, Heidelberg (2008)
2. Liu, B., Grossman, R., Zhai, Y.: Mining data records in web pages. In: Proceedings of SIGKDD 2003, Washington, USA, pp. 601–606 (2003)
3. Friburger, N., Maurel, D.: Finite-state transducer cascades to extract named entities in texts. Theoretical Computer Science 313, 93–104 (2004)
4. Hobbs, J.R., Appelt, D., Bear, J., Israel, D., Kameyama, M., Stickel, M., Tyson, M.: FASTUS: A Cascaded Finite-State Transducer for Extracting Information from Natural-Language Text. In: Roche, E., Schabes, Y. (eds.) Finite-State Language Processing, pp. 383–406. The MIT Press, Washington (1997)
5. Feng, D., Burns, G., Hovy, E.: Extracting Data Records from Unstructured Biomedical Full Text. In: Proceedings of the EMNLP Conference, Prague, Czech Republic (2007)
6. Zhong, P., Chen, J., Cook, T.: Web Information Extraction Using Generalized Hidden Markov Model. In: 1st IEEE Workshop on Hot Topics in Web Systems and Technologies (HOTWEB 2006), pp. 1–8 (2007)
7. Garrity, G.M.: Systematic Bacteriology. In: The Proteobacteria, Part C: The Alpha-, Beta-, Delta-, and Epsilonproteobacteria, Bergey's Manual Trust, Department of Microbiology and Molecular Genetics, 2nd edn. vol. 2. Michigan State University, USA (2005)
8. Jurafsky, D., Martin, J.H.: Speech and language processing. Prentice-Hall Inc. Englewood Cliffs (2000)
9. Gross, M., Perrin, D.: Electronic Dictionaries and Automata in Computational Linguistics. In: Proceedings of LITP Spring School on Theoretical Computer Science Saint-Pierre d'Oleron, France, May 25-29 (1987)
10. Aho, A.V., Hopcroft, J.E., Ullman, J.D.: The Design and Analysis of Computer Algorithms. Addison Wesley, Reading (1974)
11. Casacuberta, F., Vidal, E., Picó, D.: Inference of finite-state transducers from regular languages. Pattern Recognition 38(9), 1431–1443 (2005)
12. Kornai, A.: Extended finite state models of language. Cambridge University Press, Cambridge (1999)
13. Pajic, V.: Finite State Transducers in Web Monitoring, Master Thesys, Faculty of Mathematics, University of Belgrade, Republic of Serbia (2010)
14. Sastre, J.M., Forcada, M.: Efficient parsing using recursive transition networks with output, In: Vetulani, Z., Uszkoreit, H. (eds.) LTC 2007. LNCS, vol. 5603, pp. 280–284. Springer, Heidelberg (2009)
15. Olivier, B., Constant, M., Laporte, E.: Outilex, plate-forme logicielle de traitement de textes ecrits. In: Proceedings of TALN 2006. UCL Press, London (2006)
16. Paumier, S.: Unitex 1.2 User Manual, Universit´e de Marne-la-Vallée (2006), http://www-igm.univ-mlv.fr/~unitex/UnitexManual.pdf
17. Silberztein, M.D.: Dictionnaires ´electroniques et analyse automatique de textes. Le systeme INTEX, Paris, Masson (1993)

Experimental Study of the Shortest Reset Word
of Random Automata

Evgeny Skvortsov[1] and Evgeny Tipikin[2,*]

[1] Google Inc., USA
[2] Ural State University, Yekaterinburg, Russia
{skvortsoves,etipikin}@gmail.com

Abstract. In this paper we describe an approach to finding the shortest
reset word of a finite synchronizing automaton by using a SAT solver.
We use this approach to perform an experimental study of the length of
the shortest reset word of a finite synchronizing automaton. The largest
automata we considered had 100 states. The results of the experiments
allow us to formulate a hypothesis that the length of the shortest reset
word of a random finite automaton with n states and 2 input letters with
high probability is sublinear with respect to n and can be estimated as
$1.95n^{0.55}$.

1   Introduction

A *deterministic finite automaton* (DFA) is a triple $\mathcal{A} = (Q, \Sigma, \delta)$, where Q is a
set of states, Σ is an input alphabet, and $\delta : Q \times \Sigma \to Q$ is a transition function
defining an action of the letters in Σ on Q. We use a common concise notation
denoting $\delta(\ldots \delta(\delta(\mathbf{q}, a_0), a_1), \ldots a_k)$ by $\mathbf{q}a_0 \ldots a_k$.

A word $w \in \Sigma^*$ is said to be a *reset word* for a DFA $\mathcal{A}$ if its action leaves
$\mathcal{A}$ in one particular state no matter what state it starts at: $\mathbf{q}_1 w = \mathbf{q}_2 w$ for all
$\mathbf{q}_1, \mathbf{q}_2 \in Q$. A DFA $\mathcal{A}$ is called *synchronizing* if it possesses a reset word. In this
paper we describe results of an experimental study of the length of the shortest
reset word of random automata.

It can be easily shown that if an automaton with n states is synchronizing
then it has a reset word of length less than n^3. However, the tightness of this
bound is far from obvious. In 1964, Černý formulated a conjecture concerning
the upper bound of the length of the shortest reset word of a synchronizing
DFA [5]: the length cannot be larger than $(n-1)^2$. By now the Černý conjecture
is arguably the longest standing open problem in the combinatorial theory of
finite automata. The tightest upper bound that has been obtained so far is
$(n^3 - n)/6$; it was proved by Pin [15] in 1983.

Though no bound better than cubic has been proven for the shortest reset word, most naturally occurring automata have reset words of subquadratic

---

* The second author acknowledges support from the Ministry for Education and Science of Russia, grant 2.1.1/13995, and from the Russian Foundation for Basic Research, grants 10-01-00524.

B. Bouchou-Markhoff et al. (Eds.): CIAA 2011, LNCS 6807, pp. 290–298, 2011.

length. Automata with reset word of length $\Theta(n^2)$ are considered to be exceptional. For a long time the only infinite series of such automata was the one proposed by Černý [5]. The other substantially different ones [1,2] have only recently been constructed.

There are several theoretical and experimental results that support the statement that most synchronizing automata have a relatively short reset word. First, Higgins [11] has shown that the composition of $2n$ random mappings of a set of size n into itself *with high probability* (whp) is a mapping with an image of size 1. (By "high probability" we mean that the probability tends to 1 as n goes to infinity.) In terms of automata, Higgins's result means that a random automaton with an alphabet of size larger than $2n$ whp has a reset word of length $2n$. Indeed, if we pick an automaton uniformly at random among all automata with n states and $2n$ letters, then the action of a word composed of all the letters is identical to a mapping composed of $2n$ random mappings. Later it was shown [18] that a random automaton with n states over an alphabet of size $n^{0.5+\varepsilon}$ has a reset word of quadratic length with high probability for any $\varepsilon > 0$.

The probability distribution of the length of the shortest reset word of a random automaton can be studied experimentally for small n. It is unlikely that there is a polynomial algorithm that can find the shortest reset word in general case because the problem belongs to $FP^{NP[log]}$ [14], which means that the problem is both NP-hard and co-NP-hard. Moreover, approximating the length of the shortest reset word has also been shown to be hard [3]. Nevertheless, it is possible that the problem restricted to a certain class of automata (for instance see [10]) or to random automata is easy and can be successfully solved by an appropriate heuristic. Recently, Roman [16] has developed a genetic algorithm for finding a short reset word and in particular, applied it to random automata. In this paper we present the results of applying of SAT solvers to the problem of finding the shortest reset word.

SAT (or Boolean Satisfiability) is a combinatorial problem of finding a boolean assignment that satisfies a given boolean formula in conjunctive normal form. SAT was one of the first problems proven to be NP-complete [6]. The development of practical algorithms for solving instances of SAT (so called SAT-solvers) is an area of active research and there is a regular competition of these algorithms. These days the problems that participate in SAT competitions have hundreds of thousands of variables and millions of literals. This is especially surprising when one recalls that SAT is NP-complete. This observation does not formally contradict the NP-hardness of SAT, but shows that hard instances of SAT rarely occur in practice. There are various approaches to explaining this phenomenon in greater detail [12,7,4].

SAT is also known to be a natural language for a variety of combinatorial problems. In this paper we show that the problem of finding the shortest reset word of a finite automaton can be naturally reduced to a few SAT instances. We apply a SAT solver to those instances and recover the reset word from the resulting boolean assignment.

As mentioned, Roman [16] was using a genetic algorithm to find a reset word of random automata. Since genetic algorithms are incomplete, the results of [16] allow one to assume only an upper bound on the length of the shortest reset word. It turns out that even for an alphabet of size 2 as the number of states grows, the probability of the automata being synchronizing approaches 1. In this paper we also study automata over a 2-letter alphabet. It is easy to see that if the size of the alphabet gets larger, the length of the shortest reset word of a random automata decreases.

We were able to find the shortest reset words of randomly generated automata with up to 100 states. We argue that the results of our experiments are a reasonable basis for the hypothesis of the length of the shortest reset word of a random automaton. The hypothesis is given in the following formula:

$$\ell(n) \approx 1.95n^{0.55},$$

where n is the number of states of the random automaton and $\ell(n)$ is the length of the shortest reset word.

The rest of the paper is organized as follows. In Section 2, we describe how the problem of finding the shortest reset word can be reduced to a collection of instances of SAT. In Section 3, we formally define the notion of a random automaton. In Section 4, we present results of experiments and what we believe they mean. We conclude in Section 5 with a short discussion.

Due to space constraints we have to omit a few figures and proofs. Those can be found at the full version at the first author's website [17].

2 Solving Automata Synchronization Problem via Reduction to SAT

Given a finite automaton $\mathcal{A} = (Q, \{a, b\}, \delta)$ and an integer c, we build a 3-CNF formula $\phi_{\mathcal{A}}^c$ such that $\phi_{\mathcal{A}}^c$ is satisfiable if and only if $\mathcal{A}$ has a reset word w of length c. We denote the prefix of w of length t by $w|_{1...t}$. The formula $\phi_{\mathcal{A}}^c$ contains two types of variables:

- For each $t \in 1, \ldots, c$, we introduce a variable u_t. Setting u_t to *true* is interpreted as "the t-th letter of w is a" and setting u_t to *false* is interpreted as "the t-th letter of w is b".
- For each $\mathbf{q} \in Q$ and $t \in \{0, \ldots, c\}$, we introduce a variable $x_{\mathbf{q}t}$. A variable $x_{\mathbf{q}0}$ is used to mark whether an automaton can be initially in a state $\mathbf{q}$ or not. When $t \neq 0$, setting $x_{\mathbf{q}t}$ to *false* is interpreted as "there does not exist a state $\mathbf{u}$ such that $\mathbf{u}w|_{1...t} = \mathbf{q}$". It is convenient for us to interpret setting $x_{\mathbf{q}t}$ to *true* as "there *may* exist a state $\mathbf{u}$ such that $\mathbf{u}w|_{1...t} = \mathbf{q}$". In other words, we will enforce setting $x_{\mathbf{q}t}$ to *true* and will not enforce *false*.

There are c variables of the first type and $(c+1)n$ variables of the second type. Therefore the resulting boolean formula contains $(c+1)n + c$ boolean variables.

There are also three types of clauses in $\phi_{\mathcal{A}}^c$:

- For each $\mathbf{q} \in Q$ we assert that initially the automaton can be in this state by adding a one literal clause

$$x_{\mathbf{q}0}.$$

- For each $\mathbf{q} \in Q$ and $t \in \{0, \ldots, c-1\}$ we add the following elementary disjunctions to ϕ_A^c:

$$\neg x_{\mathbf{q}t} \vee \neg u_t \vee x_{(\mathbf{q}a)(t+1)},$$

$$x_{\mathbf{q}t} \vee u_t \vee x_{(\mathbf{q}b)(t+1)}.$$

Note that these disjunctions are equivalent to the following implications:

$$x_{\mathbf{q}t} \wedge u_t \rightarrow x_{(\mathbf{q}a)(t+1)}, \tag{1}$$

$$x_{\mathbf{q}t} \wedge \neg u_t \rightarrow x_{(\mathbf{q}b)(t+1)}. \tag{2}$$

The clauses of the first and the second types together enforce setting $x_{\mathbf{q}t}$ to *true* if and only if the state $\mathbf{q}$ can be achieved from some state of $\mathcal{A}$ by applying the prefix $w|_{1\ldots t}$.
- For each 2-element subset $\{\mathbf{p}, \mathbf{q}\} \subseteq Q$, where $\mathbf{p} \neq \mathbf{q}$ we add the following elementary disjunctions to ϕ_A^c:

$$\neg x_{\mathbf{q}c} \vee \neg x_{\mathbf{p}c}. \tag{3}$$

The clauses of the third type ensure that at most one of the variables $x_{\mathbf{q}c}$ may be true.

If w is a reset word of length c for $\mathcal{A}$, then the formula ϕ_A^c is satisfiable. Indeed the satisfying assignment is obtained as follows. Values of the variables $u_1, \ldots, u_c$ are determined by reading the word w and setting u_t to *true* or *false* according to the value of the t-th letter of w. Then we assign $x_{\mathbf{q}0} = true$ for all $\mathbf{q} \in Q$. Next, for each $t = 1, \ldots, k$ and for each $\mathbf{q} \in Q$, we assign $x_{\mathbf{q}t}$ to *true* if it must be done to satisfy some clause of type (1) or (2). Otherwise, we assign $x_{\mathbf{q}t}$ to *false*. It is easy to see that after such an assignment for any t and $\mathbf{q}$ we have $x_{\mathbf{q}t}$ equal to true if and only if $\mathbf{q} = \mathbf{u}w|_{1\ldots t}$, for some $\mathbf{u} \in Q$. Since w is a synchronizing word, all clauses of type (3) are satisfied. Analogously, if the formula ϕ_A^c is satisfiable, then the values of the variables $u_1, \ldots, u_c$ in the satisfying assignment define a word w of length c, and the fact that all clauses of ϕ_A^c are satisfied implies that w is a reset word.

There are n clauses of the first type, $2cn$ clauses of the second type and $\frac{n(n-1)}{2}$ clauses of the third type. In total we have $\frac{n(n-1)}{2} + n(2c+1)$ clauses. Clauses of the first type have one literal, clauses of the second type have three literals and clauses of the third type have two literals each. Therefore the formula ϕ_A^c contains $n^2 + 6cn$ literals in total. Thus, we can use a SAT solver to answer the question: "Can $\mathcal{A}$ be synchronized by a word of length c?"

We use MiniSAT solver [9] to find the solution to this problem. SAT algorithms development is a very active research area and each year new solvers win the competition. MiniSAT was developed in 2003 and has become a state-of-the-art algorithm since then. The algorithm is relatively simple and yet very efficient — its performance is comparable to the best present day solvers. In some years, the SAT competition has a specialized MiniSAT-hack tournament. For more details on the algorithm see [8,9].

Once we have an algorithm that can check whether there is a reset word of given length we can find the length of the shortest reset word by performing binary search. Note that there is a polynomial algorithm for checking whether $\mathcal{A}$ is synchronizing [5]. Thus, we use SAT solver only for synchronizing automata.

3   Random Automaton

In the experimental section we study the length of the shortest reset word of a random automaton over a 2-letter alphabet. Formally, Random Automaton $\mathbb{A}(n)$ with n states over an alphabet Σ can be defined as a discrete probability space $(\Omega_{\mathbb{A}}, P)$, where sample space $\Omega_{\mathbb{A}}$ is the set of all automata over Σ. To define a specific automaton $\mathcal{A} = (Q, \Sigma, \delta)$ one needs to define $\delta(\mathbf{q}, a)$, for each $\mathbf{q} \in Q$ and $a \in \Sigma$. Thus, it is easy to see that $|\Omega_{\mathbb{A}}| = n^{|\Sigma| n}$. We set the probability of all elements of the sample space to be equal, and consequently for all $\mathcal{A}$ we have $P(\mathcal{A}) = n^{-|\Sigma| n}$. We also consider a probabilistic space Random Synchronizing Automaton $\mathbb{A}'(n)$. Formally, $\mathbb{A}'(n)$ is defined as a probabilistic space induced by $\mathbb{A}(n)$ on the event "$\mathcal{A}$ is synchronizing".

The length of the shortest reset word $\ell(n)$ is a random variable over the probabilistic space $\mathbb{A}'(n)$. To study the behaviour of the random variable $\ell(n)$ as n tends to infinity we define the expectation of $\ell(n)$ by $r(n)$ and the variance of $\ell(n)$ by $d(n)$, that is $r(n) = \mathbf{E}(\ell(n)), d(n) = \mathbf{V}(\ell(n)).$

Note that while $\ell(n)$ is a random variable for each n, the functions $r(n)$ and $d(n)$ are deterministic.

4   Experimental Results

We performed a series of experiments for different n, where n is the number of states in the automaton. For a given n, the experiment consists of the following.

We generate a random automaton with n states. Then we check whether this automaton is synchronizing and if so, we find a reset word for this automaton using binary search. Then we record the result of synchronization, i.e., whether the automaton is synchronizing and the length of the shortest reset word.

For a specified number of states n, we performed a number of such experiments. The larger n is the more time it takes to solve the problem of finding the reset word, so for larger n we performed fewer experiments. For each $n \in \{1, 2, \ldots, 20, 25, 30, \ldots, 50\}$ we performed 2000 experiments, for each $n \in \{55, 60, 65, 70\}$ we performed 500 experiments and for $n \in \{75, 80, \ldots, 100\}$ we performed 200 experiments. In our experiments we used a personal computer with an Intel(R) Core(TM)2 Duo P8600 2.4GHz CPU and 4GB of RAM. The program for calculations was written in Java. The average calculation time was 2.7 seconds for $n = 50$ and 70 seconds for $n = 100$.

Thus, for each value of n participated in experiments we have an approximated probabilistic distribution of $\ell(n)$ and an estimated probability of the event "$\mathbb{A}(n)$ is synchronizing".

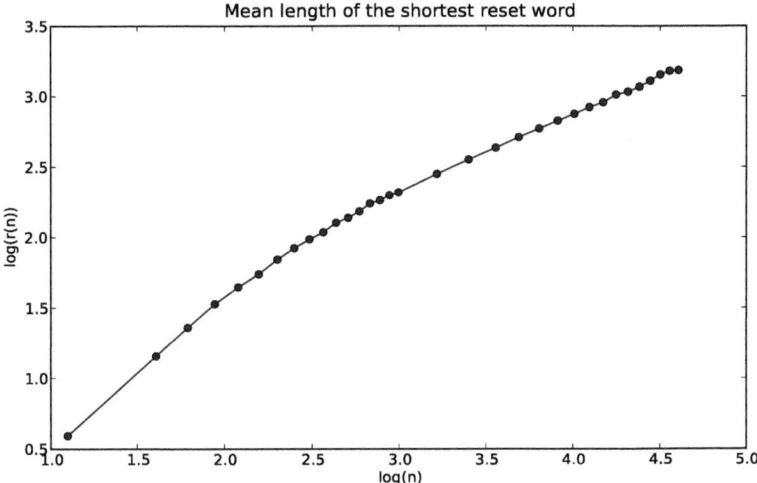

Fig. 1. The graph of the logarithm of the number of states of automata n versus the logarithm of the length of the shortest reset word r

Synchronization of $\mathbb{A}$**.** The larger n is, the larger the fraction of generated random automata that are synchronizing. For $n = 100$ only 1 out of 200 automata that we generated happened to not be synchronizing. Thus, we conclude that it is likely that $P(\text{``}\mathbb{A}\text{ is synchronizing''}) \xrightarrow[n \to \infty]{} 1$.

Expectation of $\ell(n)$**.** It appears that the function $r(n)$ follows a certain trend. To check whether the dependence of the mean value of the distribution $\ell(n)$ follows a power law, we plot the graph in log/log space in Fig. 1. From the graph we conclude that it is a combination of some effects that are present for small n and an affine function that is obeyed for large n. To extract the behaviour of $\mathbb{A}$ for large n, we ignore data points for $n < 20$. We use the least squares method to find an affine function that best reflects the dependency of $\log(r)$ on $\log(n)$. We find that $\log(r(n)) \approx 0.55 \log(n) + 0.67$ and taking the exponent of both sides we obtain the equation

$$r(n) \approx 1.95 n^{0.55} \tag{4}$$

In Fig. 2, we plot the graph of r versus n and the curve given by (4). It is interesting to note that the obtained approximation starts to fit the data at $n = 17$, approximating some data points that were not used in training.

Variance of $\ell(n)$**.** Recall that we denote variance of $\ell(n)$ by $d(n)$. Our experiments show that as n grows, $d(n)$ also grows, but function $\frac{\sqrt{d(n)}}{r(n)}$ appears to tend to 0 as n goes to infinity. It is not very hard to see that this implies that $P(\ell(n) = r(n) + o(r(n))) \longrightarrow_{n \to \infty} 1$. In other words, with high probability $\ell(n)$ is approximately equal to $r(n)$.

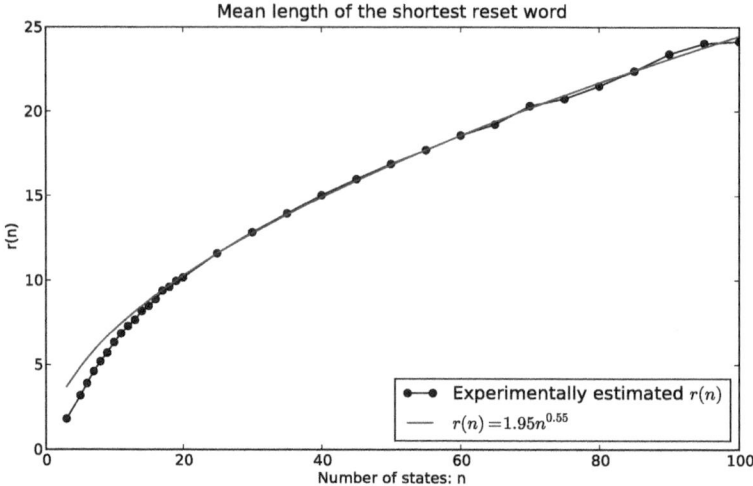

Fig. 2. The graph of the mean length of the shortest reset word versus the number of states of the random automata and a power function approximating it

5   Conclusion and Discussion

We interpret experimental results as indicating that as n goes to infinity, a random automaton is synchronizing with high probability. Also with high probability the the length of its shortest reset word can be computed as

$$\ell(n) \approx 1.95n^{0.55}. \tag{5}$$

In particular, we believe that the experimental data we obtained suggests that the length of the shortest synchronizing word of a random automaton is sublinear with respect to the number of states.

It worth noting that our conclusion (5) directly contradicts a conjecture that Roman formulated in [16]. Namely, Roman conjectured that the mean length of the shortest reset word for a random n-state synchronizing automaton is almost equal to $0.486n$. Roman's experiments with random automata consisted of two parts: for each $n = 5, 6, \ldots, 14$ one thousand random n-state automata were generated and then for each $n = 15, 16, \ldots, 100$ ten random n-state automata were generated. The linear estimate $\ell(n) \approx 0.486x + 1.654$ was suggested on the basis of the results of the first part of the experiments and then it was extrapolated even though the reported results of the second part did not really support the extrapolation. In contrast, we believe that both our and Roman's experiments with larger n indicate that a random automaton is synchronized by a word of length sublinear with respect to the number of states.

We are also aware of another series of experiments with random automata synchronization performed by Gusev (these experiments are mentioned in [1]). A direct comparison of our results with those by Gusev is impossible because he used a different random automata model. However, on a qualitative level our conclusions tend to quite agree with Gusev's.

Acknowledgement. We are grateful to Prof. M.V. Volkov for numerous productive discussions on the topic, and to the anonymous reviewers for their remarks which have helped us make the article more accurate and clear.

References

1. Ananichev, D.S., Gusev, V.V., Volkov, M.V.: Slowly synchronizing automata and digraphs. In: Hlinený and Kucera [13], pp. 55–65
2. Ananichev, D.S., Volkov, M.V., Zaks, Y.I.: Synchronizing automata with a letter of deficiency. Theor. Comput. Sci. 376(1-2), 30–41 (2007)
3. Berlinkov, M.V.: Approximating the Minimum Length of Synchronizing Words Is Hard. In: Ablayev, F., Mayr, E.W. (eds.) CSR 2010. LNCS, vol. 6072, pp. 37–47. Springer, Heidelberg (2010)
4. Bulatov, A.A., Skvortsov, E.S.: Phase transition for local search on planted SAT. CoRR, abs/0811.2546 (2008)
5. černy, J.: Poznámka k homogénnym eksperimentom s konečnými automatami. Matematicko-fyzikalny Časopis Slovensk. Akad. Vied 14, 208–216 (1964)
6. Cook, S.A.: The complexity of theorem-proving procedures. In: STOC, pp. 151–158 (1971)
7. Dantsin, E., Hirsch, E.A., Wolpert, A.: Clause shortening combined with pruning yields a new upper bound for deterministic SAT algorithms. In: Calamoneri, T., Finocchi, I., Italiano, G.F. (eds.) CIAC 2006. LNCS, vol. 3998, pp. 60–68. Springer, Heidelberg (2006)
8. Eén, N., Biere, A.: Effective Preprocessing in SAT Through Variable and Clause Elimination. In: Bacchus, F., Walsh, T. (eds.) SAT 2005. LNCS, vol. 3569, pp. 61–75. Springer, Heidelberg (2005)
9. Eén, N., Sörensson, N.: An Extensible SAT-solver. In: Giunchiglia, E., Tacchella, A. (eds.) SAT 2003. LNCS, vol. 2919, pp. 502–518. Springer, Heidelberg (2004)
10. Eppstein, D.: Reset sequences for monotonic automata. SIAM J. Comput. 19, 500–510 (1990)
11. Higgins, P.: The range order of a product of i transformations from a finite full transformation semigroup. Semigroup Forum 37, 31–36 (1988), doi:10.1007/BF02573120
12. Hirsch, E.A.: Sat local search algorithms: Worst-case study. J. Autom. Reasoning 24(1/2), 127–143 (2000)
13. Hlinĕný, P., Kučera, A. (eds.): MFCS 2010. LNCS, vol. 6281. Springer, Heidelberg (2010)
14. Olschewski, J., Ummels, M.: The complexity of finding reset words in finite automata. In: Hlinený and Kucera [13], pp. 568–579
15. Pin, J.-E.: On two combinatorial problems arising from automata theory. In: Berge, C., Bresson, D., Camion, P., Maurras, J.F., Sterboul, F. (eds.) Proceedings of the International Colloquium on Graph Theory and Combinatorics, Combinatorial Mathematics, vol. 75, pp. 535–548. North-Holland, Amsterdam (1983)

16. Roman, A.: Genetic Algorithm for Synchronization. In: Dediu, A.H., Ionescu, A.M., Martín-Vide, C. (eds.) LATA 2009. LNCS, vol. 5457, pp. 684–695. Springer, Heidelberg (2009)
17. Evgeny Skvortsov. Google profile,
    `http://profiles.google.com/u/0/108501114510819324139/about`
18. Skvortsov, E.S., Zaks, Y.: Synchronizing random automata. Discrete Mathematics & Theoretical Computer Science 12(4), 95–108 (2010)

Author Index